Sustainable Development and Biodiversity

Volume 39

Series Editor

Kishan Gopal Ramawat, Botany Department
Mohanlal Sukhadia University
Udaipur, India

Sustainable Development Goals are best achieved by mechanisms such as research, innovation, and knowledge sharing. This book series aims to help researchers by reporting recent progress and providing complete, comprehensive, and broad subject-based reviews about all aspects of sustainable development and ecological biodiversity.

The series explores linkages of biodiversity with delivery of various ecosystem services and it offers a discourse in understanding the biotic and abiotic interactions, ecosystem dynamics, biological invasion, ecological restoration and remediation, diversity of habitats and conservation strategies. It is a broad scoped collection of volumes, addressing relationship between ecosystem processes and biodiversity. It aims to support the global efforts towards achieving sustainable development goals by enriching the scientific literature. The books in the series brings out the latest reading material for botanists, environmentalists, marine biologists, conservationists, policy makers and NGOs working for environment protection.

We welcome volumes on the themes --

Agroecosystems, Agroforestry, Biodiversity, Biodiversity conservation, Conservation of ecosystem, Ecosystem, Endangered species, Forest conservation, Genetic diversity, Global climate change, Hotspots, Impact assessment, Invasive species, Livelihood of people, Plant biotechnology, Plant resource utilization, Sustainability of the environment, Sustainable management of forests, Sustainable use of terrestrial ecosystems and plants, Traditional methods, Urban horticulture.

Nabarun Ghosh • Debajyoti Ghosh
Shaily Goyal
Editors

Nanotechnology in Air Quality Management

For Sustainable Environment with high interest

Springer

Editors
Nabarun Ghosh
Department of Life, Earth & Environmental Sciences
West Texas A&M University
Cayon, Texas, USA

Shaily Goyal
Department of Life, Earth & Environmental Sciences
West Texas A&M University
Canyon, Texas, USA

Debajyoti Ghosh
Division of Rheumatology
Allergy, and Immunology
Department of Internal Medicine
University of Cincinnati College of Medicine
Cincinnati, OH, USA

ISSN 2352-474X ISSN 2352-4758 (electronic)
Sustainable Development and Biodiversity
ISBN 978-981-96-8338-3 ISBN 978-981-96-8339-0 (eBook)
https://doi.org/10.1007/978-981-96-8339-0

This Springer imprint is published by the registered company Springer Nature Singapore Pte Ltd.
The registered company address is: 152 Beach Road, #21-01/04 Gateway East, Singapore 189721, Singapore

Preface

Air pollution is a leading global cause of death and disease, responsible for an estimated 4.2 million fatalities annually from stroke, heart disease, lung cancer, and chronic respiratory illnesses. The escalating levels of particulate matter (PM2.5), aeroallergens like pollen and mold spores, and unchecked emissions of volatile organic compounds (VOCs) have significantly worsened air quality worldwide. While traditional methods for controlling pollution remain crucial, they often come with limitations. In this context, nanotechnology has emerged as a promising solution, offering innovative approaches to combat air pollution through the exceptional properties of nanomaterials.

In recent years, nanotechnology has emerged as a groundbreaking field with the potential to revolutionize pollution control. Its ability to manipulate materials at the nanoscale enables the creation of advanced solutions that can detect, capture, and neutralize contaminants with unparalleled precision. This book explores the transformative power of nanotechnology in addressing air pollution and its broader environmental applications.

This book offers an in-depth examination of how nanotechnology can provide sustainable, effective solutions to these challenges. It delves into the versatility of nanotechnology and its ability to address environmental pollution comprehensively. Through a combination of cutting-edge research, real-world applications, and future directions, the book highlights the critical role of nanotechnology in developing sustainable strategies for environmental health.

The contents of this book are organized to provide a holistic understanding of nanotechnology's potential. Readers will discover the principles of nanotechnology, the synthesis and characterization of nanomaterials, and their diverse applications in environmental remediation. From nanoadsorbents and nanocatalysts to nanofilters and nanosensors, the capabilities of these materials are explored in depth, emphasizing their role in creating a cleaner and healthier world.

The aim of this book is to inspire researchers, industry leaders, and policymakers to harness the potential of nanotechnology in the fight against pollution. By fostering collaboration and encouraging innovation, it is our hope that this book will contribute to the creation of a future where technology and nature coexist harmoniously. A future where we can breathe cleaner air and work toward restoring the balance between human activity and the environment.

This preface is an invitation to embark on a journey of discovery and exploration. As you delve into the pages ahead, we hope you will find not only the knowledge and insights needed to address the pressing environmental challenges of our time but also the inspiration to be part of a global movement toward a sustainable and healthier planet.

Cayon, Texas, USA — Nabarun Ghosh
Cincinnati, OH, USA — Debajyoti Ghosh
Cayon, Texas, USA — Shaily Goyal

Acknowledgments

This book would not have been possible without the love, support, and encouragement of the incredible people in my life.

First and foremost, I am deeply thankful to my father, Dr. B.M. Goyal, for his continuous motivation and unwavering belief in me and to my mother Kanta Goyal, whose unconditional love has been my foundation.

I am profoundly grateful to my husband, Dr. Komal Sharma, for his unwavering support, and patience throughout this journey. Your encouragement and understanding have been my greatest comfort as I immersed myself in this process.

A special mention to my two beautiful daughters, Shireen and Krisha—you are my greatest joys and constant inspirations. Your smiles and love light up my world and keep me moving forward.

I also want to thank Prof. K.G. Ramawat for his support, encouragement, and belief in my vision. I am eternally grateful for your guidance and the inspiration you have provided throughout this journey.

Finally, to you, the reader—thank you for taking the time to engage with this book.

With all my gratitude,
Shaily Goyal

I am thankful to my wife, Mrs. Parama Ghosh MBA, my daughters, Monica Ghosh MS and Olie Ghosh BS for their constant inspiration, love and affection. I appreciate their constant support and help towards the completion of this book.

I am also thankful to Dr. Jeff Bennert, PhD, CTN, former President and CEO of Air Oasis, Dr. Constantine Saadeh, MD, FICP of Allergy ARTS, USA, and Neil Vitale of Air For Life, UK, for their invaluable contributions to the research and development in nanotechnology and air purification.

Publication of this book would not be possible without the constant effort and hard work by Dr. Shaily Goyal, an eminent researcher and author of many books and articles.

Nabarun Ghosh

Contents

Editors and Contributors

About the Editors

Nabarun Ghosh (Arun) is a former Professor of Biology and Advisor for the Biology Education and Biotech Program at the Department of Life, Earth, and Environmental Sciences at West Texas A&M University (2000–2024). He has taught at several institutions, including the University of North Texas (1992–98), Richland College (1997), and Texas Tech University (Adjunct). He holds two PhD degrees: one in Biology from the University of North Texas, USA, in 1998 and another in Cytogenetics from the University of Calcutta, India, in 1995. He completed a post-doctoral fellowship in Neuroscience at Baylor College of Medicine (1999–2000). He was a CSIR Fellow (Council of Scientific & Industrial Research) and DNES Fellow (Department of Non-Conventional Energy Sources) at the Center of Advanced Study in Cell and Chromosome Research, University of Calcutta, India (1988–1993). His research interests include Nanotechnology, air and water quality, Biotechnology, and Electron Microscopy. He has collaborated for Nanotechnology with Air For Life (UK) and Air Oasis (USA). He also collaborated with USDA; Texas Tech University; Allergy ARTS Clinic, Texas; and scientists from all over India. He has visited and delivered lectures on Nanotechnology at many universities all over the world: USA, UK, UAE, China, India, Italy, and at the NANOTEC, Science Park, Thailand, etc.

He has published over 200 research articles and authored a book titled "What's in Your Air?". He is an active member of several academic bodies and editorial boards, including the European Academy of Allergy and Clinical Immunology (EAACI) and the American Academy of Allergy, Asthma, and Immunology (AAAAI). He has received numerous awards from WTAMU, Texas A&M University system, Texas Society for Microscopy, and Phytopathological Society, India. He has served as the President of the Texas Society for Microscopy. Throughout his career, he has supervised many graduate students and examined over 20 PhD theses as a foreign expert. His work in air quality, allergy, and asthma research has garnered attention from various media outlets, TV stations, newspapers, and journals including *Scientific American* in 2024.

Debajyoti Ghosh (Deb) is a distinguished biomedical researcher and educator with a strong track record of academic and research excellence. He holds a PhD in Allergy/Immunology and has conducted postdoctoral research in France and the United States, focusing on the pathogenesis of complex allergic inflammatory diseases. He is currently a Research Assistant Professor at the University of Cincinnati, Ohio, where his research is supported by US federal and industry-sponsored grants.

Throughout his career, Deb has been recognized for his significant contributions to science, including being elected a Fellow of the American Academy of Allergy, Asthma and Immunology (AAAAI). He has received multiple travel grants and invitations to present in prestigious national and international congresses. Other honors include the DAAD fellowship and several AAAAI Featured Abstract and Travel Grant awards.

Deb's current research leverages omics data analysis to identify gene expression signatures of complex inflammatory diseases, resulting in numerous peer-reviewed publications and a patent granted in 2022. In addition, his team is actively working to identify omics signatures associated with hazardous environmental and occupational exposures, aiming to advance precision medicine.

Shaily Goyal earned her PhD in Plant Biotechnology from Mohanlal Sukhadia University, India, where she contributed as a Research Associate in the Laboratory of Biomolecular Technology, funded by CSIR, India. During her time there, she also served as a Teaching Assistant, delivering lectures and assisting in courses across Biology, Botany, and Biotechnology.She further advanced her academic career by earning a second MS degree from West Texas A&M University, USA. Her research since then has concentrated on air quality, with a particular focus on PM 2.5, aeroallergens and Nanotechnology. She has led and completed numerous air quality analysis projects at West Texas A&M University.Her contributions to academia include more than 40 research articles published in peer-reviewed journals and book chapters. Additionally, she has co-authored two textbooks tailored for graduate and postgraduate students and has served as the editor of a research-focused book.

Contributors

Tri Partono Adhi Department of Chemical Engineering, Institut Teknologi Bandung, Bandung, Indonesia

D. K. Aswal Health, Safety and Environment Group, Bhabha Atomic Research Centre, Mumbai, India

Prabir K. Banerjee Department of Gastroenterology, RN Tagore Hospital, Kolkata, India

Marcio Peron Franco de Godoy Universidade Federal de São Carlos – UFSCar, São Carlos, Brazil

Julia Eilers University of Cincinnati College of Medicine, Cincinnati, OH, USA

Marwa Farouk Elkady Chemical and Petrochemical Engineering Department, Egypt-Japan University for Science and Technology (E-JUST), New Borg El-Arab City, Alexandria, Egypt

Ahmed M. El-Khawaga Department of Basic Medical Sciences, Faculty of Medicine, Galala University, Suez, Egypt

Yosri A. Fahim Department of Basic Medical Sciences, Faculty of Medicine, Galala University, Suez, Egypt

Debajyoti Ghosh Division of Rheumatology, Allergy, and Immunology, Department of Internal Medicine, University of Cincinnati College of Medicine, Cincinnati, OH, USA

Nabarun Ghosh Department of Life, Earth & Environmental Sciences, West Texas A&M University, Canyon, Texas, USA

Shaily Goyal Department of Life, Earth & Environmental Sciences, West Texas A&M University, Canyon, Texas, USA

Jayanta Gupta Department of Health Sciences, Marieb College of Health & Human Services, Florida Gulf Coast University, Fort Myers, FL, USA

Ibrahim W. Hasani Department of Pharmaceutics, Faculty of Pharmacy, Al-Shamal Private University, Idlib, Syria

Hassan Shokry Hassan Environmental Engineering Department, Egypt-Japan University for Science and Technology (E-JUST), New Borg El-Arab City, Alexandria, Egypt

Yong He Key Laboratory of Optoelectronic Technology and Systems of the Education Ministry of China, College of Optoelectronic Engineering, Chongqing University, Chongqing, China

State Key Laboratory of Coal Mine Disaster Dynamic and Control, Chongqing University, Chongqing, China

Aubrey D. Howard Science, Roswell High School, Roswell, NM, USA

Chun Huang Key Laboratory of Optoelectronic Technology and Systems of the Education Ministry of China, College of Optoelectronic Engineering, Chongqing University, Chongqing, China

Wen-Min Huang Department of Physics, National Chung Hsing University, Taichung, Taiwan

Antonius Indarto Department of Chemical Engineering, Institut Teknologi Bandung, Bandung, Indonesia

Department of Bioenergy Engineering and Chemurgy, Institut Teknologi Bandung, Sumedang, Indonesia

Manish Joshi Health, Safety and Environment Group, Bhabha Atomic Research Centre, Mumbai, India

Utkarsh Kumar Department of Physics, National Chung Hsing University, Taichung, Taiwan

Hangyu Liu Key Laboratory of Optoelectronic Technology and Systems of the Education Ministry of China, College of Optoelectronic Engineering, Chongqing University, Chongqing, China

Wan Hanna Melini Wan Mohtar Department of Civil Engineering, Faculty of Engineering and Built Environment, Universiti Kebangsaan Malaysia, Bangi, Selangor, Malaysia

Environmental Management Center, Institute of Climate Change, Universiti Kebangsaan Malaysia, Bangi, Selangor, Malaysia

Marwa Nabil Advanced Technology and New Materials Research Institute, City of Scientific Research and Technology Applications (SRTA-City), New Borg El-Arab City, Alexandria, Egypt

Yina Julieth Onofre Universidade Federal de São Carlos – UFSCar, São Carlos, Brazil

Saurabh Kumar Pandey Sensors and Optoelectronics Research Group (SORG), Department of Electrical Engineering, Indian Institute of Technology Patna, Patna, Bihar, India

Yano Surya Pradana Department of Chemical Engineering, Institut Teknologi Bandung, Bandung, Indonesia

Department of Chemical Engineering, Faculty of Engineering, Universitas Gadjah Mada, Yogyakarta, Indonesia

Deepak Punetha Department of Electronics and Communication Engineering, Motilal Nehru National Institute of Technology, Allahabad, Prayagraj, Uttar Pradesh, India

Raghvendra Department of Electronics and Communication Engineering, SRM University, Amaravati, Andhra Pradesh, India

Mahmoud Samy Public Works Engineering Department, Faculty of Engineering, Mansoura University, Mansoura, Egypt

B. K. Sapra Health, Safety and Environment Group, Bhabha Atomic Research Centre, Mumbai, India

Vivek Pratap Singh Sensors and Optoelectronics Research Group (SORG), Department of Electrical Engineering, Indian Institute of Technology Patna, Patna, Bihar, India

Yohanes Andre Situmorang Department of Chemical Engineering, Parahyangan Catholic University, Bandung, Indonesia

Jay Vitale Air for Life UK LTD, Milton Keynes Business Centre, Milton Keynes, UK

Chiu-Hsien Wu Department of Physics, National Chung Hsing University, Taichung, Taiwan

Institute of Nanoscience, National Chung Hsing University, Taichung, Taiwan

B. C. Yadav Nanomaterials and Sensor Research Laboratory, Department of Physics, Babasaheb Bhimrao Ambedkar University, Lucknow, India

Brijesh Yadav Pulmonary Pathogenesis and Immunotoxicology Laboratory, Department of Environmental and Public Health Sciences, Division of Environmental Genetics and Molecular Toxicology, University of Cincinnati College of Medicine, Cincinnati, OH, USA

Department of Nephrology and Renal Transplantation, Sanjay Gandhi Postgraduate Institute of Medical Sciences, Lucknow, Uttar Pradesh, India

Jagjit S. Yadav Pulmonary Pathogenesis and Immunotoxicology Laboratory, Department of Environmental and Public Health Sciences, Division of Environmental Genetics and Molecular Toxicology, University of Cincinnati College of Medicine, Cincinnati, OH, USA

Yuanting Yu Key Laboratory of Optoelectronic Technology and Systems of the Education Ministry of China, College of Optoelectronic Engineering, Chongqing University, Chongqing, China

Chapter 1
Nanotechnology: An Emerging Field for Air Quality Sustainability

Chun Huang, Hangyu Liu, Yuanting Yu, and Yong He

Abstract Air quality is closely related to the concentration of pollutant gases (NO_2, SO_2, CO, O_3, etc.) in the environment. Excessive pollutant gases can adversely affect both the environment and human health, so efficient gas detection is crucial for human health and supporting air quality sustainability. Conventional gas detection methods generally suffer from various shortcomings, such as susceptibility to environmental factors, high operating temperatures, and low selectivity, hindering the further development of gas detection technology. With the emergence of nanotechnology, the construction of nano gas sensors is considered a promising solution to address the limitations of conventional detection techniques. The research on nano gas sensors focuses on the design and synthesis of nanomaterials, which are expected to exhibit superior gas detection capabilities owing to their distinctive structural characteristics and enhanced surface reactivity at the nanoscale. In general, the involvement of nanomaterials in gas detection includes nanosorbents for the sampling process and nano-gas-sensing materials for the sensing process. In this chapter, we briefly introduce the hazards and sources of environmental pollutant gases, summarize the nanosorbents and nano-sensing materials used for gas detection, and the synthesis methods of nanomaterials.

Keywords Nanomaterial · Gas sampling · Gas sensing · Nanomaterial synthesis · Environmentally hazardous gases

C. Huang · H. Liu · Y. Yu
Key Laboratory of Optoelectronic Technology and Systems of the Education Ministry of China, College of Optoelectronic Engineering, Chongqing University, Chongqing, China

Y. He (✉)
Key Laboratory of Optoelectronic Technology and Systems of the Education Ministry of China, College of Optoelectronic Engineering, Chongqing University, Chongqing, China

State Key Laboratory of Coal Mine Disaster Dynamic and Control, Chongqing University, Chongqing, China
e-mail: yonghe@cqu.edu.cn

N. Ghosh et al. (eds.), *Nanotechnology in Air Quality Management*, Sustainable Development and Biodiversity 39, https://doi.org/10.1007/978-981-96-8339-0_1

1.1 Introduction

With the improvement of people's standard of living, air quality has been widely concerned because of the great influence of polluting gases on the environment and human health. Air quality index (AQI) is commonly used to reflect air quality conditions by simplifying the concentrations of several pollutants into a conceptual exponential value. Several pollutants that are used to gain an AQI include the fine particulate matter (PM2.5), respirable particulate matter (PM10), nitrogen dioxide (NO_2), sulfur dioxide (SO_2), carbon monoxide (CO), and ozone (O_3) (Yang et al. 2017). Furthermore, comprehensive environmental monitoring also requires the detection of some other atmospheric pollutants, including harmful gases like hydrogen sulfide (H_2S) and ammonia (NH_3), greenhouse gases such as carbon dioxide (CO_2) and methane (CH_4), and volatile organic compounds (VOCs), all of which significantly harm the environment. (Zhou et al. 2023). The emission of environmentally detrimental gases originates from both anthropogenic sources and natural processes, posing significant threats to ecological systems and public health.

Consequently, atmospheric monitoring of hazardous gas concentrations provides a critical metric for evaluating air quality standards and assessing associated health risks. Existing gas detection methods have various limitations that hinder their wide application in the field of hazardous gas detection. The emergence of nanotechnology provides a solution to overcome these limitations. The key to nanotechnology lies in the study and preparation of nanomaterials, which exhibit unique physicochemical properties, including mechanical, chemical, electrical, and mechanical properties that have great potential for gas detection. This chapter focuses on the application of nanomaterials in detecting hazardous atmospheric gases and their mechanisms. It summarizes the synthesis methods for nanomaterials (Fig. 1.1) and analyzes the opportunities and challenges of nanotechnology in promoting sustainable air quality.

1.1.1 Environmentally Hazardous Gases

Various types of environmentally hazardous gases are the main causes of air quality deterioration. These gases originate from a variety of sources and cause a wide range of environmental and human health problems (Fig. 1.2). The main environmentally hazardous gases are nitrogen oxides (NO_x), carbon oxides (CO_x), SO_2, H_2S, NH_3, and VOCs. These dangerous gases are emitted into the atmosphere naturally or anthropogenically. In addition to endangering the environment and the human body as primary pollutants, they also react with other substances to produce secondary pollutants such as O_3 (under the conditions of light, oxygen, etc.), which also adversely affect the environment and human health.

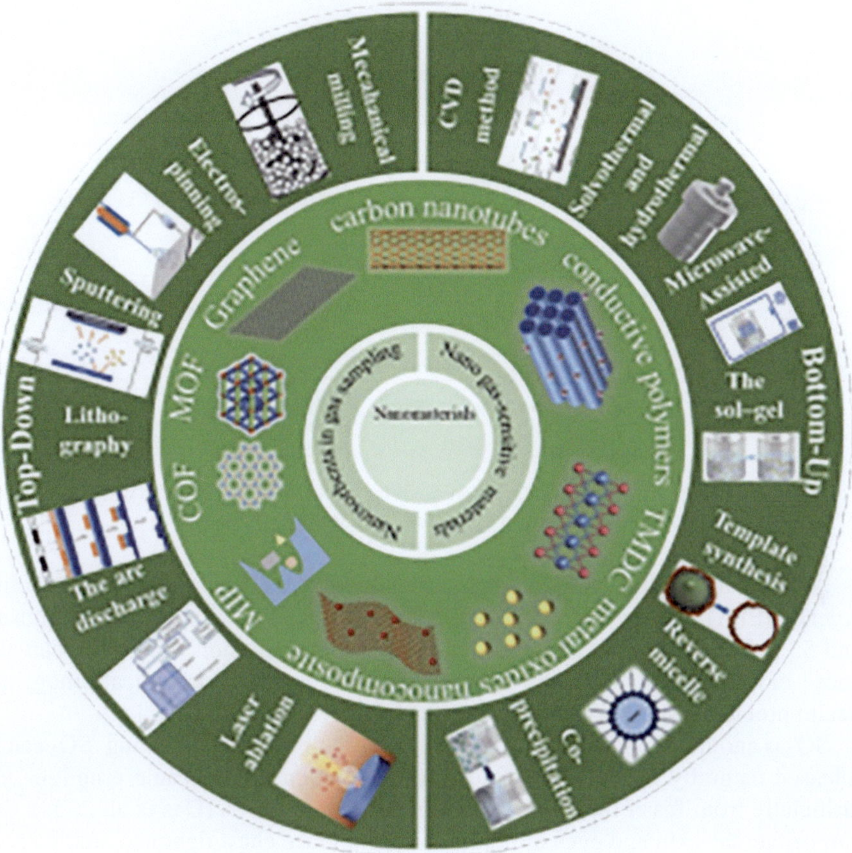

Fig. 1.1 Schematics illustrating nanomaterials in gas sampling and detection and their different synthesis methods

NO_x are common toxic gases mainly from human activities, such as industrial and automotive emissions, but are also released into the air from natural sources, such as forest fires caused by high temperatures or lightning (Atkinson 2000). NO_2, the predominant NO_x species, is a significant atmospheric pollutant with substantial environmental and public health implications. Excessive emissions of NO_2 contribute to multiple environmental phenomena, including photochemical smog, acid rain, and the formation of ground-level ozone, thereby inducing secondary pollution (Ahmadi et al. 2015). From a public health perspective, NO_2 exposure adversely affects multiple physiological systems, particularly causing dermatological, ocular, and respiratory complications in humans. Chronic exposure to concentrations as low as 1 ppm can induce pulmonary tissue damage, potentially leading to severe outcomes such as lung lesions, pulmonary edema, and mortality (Wang et al. 2021).

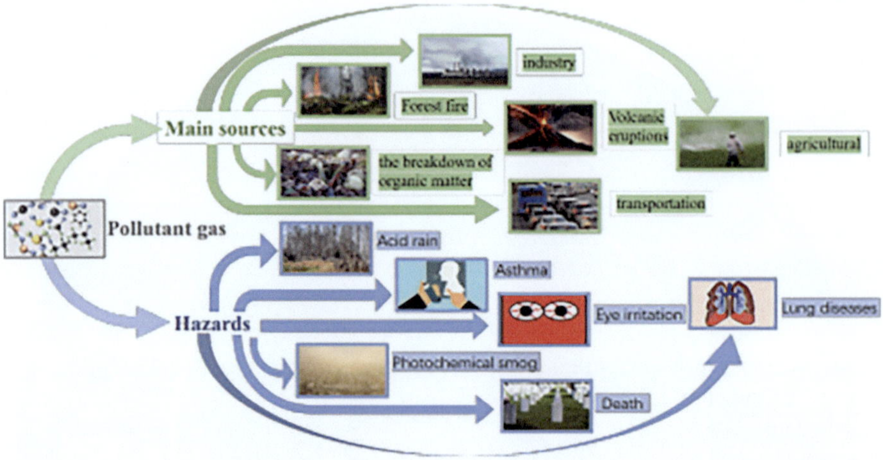

Fig. 1.2 The main sources of air-polluting gas and their hazards. The schematic was created using public domain images obtained from an open-source database

Ambient air quality standards established by the U.S. Environmental Protection Agency (EPA) specify maximum NO_2 concentrations of 100 ppb for 1-h exposure and 53 ppb for annual mean levels. In addition, nitrogen monoxide (NO) can also cause environmental problem, such as acid rain and photochemical smog, and health problems, such as asthma and heart failure (Taufik et al. 2021).

SO_2 is another representative toxic gas that is colorless and irritating. SO_2 can be released naturally from the breakdown of vegetation and volcanic eruptions and artificially from fossil fuel combustion, metal smelting, etc. (Das et al. 2008). SO_2 can irritate the skin and eyes of human beings, induce chest tightness, and damage the respiratory system, and the exposure limit of SO_2 to the human respiratory system is 5 ppm (Lee et al. 2011). In addition to health concerns, SO_2 can also affect the environment via the formation of acid rain and soil acidification. The existing SO_2 standard established by the U.S. EPA is 75 ppb. Also, SO_2 serves as a primary precursor for sulfate aerosols, constituting a significant fraction of respirable particulate particles in the atmosphere (Wetchakun et al. 2011).

H_2S is also a colorless environmental pollutant with a rotten egg odor, originating from both natural and anthropogenic sources. Natural emissions primarily derive from geothermal activities (hot springs and volcanic emissions), hydrocarbon reservoirs (crude petroleum and natural gas), and microbial decomposition of organic sulfur compounds. Industrial sources include food processing, craft paper mills and petroleum refineries (Gong et al. 2006). H_2S is a highly corrosive and toxic gas that can irritate human skin and eyes even at a low concentration. When the concentration of H_2S exceeds the short-term exposure limit value of 15 ppm, the human body suffers from olfactory fatigue, and the ability of the human nose to detect H_2S decreases (Ramgir et al. 2013). Besides, as a chemical asphyxiant, H_2S can endanger the nervous and metabolic systems of the human body, causing people to lose consciousness at a very low concentration.

NH_3, a colorless air pollutant with a pungent odor, primarily originates from the breakdown of organic matter. The presence of excess NH_3 in the atmosphere poses significant environmental and public health risks to humans. Human exposure to concentrations exceeding 25 ppm can cause immediate eye irritation, repeated coughing, acute poisoning, and even death (Gangopadhyay and Das 2008).

CO_x are also an important class of environmentally hazardous gases, including CO_2 and CO. CO_2 is colorless, odorless, and the most important greenhouse gas. Sources of carbon dioxide include the respiration of creatures and the burning of fossil fuels. Excessive carbon dioxide can lead to oxygen deprivation, respiratory distress, and even asphyxiation (Wetchakun et al. 2011). Growing levels of carbon dioxide aggravate the greenhouse effect and cause global warming. CO is a toxic, colorless, and odorless gaseous byproduct of the incomplete combustion of carbon-containing substances. It prevents the bloodstream from absorbing oxygen and causes damage to the body by reducing cellular respiration, and high levels of carbon monoxide can lead to death. The exposure limit for CO is 10 ppm.

O_3 in the context of environmental pollutants refers to tropospheric O_3, also known as ground-level O_3, which is typically a photochemical reaction of VOCs with NO_x (Ziegler et al. 2018). High concentrations of ozone can inhibit photosynthesis, affect plant growth, and alter plant cover and health (Singh et al. 2009). O_3 can also cause eye irritation and damage the human respiratory system (Rowe et al. 1983). The newest standard for O_3, established by the U.S. EPA is 70 ppb.

VOCs comprise a wide range of chemicals. Typical VOCs include ethanol, formaldehyde, benzene, and acetone. Various activities in daily life contribute to VOCs' emissions, such as tobacco production, printing, photocopying, use of paints, detergents, etc. Exposure to VOCs can irritate the eyes, nose, and throat and cause breathing difficulties. When inhaled, VOCs can damage the central nervous system and other organs of the body. Some highly toxic VOCs can also cause cancer (Popoola et al. 2023).

The presence of these environmentally hazardous gases presents significant dual threats to public health and environmental quality. Accurate and efficient detection of environmentally hazardous gases is an important part of maintaining sustainable air quality.

1.1.2 Gas Detection Technologies Based on Sensors

As one of the gas detection technologies, sensors offer advantages such as speed, portability, selectivity, and sensitivity compared to conventional analytical methods such as mass spectrometry and chemical analysis. In order to achieve effective detection of environmentally hazardous gases, the gas-sensitive properties of a variety of gas sensors have been investigated, ranging from optical to electrochemical sensors. Among optical sensors, fiber optic and photonic crystal-based devices have demonstrated particular promise. These sensors operate based on light propagation modulation principles (Bogue 2015). Fiber optic-based sensors use optical fibers to measure changes in optical properties (i.e., wavelength) and detect analytes on the sensing layer. This configuration offers excellent sensitivity, environmental stability, and an extended operational lifespan, although miniaturization presents technical challenges. Photonic crystal sensors employ precisely engineered periodic structures of dielectric materials with varying refractive indices, offering unique light manipulation capabilities for gas detection applications (Li et al. 2023).

In acoustic gas-sensing applications, two types of acoustic devices are commonly used (Kumar and Prajesh 2022). Bulk acoustic wave (BAW) devices operate through acoustic wave propagation within bulk materials, including quartz crystal microbalance (QCM) and thin-film bulk acoustic wave resonators (FBAR). In contrast, surface acoustic wave (SAW) devices confine wave propagation to the piezoelectric material's surface. QCM sensors exploit the piezoelectric properties of quartz crystals, where the applied voltage induces resonant frequency shifts proportional to the surface mass changes resulting from gas molecule adsorption/desorption on functionalized surfaces. The sensor's performance is predominantly determined by the physicochemical characteristics of the sensing materials deposited on the active electrodes (Fauzi et al. 2021). This operational principle enables QCM's widespread application in selective gas detection across diverse target analytes. FBAR devices share fundamental operational similarities with QCM, but differ structurally through the implementation of piezoelectric thin films instead of quartz crystals. This architectural modification permits operation at significantly higher resonance frequencies (Zhang et al. 2018b). Both technologies rely heavily on surface adsorption phenomena, with the sensor response being largely dependent on gas-surface interactions. SAW devices function through surface-confined acoustic wave propagation in elastic substrates, exhibiting a characteristic exponential amplitude decay with increasing depth. This phenomenon creates a high surface energy density, enabling gas detection through modifications in wave propagation boundary conditions. The deposition of thin, gas-sensitive films facilitates selective molecular adsorption, thereby altering the surface wave characteristics (Jakubik 2011).

Calorimetric gas sensors use calorimetry as a transducing principle to measure the change in heat caused by the combustion of combustible gases in a catalytic layer. This class of devices is also called "pellistor," "catalytic bead," "catalytic gas sensor," "combustible gas sensor," or "thermometric gas sensor". The catalytic layer's activity, which facilitates efficient analyte combustion, serves as the primary determinant of the sensor performance characteristics (Cavicchi 2011).

Electrochemical gas sensors function by measuring the electrical current variations induced by the oxidation or reduction reactions of the target gas (Ali et al. 2022). The working electrode, serving as the primary reaction interface, represents the most critical component of the sensor architecture. To enhance the sensor performance parameters, particularly selectivity and sensitivity, the electrode surface is typically functionalized with a chemically stable material (Saputra 2023).

Resistive gas sensors operate through measurable changes in electrical resistance resulting from surface interactions between the sensing material and target gas molecules (Naganaboina and Singh 2023). The sensing material is the key element of this type of sensor, and in this regard, researchers worldwide are working on the adaptation of existing materials and the development of new sensing materials.

Various types of gas sensors exhibit distinct characteristics in terms of their performance and limitations. While optica-l and acoustic-based sensors demonstrate superior sensitivity and electromagnetic interference resistance, their performance is significantly influenced by environmental conditions, particularly temperature and humidity fluctuations, coupled with higher implementation costs. Calorimetric gas, though characterized by operational stability and extended service life, is restricted to combustible gas detection with relatively low precision. Electrochemical and chemical gas sensors have high sensitivity and a wide linear range; however, they have serious cross-sensitivity problems. These inherent limitations across sensor categories have substantially constrained the advancement of gas detection methodologies. The emergence of nanotechnology offers promising avenues for sensor optimization through atomic-level material manipulation, potentially addressing these persistent challenges.

1.1.3 Nanotechnology for Gas Detection

The term "nanotechnology" was formally introduced by Professor Norio Taniguchi of Tokyo University of Science in his seminal 1974 paper titled "On the Basic Concept of 'Nanotechnology'," presented at the Japan Society for Precision Engineering conference. Taniguchi defined nanotechnology as "the processing of materials through the separation, consolidation, and deformation of individual atoms or molecules" (Sandhu 2006). However, the conceptual foundations of nanotechnology can be traced back to 1959, when Richard Feynman, an American physicist and Nobel Laureate, presented his visionary lecture "Enough room at the bottom" at the California Institute of Technology during an American Physical Society meeting. In this presentation, Feynman posed the hypothesis "Why cannot we write the entire 24 volumes of the Encyclopedia Britannica on the head of a pin?" and described the idea of constructing smaller machines down to the molecular level (Feynman 2011). Although he does not explicitly use the term nanotechnology, he discusses methods for manipulating and controlling atoms and molecules in detail.

The modern era of nanotechnology began in 1981 with the development of scanning tunneling electron microscopy (STEM), which enabled the manipulation of individual atoms at the nanoscale. This revolutionary invention earned IBM scientists Gerd Binnig and Heinrich Rohrer the 1986 Nobel Prize in Physics (Prize 2013). Furthermore, people have been using nanomaterials as far back as the fourth century, when they added gold and silver to glass to create amazing effects (Bayda et al. 2020). Another notable example is the Lycurgus Cup, which exhibits color-changing properties depending on the light source's position (Arrigan 2018). Systematic research on nanoscale manipulation emerged in the late 1960s, facilitated by advancements in micro- and nano-scale instrumentation.

So far, nanotechnology has been involved in the further development of all areas of life, including information technology, energy, medicine, food safety, etc. Similarly, nanotechnology provides solutions for the development of more advanced gas sensors from gas sampling and sensing. Research on nano-gas sensors focuses on the discovery and synthesis of nanomaterials, and the use of nanotechnology enables the design and modulation of material properties at the atomic scale, which makes nanomaterials exhibit unique physical and chemical properties that traditional materials do not have, which may be more conducive to gas detection. In addition to being used as gas-sensitive materials for the direct detection of target gases, nanomaterials can also be used as nano-adsorbents for gas sampling, a pretreatment process for gases, which is also a way to enhance the performance of gas detection. Compared to conventional gas sensors, nano-gas sensors offer a number of key advantages, including low power consumption, high selectivity, low detection limits, ease of integration, and cost efficiency unmatched by other technologies.

1.2 Gas Sampling Based on Nanosorbents

1.2.1 Gas Sampling for Gas Detection

Although simpler and faster gas detection is desirable, sample handling is required in practical environments due to the following facts (Burlachenko et al. 2016): (a) the detection target is always present in the mixture, (b) low concentration of the target molecule, and (c) presence of dominant components (e.g., water) in the mixture. Therefore, in the area of gas detection, optimal sample handling technology is of great importance for high reliability gas detection, in particular, identifying target molecules in multicomponent mixtures. On the one hand, it is conducive to reproducing the experimental conditions, which is a prerequisite for obtaining accurate detection results; on the other hand, it provides a relatively safe working environment for the sensor (avoiding damage by non-target gases), helping to prolong the life of the sensor elements.

There are two principal methods of gas sampling: sampling with or without enrichment (Hebisch et al. 2009). The gas bags, canisters, or cylinders are used in sampling without enrichment (Arrhenius et al. 2016). But it can only collect a limited volume of samples and may not effectively collect enough trace compounds for reliable detection. Moreover, the loss of gas samples during storage and transportation may lead to a change in the gas composition, causing distortion of the detection results. Direct onsite enrichment sampling overcomes these disadvantages. Sampling with enrichment involves adsorbing the target substance onto a solid or absorbing it into a solution to support subsequent detection and analysis (Lecharlier et al. 2022).

1.2.2 Nanosorbent-Based Solid-Phase Microextraction (SPME)

SPME is a typical gas enrichment technique developed in 1990 (Arthur and Janusz 1990). SPME procedure involves the adsorption of the target analyte onto the surface of the coated fiber. After extraction, the adsorbed analytes are desorbed for analysis and determination (Murtada 2020). SPME has become one of the most popular sample enrichment techniques because of its rapidity, simplicity, greenness (no additional solvents are required), and convenience of integration into portable instruments. The coating material of SPME is a key component of extraction efficiency and selectivity. The coating materials that are currently commercialized are polydimethylsiloxane (PDMS), polyacrylate (PA), polydimethylsiloxane divinylbenzene (PDMS/DVB), carboxen-polydimethylsiloxane (CAR/PDMS), carbowax-polyethylene glycol (CW/PEG), etc. (Portillo-Castillo et al. 2018). However, they generally have the limitations of a relatively low operating temperature range, easily depleted coatings, and a short service life.

The enhanced extraction efficiency can be attributed to the high surface area and porosity of the nanomaterials. These materials demonstrate exceptional mechanical, chemical, and thermal stability owing to their unique structural properties, making them ideal candidates for SPME coatings (Jagirani and Soylak 2020). The development of advanced SPME devices utilizing nanosorbents is particularly promising, as they exhibit superior extraction capabilities compared to conventional materials.

Headspace SPME (HS-SPME) is one of the most widely used modes of SPME, in which analytes from the HS phase equilibrate with the sample. This approach offers distinct advantages by minimizing matrix interference from non-volatile impurities while simultaneously preserving the integrity of the extraction coatings. Particularly effective for volatile (VOC) and semi-volatile organic compound (SVOC) analysis, HS-SPME demonstrates exceptional detection sensitivity (Lancioni et al. 2022). Technically, sampling from a gaseous matrix is very similar to sampling from the headspace. The analytical performance of HS-SPME is predominantly governed by the physicochemical properties of the fiber's extraction phase. The main nanosorbent materials applied to HS-SPME are carbon-based nanomaterials, metal–organic frameworks (MOFs), covalent organic frameworks (COFs), molecularly imprinted polymers (MIPs), and their complexes. Table 1.1 summarizes the application of nanomaterials in HS-SPME over the last five years (2019–2023).

1.2.2.1 Carbon-Based Nanosorbents

Carbon-based nanomaterials, including carbon, carbon nanotubes (CNTs), graphene, and their functionalized derivatives, have been extensively utilized as sorbents in solid-phase microextraction (SPME). These materials are employed independently or in composite forms with other materials (Ahmadi et al. 2017; Jing

Table 1.1 Recent applications of nanomaterials for HS-SPME

Material	Analyte	Linear range	LOD	RSD (%)		References
				Repeatability	Reproducibility	
TAPT-OMe-PDA	PAHs	2–5000 ng L^{-1}	0.11–1.6 ng L^{-1}	7.1–9.8	3.3–6.1	Liu et al. (2024)
Aniline-modulated TAPB-BTCA-COFs	PAHs	0.05–200 ng g^{-1}	0.06–0.17 ng g^{-1}	3.2–5.0	4.4–8.4	Yang et al. (2020)
DB-modulated TAPB-DMTP-COF	PBBs	0.25–5000 ng L^{-1}	0.04–0.28 ng L^{-1}	1.3–7.7	6.1–7.8	Zhou et al. (2023a)
TFPB-BD	PCBs	1–1000 ng L^{-1}	0.07–0.35 ng L^{-1}	3.8–8.8	5.3–9.8	Guo et al. (2019b)
CF–COF	PCBs	0.01–1000 g L^{-1}	1.5–8.8 ng L^{-1}	3.94–9.26	1.32–9.33	Su et al. (2021)
OH-TPB-COFs	PAEs	1–100μg L^{-1}	32–451 ng L^{-1}	0.83–4.67	3.08–9.73	Guo et al. (2019a)
CC-QP	PAHs	0.1–100μg L^{-1}	0.03–0.19μg L^{-1}	3.3–7.4	5.3–9.9	Wang et al. (2020)
TPT-COF	PAEs	0.01–100μg L^{-1}	0.01–0.1μg L^{-1}	0.83–4.72	3.08–10.9	Guo et al. (2019)
TAPB-TMC-COF	PAHs	0.2–2000 mg L^{-1}	0.29–0.94μg L^{-1}	4.3–8.4	8.5–11.0%	Yang et al. (2019)
TpBD-COFs	PAEs	0.004–50μg L^{-1}	1–430 ng L^{-1}	4.1–9.4	6.8–10.8	Yue et al. (2020)
TFPA-TAPP-COF	PAHs	0.1–50μg L^{-1}	6–24 ng L^{-1}	3.1–8.50	5.79–9.98	Tian et al. (2020)
GO/SNW-1	PAEs	0.05–100μg L^{-1}	0.01–0.50μg L^{-1}	5.9–8.3	8.2–10.8	Mohammad Mahdi et al. (2020)
IRMOF-8	TATP	50–5000 ng mL^{-1}	13 ng mL^{-1}	4.1	8.0	Xie et al. (2023)
ZNPC	Phenols	5–5000 ng g^{-1}	0.73–2.3 ng g^{-1}	2.8–7.3	9.7–11.7	Yan et al. (2021)

MIL-101	cVMS	0.5–60 ng mL^{-1}	4–6 pg mL^{-1}	–	–	Zhang et al. (2020b)
MIL-101-NH_2-derived NPC	BTEX	5–5000 ng L^{-1}	6–52 pg L^{-1}			
SOM-ZIF-8	BTEX	0.004–50μg L^{-1}	1.0–12 ng L^{-1}	–	5.6–6.7	Maya and Ghani et al. (2019)
MOF/COF composite	PAHs	0.2–50.0 ng mL^{-1}	0.02–0.07 ng mL^{-1}	2.25–6.69	3.82–7.05	Yu et al. (2023)
CIM-80(Al)	Cyclic methylsiloxanes	2–200μg·L^{-1}	–	–	–	González-Hernández et al. (2021)
	Musk fragrances	0.1–20μg·L^{-1}				
SiO_2@MOF-199	PAEs	5–1000 ng L^{-1}	0.3–19.8 ng L^{-1}	4.7–8.5	7.2–11.3	Jiang et al. (2022)
UiO-66/MoS_2	PAHs	0.5–10,000 ng kg^{-1}	0.11–1.40 ng kg^{-1}	3.1–8.2	4.4–8.5	Yuan et al. (2019)
Hollow ZIF-7	Volatile biomarkers	1–500μg L^{-1}	0.07–0.53μg L^{-1}	2.8–7.5	4.2–14.6	Jiang et al. (2021)
NH_2-UiO-66(Zr)-shp	Organic ultraviolet filters	10–50,000 ng g^{-1}	0.2–2.1 ng g^{-1}	2.3–6.0	1.8–5.8	Liu et al. (2020b)
MAFs	PAHs	0.01–100μg L^{-1}	0.1–7.5 ng L^{-1}	0.23–4.2	2.9–8.3	Liu et al. (2019)
PEDOT/UiO-66	VOCs	0.09–100 ng mL^{-1}	0.03–0.06 ng mL^{-1}	2.6–12.0	3.4–12.3	Zhang et al. (2020)
DES–GA	PCNs	1–1000 pg g^{-1}	0.00983–0.0661 pg g^{-1}	9.51–15.8	15.7–20.5	Wang et al. (2022)

(continued)

Table 1.1 (continued)

Material	Analyte	Linear range	LOD	RSD (%)		References
				Repeatability	Reproducibility	
$MWCNTs/MnO_2$/ PEDOT	PAHs	0.5–250 ng·g^{-1}	0.1–0.8 ng·g^{-1}	2.0–8.9	6.0–10	Ma et al. (2019)
nHA/mCNTs	PAHs	0.01–0.5μg ml^{-1}	2.6–2.9 ng ml^{-1}	–	–	Ghorbani et al. (2023)
Carbon nanospheres (CNs)	PAHs from water sample	0.5–300 ng mL^{-1}	0.12–0.45 ng mL^{-1}	6.9–9.1	6.9–11.1	Chang et al. (2020)
	PAHs from solid sample	6.0–2700 ng g^{-1}	0.12–0.45 ng mL^{-1}	5.8–10.2	7.9–11.2	
$CoNi_2S_4$/GPL	PAHs	00001–10μg L^{-1}	0.1–30 mg L^{-1}	3.8–9.1	8.3–13.1	Piryaei et al. (2023)
BN@rGO	PAHs from water sample	0.25–50 ng mL^{-1}	0.05–0.15 ng mL^{-1}	6.6–7.3	8.7–11	Li et al. (2021)
	PAHs from solid sample	1.0–400 ng g^{-1}	0.3–0.5 ng g^{-1}	4.8–7.2	7.8–11	

TAPT 2,4,6-tri(4-aminophenyl)-1,3,5-triazine, *OMe-PDA* 2,5-dimethoxy-phenyl-1,4-diformaldehyde, *PAHs* polycyclic aromatic hydrocarbons, *TAPB* 1,3,5-Tris (4-aminophenyl) benzene, *BTCA* benzene-1,3,5-tricarbaldehyde, *DB* 2,5-dimethoxybenzaldehyde, *DMTP* 2,5-dimethoxyterephaldehyde, *PBBs* polybrominated biphenyls, *TFPB* 1,3,5-tris-(4-formylphenyl)benzene, *BD* benzidine, *PCBs* polychlorinated biphenyls, *TPB* 2,4,6-triphenoxy-1,3,5-benzene, *PAES* phthalate esters, *CC-QP* cyanuric chloride-*p*-quaterphenyl, *ZNPC* MOF-derived zinc and nitrogen co-doped porous carbon, *PANI* polyaniline, *BTEX* benzene series compounds, *NPC* nanoporous carbon, CF–*COF* chlorine-functionalized covalent organic framework, *PAEs* phthalic acid esters, *BaP* benzo[a] pyrene, *DES-GA* graphene aerogel modified with DES, *PCNs* polychlorinated naphthalenes, *Shp* superhydrophobic, *MAFs* metal azolate frameworks, *cVMs* cyclic volatile methylsiloxanes, *MWCNTs/*MnO_2 multi-walled carbon nanotubes/manganese dioxide, *GPL* graphenized pencil lead, *BN@rGO* h-BN functionalized reduced graphene oxide material, *ANOCs* aromatic and organophosphorus compounds fet al. 2021). Graphene, a two-dimensional carbon allotrope, consists of a single layer of sp^2-hybridized carbon atoms arranged in a hexagonal lattice, serving as the fundamental structural unit for all carbon-based nanomaterials. CNTs are rolled graphene sheets, including single-wall CNTs (SWCNTs) and multiple-wall CNTs (MWCNTs), formed by curling single-layer and multilayer graphene, respectively. Both CNTs and graphene exhibit exceptional properties for SPME applications, including a large surface area, versatile functionalization capabilities, delocalized π-electron system, and superior mechanical and chemical stabilities. However, the adsorption capacity of CNTs is limited by steric hindrance, which prevents the inner layers from participating in the adsorption process (Liu et al. 2012). Despite this limitation, CNT-based composites demonstrate enhanced an extraction efficiency. For instance, a nanocomposite comprising hydroxyapatite nanoparticles (nHA) and modified carbon nanotubes (mCNTs) showed remarkable thermal stability and an extended lifespan in HS-SPME (Ghorbani et al. 2023). Similarly, a porous nano-composite coating of multi-walled carbon nanotubes/manganese dioxide/poly (3,4-ethylenedioxythiophene) (MWCNTs/MnO2/PEDOT) exhibited superior extraction capability due to its large specific surface area (Ma et al. 2019).

Unlike CNTs, graphene offers superior adsorption capacity as both surfaces are available for adsorption processes. The accessibility of adsorption sites on both sides of graphene layers enables a higher degree of functionalization. However, the practical application of pristine graphene in solid-phase microextraction (SPME) is limited by its poor dispersibility in common solvents, which results from strong van der Waals interactions and pronounced hydrophobic effects between its carbon layers. In contrast, graphene derivatives, particularly graphene oxide (GO) and reduced graphene oxides (rGO), have gained greater popularity in SPME applications. GO is synthesized through graphite oxidation and retains the fundamental structure of graphene while incorporating numerous reactive oxygen-containing functional groups. These oxygen functional moieties provide GO with many additional useful properties, making it an ideal adsorbent for SPME. rGO is prepared by reducing graphene oxide, and upon reduction, most of these groups disappear, making rGO an ideal adsorbent for nonpolar analytes. Graphene and graphene derivatives are often modified to increase selectivity for analytes and thus improve extraction efficiency (Kori et al. 2023). rGO was modified by hexagonal boron nitride to form a loose porous 3D frame structure that is conducive to improving its adsorption capability (Li et al. 2021). Similarly, GO modified with deep eutectic solvents develops a three-dimensional porous structure with high enrichment factors and thermal stability, demonstrating excellent suitability for SPME applications (Wang et al. 2022).

Carbon nanomaterials exhibit significantly distinct properties depending on their nanostructural configurations. Among these, carbon nanospheres (CNs) have attracted substantial research interest owing to their tunable size distribution and exceptional monodispersity. When employed as solid-phase microextraction (SPME) coating materials, CNs demonstrate superior analytical performance, including enhanced sensitivity, broad linear dynamic range, excellent extraction efficiency, and remarkable thermal stability. (Chang et al. 2020).

1.2.2.2 MOFs-Based Nanosorbents

MOFs are crystalline porous materials characterized by ultra-high porosity and exceptionally large specific surface areas. These structures are formed through the coordination bonding between inorganic metal ions (or clusters) and organic linkers. In MOF architecture, metal ions serve as nodes or coordination centers, while organic ligands function as bridging connectors, collectively forming intricate two-dimensional (2D) or three-dimensional (3D) networks (Gutiérrez-Serpa et al. 2019). MOFs exhibit the highest known specific surface areas among porous materials, ranging from approximately 150–7000 $m^2\ g^{-1}$ (Ding and Yazaydin 2013), making them exceptional candidates for solid-phase microextraction (SPME) adsorbents. SPME devices with MOF coating possess higher sensitivity and linear range, performing higher extraction efficiency than commercial coatings. A notable application was demonstrated by González-Hernández et al., who successfully employed the aluminum-based MOF CIM-80(Al) as an extraction phase for the simultaneous targeted extraction and determination of volatile methylsiloxanes and musk fragrances from aqueous samples (González-Hernández et al. 2021). Experimental results demonstrated superior extraction performance of the MOF-based coating in volatile methylsiloxanes analysis. Comparative analysis with commercial coatings showed that the MOF-based coating achieved lower limits of detection and demonstrated improved precision in analytical measurements.

A novel MOF-199-coated fiber was fabricated through an in situ solvothermal approach on etched stainless-steel substrates, as illustrated in Fig. 1.3a (Omarova

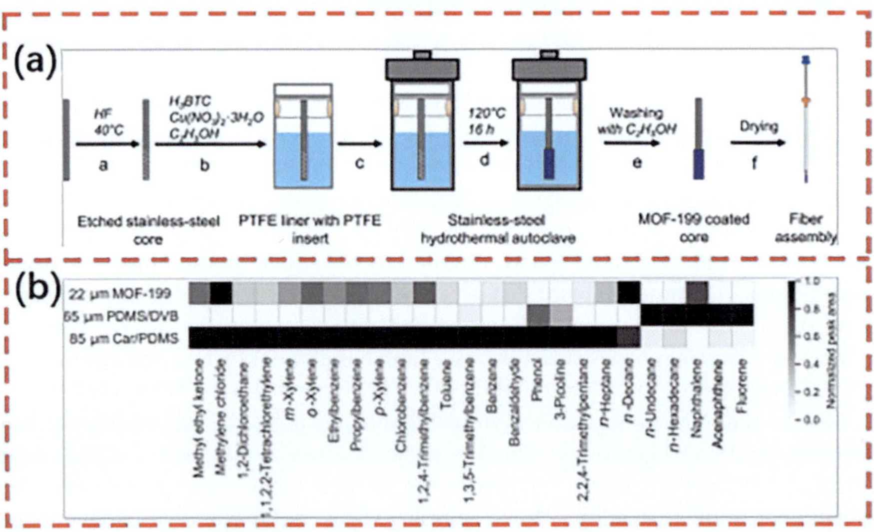

Fig. 1.3 (**a**) Schematic illustration of the in situ synthesis method applied to the manufacture of a MOF-199-based SPME fiber. (**b**) Normalized responses of 25 VOCs were obtained using the new MOF 199 (in situ, 16 h), PDMS/DVB, and Car/PDMS fibers for air samples spiked with analytes. (Reprinted with permission from Omarova et al. (2021))

et al. 2021). The developed coating demonstrated enhanced molecular recognition capabilities, exhibiting superior selectivity for low molecular weight alkanes compared to their higher molecular weight counterparts (Fig. 1.3b).

In addition, the functionalization strategies of MOF can further improve its performance. A recent study developed a superhydrophobic MOF by modifying amino-functionalized UiO-66(Zr) with phenylsilane groups (Liu et al. 2020b). This modification resulted in superior extraction capabilities for five semi-volatile UV filters, as demonstrated in Fig. 1.4a. Remarkably, the phenylsilane-modified coating maintained exceptional thermal stability when exposed to 100 °C water vapor for 21 h, showing negligible efficiency reduction compared to unmodified NH2-UiO-66(Zr) (Fig. 1.4b). This stability enhancement originates from the hydrophobic protective layer that effectively blocks water molecule penetration, thereby preserving the MOF's structural integrity (Fig. 1.4c).

The integration of MOFs with complementary materials represents another promising strategy for enhancing extraction performance. Recent advancements in hybrid material development include the synthesis of Zn-MOF/COF composites through the combination of zinc-based MOFs and melamine-based covalent organic

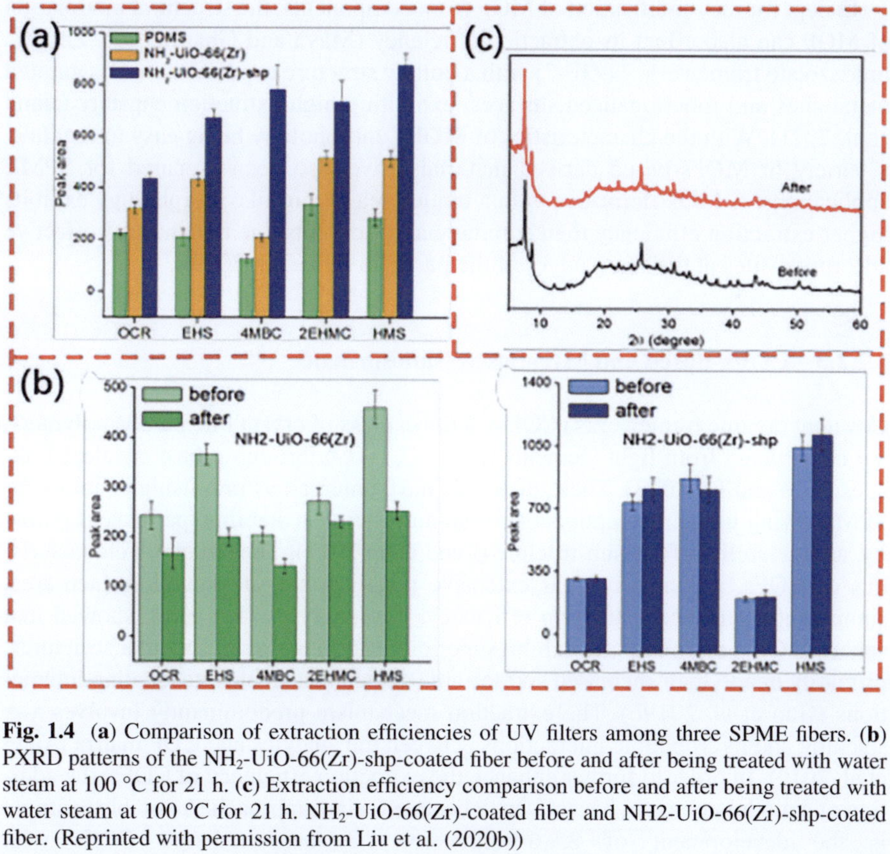

Fig. 1.4 (**a**) Comparison of extraction efficiencies of UV filters among three SPME fibers. (**b**) PXRD patterns of the NH_2-UiO-66(Zr)-shp-coated fiber before and after being treated with water steam at 100 °C for 21 h. (**c**) Extraction efficiency comparison before and after being treated with water steam at 100 °C for 21 h. NH_2-UiO-66(Zr)-coated fiber and NH2-UiO-66(Zr)-shp-coated fiber. (Reprinted with permission from Liu et al. (2020b))

frameworks (Koonani and Ghiasvand 2024). These hybrid materials demonstrate superior polycyclic aromatic hydrocarbon (PAH) extraction efficiency in soil samples compared to their components, owing to synergistic porosity effects. Similar enhancement trends were observed in food safety applications, where MOF/COF composites exhibited remarkable PAH extraction capabilities in milk samples (Yu et al. 2023). Further innovation in this field led to the development of UiO-66/MoS_2-coated SPME arrows, which outperformed conventional PDMS/CARS/DVB fibers in both adsorption capacity and PAH affinity.

Further innovation in this field led to the development of UiO-66/MoS_2-coated SPME arrows, which outperformed conventional PDMS/CARS/DVB fibers in both adsorption capacity and PAH affinity (Yuan et al. 2019). The enhanced performance stems from three key factors: (1) substantially increased specific surface area providing more binding sites, (2) effective π–π interactions between PAH aromatic rings and composite components, and (3) hierarchical pore structures accommodating diverse PAH molecular sizes. Beyond these developments, research has extended to silica-MOF and conductive polymer-MOF composites, consistently demonstrating superior extraction efficiency compared to single-component materials (Zhang et al. 2020; Jiang et al. 2022).

Except for the modification of MOF from components, the structural modulation of MOF can also affect its extraction efficiency (Maya and Ghani 2019). Zeolitic imidazolate framework-7 (ZIF-7) with a hollow structure exposes more unsaturated metal sites and functionalized surfaces, exhibiting high extraction capacity (Jiang et al. 2021). With the characteristics of MOFs' morphology being easy to regulate, a variety of MOFs-based derived materials have also been prepared for SPME applications. A MOF derivative with a unique sea urchin-like morphology exhibits higher extraction efficiency than commercial adsorbents due to enhanced effective adsorption sites at the sharp edges of the particles (Hu et al. 2019).

1.2.2.3 COFs-Based and MIP-Based Nanosorbents

Covalent organic frameworks (COFs), a novel class of crystalline porous polymers, are constructed from light elements (H, B, C, N, O) through robust covalent linkages (Gao and Lu 2023). These materials have emerged as promising coatings for SPME owing to their exceptional thermal and chemical stability, permanent porosity, and tunability of the substructure (Feng et al. 2021). The structural characteristics of COFs, particularly their extensive porosity and substantial surface area, significantly enhance extraction efficiency. The study of Guo et al. showed that spherical COF morphologies exhibit superior performance over irregular structures, primarily due to their increased surface area and enhanced analyte–coating interactions (Guo et al. 2019b). The extraction mechanism predominantly involves π–π stacking and hydrophobic interactions between the analytes and COF matrix (Yang et al. 2019). In order to further enhance the extraction efficiency of COFs, the addition of a modulator agent has proven to be an effective approach. A notable example is the development of 2,4,6-triphenoxy 1,3,5-benzene-based COFs with

4-hydroxybenzhydrazide (OH-TPB-COFs), where OH-TPB-COFs serves as a modulator, demonstrating enhanced extraction capabilities compared to conventional PDMS fibers (Guo et al. 2019a), as shown in Fig. 1.5a. Furthermore, the integration of COFs with nanomaterials has yielded promising results, as evidenced by the superior performance of rGO/SNW-1(Schiff base network-1, a new generation of COFs) nanocomposites for HS-SPME of PhEs from water samples.

The efficiency of the composite was distinctly far higher than that of their pure forms, as shown in Fig. 1.5b. The fabrication method of the COF coating can also have an impact on the extraction efficiency (Mokhtari et al. 2020). Current research emphasizes the importance of fabrication methodologies, particularly in situ growth techniques, which enable precise control over coating thickness and stability. Yue et al. developed a TpBD-TiO_2-coated titanium wire for the extraction of 11 PAEs (Yue et al. 2020). The COF of TpBD was bonded in situ to the titanium wire via TiO_2 nanotube arrays, as shown in Fig. 1.5c.

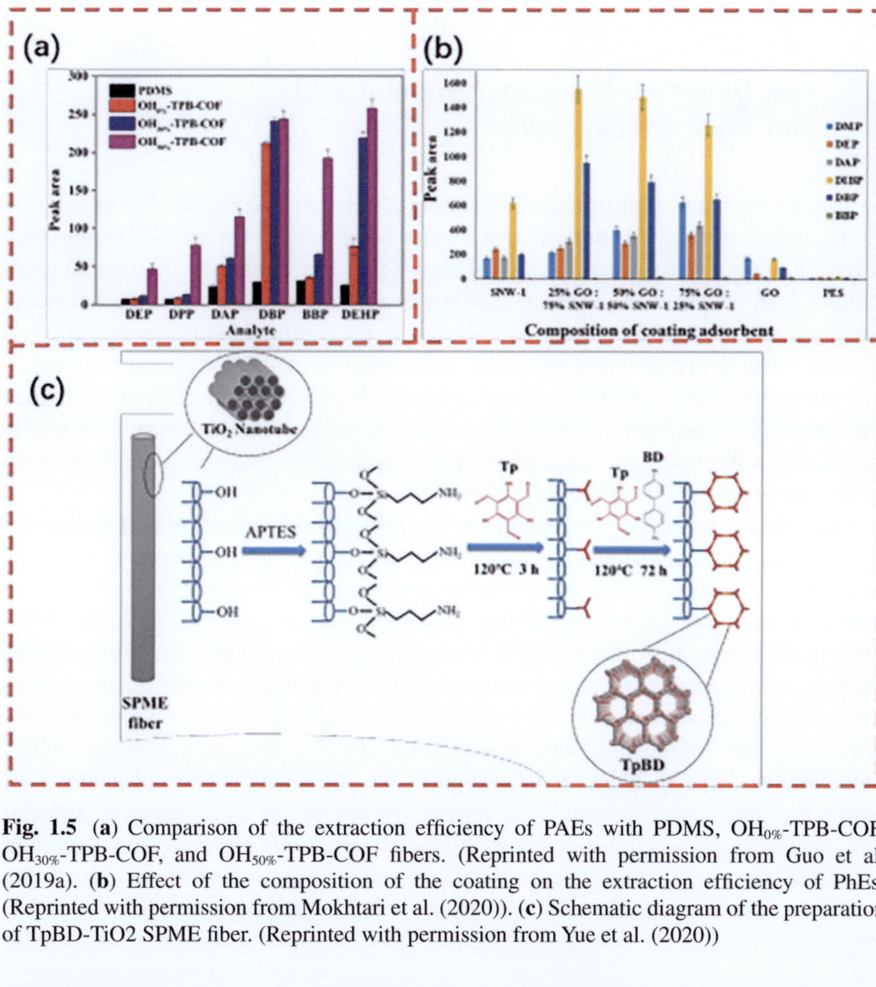

Fig. 1.5 (**a**) Comparison of the extraction efficiency of PAEs with PDMS, $OH_{0\%}$-TPB-COF, $OH_{30\%}$-TPB-COF, and $OH_{50\%}$-TPB-COF fibers. (Reprinted with permission from Guo et al. (2019a). (**b**) Effect of the composition of the coating on the extraction efficiency of PhEs. (Reprinted with permission from Mokhtari et al. (2020)). (**c**) Schematic diagram of the preparation of TpBD-TiO2 SPME fiber. (Reprinted with permission from Yue et al. (2020))

Molecularly imprinted polymers (MIPs) represent a class of synthetic nanosorbents characterized by artificially created recognition sites that exhibit specific binding affinity toward target molecules (Sereshti et al. 2020). MIPs resemble naturally occurring antibodies with the property of selectively recognizing a target molecule (Azizi and Bottaro 2020). The fabrication of MIPs typically involves three sequential steps: (1) complex formation between template molecules (target analytes or their structural analogs) and functional monomers through physical or chemical interactions; (2) polymerization of the monomer–template complex in the presence of cross-linking agents; and (3) template removal to generate three-dimensional cavities with complementary size, shape, and functional group orientation to the target molecule (Lancioni et al. 2022). Due to the superior selectivity of MIP adsorbents, the problem that most adsorbents enrich more species of molecules can be reduced. The MIP-coated fibers used for HS-SPME achieved lower limits of detection (LODs) and enhanced adsorption/desorption kinetics compared to commercial PA fibers (Hashemi-Moghaddam et al. 2017).

1.3 Gas Detection Based on Nanomaterials and Their Mechanisms

As urbanization and industrialization continue to expand, compounded by a rising global population, air pollution, and climate change are becoming increasingly severe, thereby intensifying a range of significant health and environmental challenges (Aldy 2017). Air pollution stands out as a major environmental hazard in contemporary times, intricately connected to escalating healthcare expenses, a marked increase in premature mortality, and a reduction in global economic productivity (Barnett 1974). Moreover, breathing in high concentrations of polluted gases and prolonged exposure to such pollutants can trigger various hazardous health issues, including asthma, lung infections, respiratory ailments, and, in extreme cases, even fatalities (Haghi et al. 2021).

Gas sensors have now emerged as essential tools for detecting and measuring diverse gases across various settings (John et al. 2021), and they play a crucial role in multiple domains, including industrial applications (Tian et al. 2021), environmental surveillance (Zhou and Zhang 2021), and public safety (Goel et al. 2023). Their significance is on the rise due to increasing attention to air quality, climate change, and occupational safety (Gai et al. 2022). Recent advancements in gas sensor technology have enhanced the precision and reliability of gas detection, thereby improving the monitoring and management of gas emissions (Wang 2020). Typically, gas sensors identify specific gases via chemical or physical interactions within the sensing component. This component comprises a gas-sensitive material that undergoes property changes upon contact with a target gas molecule (Robbiani et al. 2023).

As the core functional component in detection devices, gas-sensitive materials play a pivotal role in determining sensor performance through their molecular interaction capabilities and signal transduction mechanisms. The material's intrinsic characteristics not only govern gas molecule adsorption dynamics but also dictate interfacial compatibility with adjacent sensor components, thereby influencing device integration and operational stability. Recent breakthroughs in nanoscale engineering have revolutionized material design paradigms, enabling the synthesis of advanced nanostructures with tailored properties for environmental remediation applications. These technological advancements have catalyzed the development of novel mitigation approaches, particularly in atmospheric quality management systems (Guerra et al. 2018).

Nanoparticles are now engineered with high precision and specificity to target particular pollutants. This targeted approach enhances the absorption of pollutants and effectively reduces their release into the environment (Olyaie and Banejad 2012). The high surface-to-volume ratio of these nanoparticles endows them with exceptional sensitivity for detecting even trace amounts of pollutants (Das et al. 2015). Additionally, these advanced nanomaterials enable the creation of highly miniaturized and responsive pollution monitoring devices. As a result, nanomaterial-based sensors hold immense potential for the detection and management of air pollutants (Wang et al. 2023).

Contemporary industrial applications increasingly leverage advanced nanostructured materials for real-time monitoring of atmospheric contaminants, including sulfur dioxide, ammonia, and nitrogen oxides, at emission sources (Ouyang et al. 2022). The design and synthesis of these nanostructured sensing elements can be precisely engineered to accommodate distinct pollutant characteristics, encompassing chemical composition, physical properties, and concentration ranges (Zheng et al. 2021). This review systematically examines the principal categories of nanostructured sensing materials employed in atmospheric monitoring devices, focusing on their structural configurations and functional mechanisms. The discussion encompasses polymeric conductors, nanocarbon architectures, layered transition metal compounds, metallic oxides, and composite nanomaterials, each demonstrating distinctive advantages in environmental sensing applications.

1.3.1 Conductive Polymers (CPs)-Based Gas Sensor

Conjugated polymeric materials have emerged as promising candidates for atmospheric monitoring applications, exhibiting three distinctive advantages: (1) tunable electrical conductivity, (2) operational efficiency at ambient temperatures, and (3) mechanical resilience under deformation stress. The extended π-electron conjugation system inherent in these materials endows them with exceptional charge transport properties, making them versatile for multiple technological applications ranging from photovoltaic devices to energy storage systems (Aldy 2017). As a rapidly developing material class, polymeric conductors are anticipated to drive significant innovations in environmental monitoring technologies.

The distinctive electrochemical properties of conjugated polymers, particularly their reversible oxidation–reduction characteristics and versatile composite formation capabilities with metallic nanostructures and carbon-based materials, render them particularly advantageous for next-generation technological applications, including bioelectronic interfaces and advanced energy storage devices (Kumar and Sharma 1998). Among the extensively studied polymeric systems, polypyrrole (PPy), polyaniline (PANI), polyacrylonitrile (PAN), and the poly(3,4-ethylenedioxythiophene): poly (styrenesulfonate) (PEDOT: PSS) complex have demonstrated exceptional performance in atmospheric monitoring applications.

These polymeric systems predominantly display p-type semiconducting behavior, where exposure to oxidizing agents induces the formation of a hole-enriched surface layer, resulting in enhanced electrical conduction. In contrast, reducing gases typically cause a depletion of charge carriers, leading to decreased conductivity. A significant advancement in sensor fabrication was achieved through the development of a core-sheath polymeric architecture, as reported by Wu et al. (2017). This innovative design utilized aligned nanofibrous structures created through an integrated manufacturing approach combining electrospinning with in situ polymerization (Fig. 1.6a). The resulting uniaxially aligned coaxial nanofiber yarn (UACNY) composed of PANI/PAN demonstrated remarkable sensitivity in ammonia detection applications, with systematic evaluation of its gas-sensing performance parameters.

The engineered coaxial fiber architecture synergistically combines the structural integrity of polyacrylonitrile with the exceptional charge transport properties of

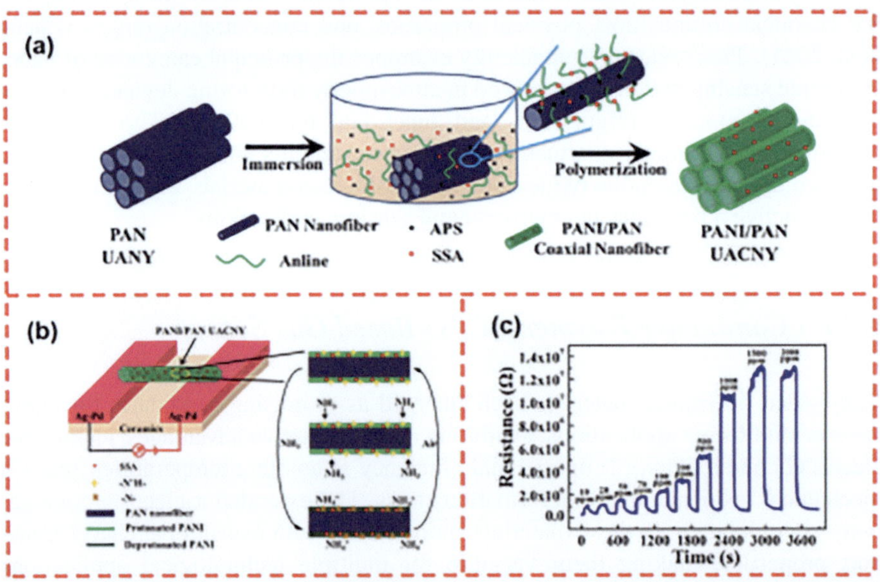

Fig. 1.6 (**a**) Schematic of UACNY fabrication from core-sheath PANI/PAN nanofibers. (**b**) PANI/PAN UACNY sensor sensitization mechanism schematic model. (**c**) PANI/PAN UACNY sensor resistance transients at 10–2000 ppm NH_3 at room temperature. (Reprinted with permission from Wu et al. (2017))

polyaniline. This unique configuration provides an extensive interfacial area that promotes efficient ammonia molecule permeation and interaction. The aligned nanostructure morphology, illustrated in Fig. 1.6b, establishes preferential pathways for charge carrier migration, resulting in anisotropic electrical conduction with enhanced signal transmission efficiency.

Experimental characterization revealed that the developed yarn-based sensing platform achieved remarkable detection capabilities across a broad concentration spectrum (10–2000 ppm), as quantitatively demonstrated in Fig. 1.6c. The device maintained consistent performance metrics, including response stability, measurement reproducibility, and signal reversibility, establishing its viability as a reliable detection medium for hazardous atmospheric constituents, particularly ammonia vapors.

In a similar vein, to enhance the gas sensitivity of polyaniline-based sensors, Guo et al. developed a flexible and transparent nanocomposite film resembling an artificial neural network. This room-temperature electronic sensor was fabricated by chemically oxidizing polyaniline in a reduced graphene oxide (rGO) solution and simultaneously depositing it onto a flexible polyethylene terephthalate (PET) substrate (Guo et al. 2016).

The innovative biomimetic architecture successfully integrated two distinct functional components: a PPANI/rGO-based central unit and FPANI-derived conductive pathways, which were co-deposited onto flexible polyethylene terephthalate substrates. This neural network-inspired configuration capitalizes on the complementary properties of its constituent materials, yielding a nanocomposite film with an optimized surface area of 47.896 m^2/g, efficient charge transport pathways, and maximized active site accessibility. The resulting room-temperature operational sensor demonstrates exceptional performance metrics for ammonia detection across a wide concentration range (100 ppb to 100 ppm), characterized by optimized response kinetics (36 s response/18 s recovery), outstanding gas discrimination capability, and robust mechanical properties.

Furthermore, the nanocomposite membrane maintains superior optical transparency (90.3% at 550 nm wavelength) while preserving structural integrity under mechanical stress. Conjugated polymeric systems maintain remarkable gas detection capabilities under ambient conditions, exhibiting wide operational detection ranges. Nevertheless, these materials face persistent challenges related to signal intensity modulation and temporal response characteristics. Recent advancements in material engineering have demonstrated that these limitations can be effectively addressed through strategic hybridization approaches, incorporating metallic components, nanostructured carbon materials, metal oxides, or complementary polymeric systems (Bandgar et al. 2015). The interfacial interactions between the polymeric matrix and these functional additives significantly enhance the hybrid material's performance, particularly in terms of electrical resistance modulation upon gas molecule adsorption. A comprehensive comparison of performance metrics for polymeric-based gas detection systems is presented in Table 1.2.

Table 1.2 Related parameters of CPs-based gas sensors

Sensing material	Target gas	Conc. (ppm)	T (°C)	Res. (%)	Res./Rec. Time(s)	References
PANI	NH_3	100	20	1640	13	Wang et al. (2017)
PANI/PAN	NH_3	200	25	1060	5/35	Wu et al. (2017)
PANI-graphene	CO_2	60	25	75	65	Wu et al. (2017)
PCDTBT	CO_2	550	25	275	20	Abdali et al. (2019)
	NO	100	25	80.6	300/2100	Gusain et al. (2017)

1.3.2 Carbon Nanomaterials-Based Gas Sensor

Carbon-based materials have garnered extensive research attention and are widely utilized in nanotechnology due to their remarkable properties. Compared to conventional materials, carbonaceous structures offer numerous advantages, especially in terms of their exceptional physicochemical characteristics (Bezzon et al. 2019). Their fabrication processes are often simplified, yielding substantial quantities of material with minimal densification defects. Moreover, carbon-based materials provide a cost-effective alternative to expensive electronic compounds, while retaining excellent performance and environmental sustainability (Peng et al. 2014). As a result, carbon nanostructures have been extensively explored as highly efficient sensor devices, thanks to their superior physical and chemical attributes, which enable outstanding sensing capabilities.

1.3.2.1 Carbon Nanotubes (CNTs)-Based Gas Sensor

CNTs are a form of carbon characterized by their nanometer-scale diameter and micrometer-scale length, with an aspect ratio exceeding 1000. Their atomic arrangement consists of hexagons, similar to graphite. Structurally, CNTs are composed of a cylindrical graphene sheet rolled into a seamless nanoscale cylinder (Aqel et al. 2012). CNTs are categorized into two main types: single-walled CNTs (SWCNTs), which have a single layer of graphite with a diameter of approximately 15 nm, and multi-walled CNTs (MWCNTs), which feature multiple concentric layers of graphite.

CNTs have gained significant attention in gas detection applications owing to their distinctive electronic characteristics, ambient temperature functionality, nanoscale dimensionality, environmental resilience, detection sensitivity, and mechanical durability (Tang et al. 2017). The structural asymmetry in CNTs induces electron density redistribution, generating extensive π-electron delocalization along their surface. This electronic configuration contributes to their remarkable charge transport properties (Zhang et al. 2008). Furthermore, their nanoscale architecture, characterized by high surface accessibility and tubular morphology, facilitates extensive gas molecule adsorption through electronic interaction mechanisms.

The unique combination of intrinsic charge carrier mobility, electrical conduction efficiency, and structural adaptability has positioned CNTs as versatile components in flexible electronic systems, functioning simultaneously as detection elements and conductive interfaces (Park et al. 2013). A notable advancement was demonstrated by Guo et al. 2019e), who developed transparent hydrogen detection devices using single-walled carbon nanotube (SWCNT) networks synthesized through floating catalyst chemical vapor deposition (FCCVD). The deposition of these nanostructures onto polyethylene terephthalate substrates yielded detection systems with sub-minute response kinetics and rapid recovery profiles, maintaining sensitivity at hydrogen concentrations as low as 0.1% by volume.

Complementary research by Lee et al. established a scalable fabrication methodology for transparent chemical sensors utilizing surface-modified SWCNTs (Lee et al. 2013). Their approach involved the aerosol deposition of nanotube films onto flexible polymeric substrates, followed by gold nanoparticle functionalization. The resulting hybrid nanostructures were systematically evaluated for ammonia detection capabilities, demonstrating the potential for cost-effective sensor production.

Compared to pristine SWCNTs, the Au NP-modified SWCNT sensor exhibited a substantial increase in film resistance upon exposure to NH_3, as illustrated in Fig. 1.7a. The incorporation of Au NPs enhanced the sensor's sensitivity to 255 ppb NH_3, primarily due to the formation of additional local depletion regions and changes in the barrier characteristics at the metal/carbon nanotube interface.

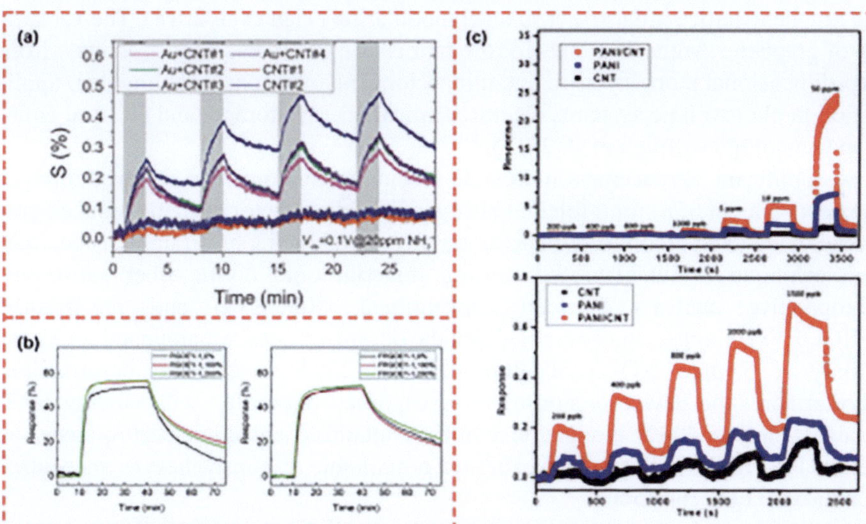

Fig. 1.7 (**a**) SWCNT film and AuNP decorated SWCNT film response curves for 20 ppm NH_3 gas. (Reprinted with permission from Lee et al. (2013)). (**b**) Gas-sensing performance of RGOEY-1 and FRGOEY-1 when exposed to 5.0 ppm NO_2 at 0%, 100%, and 200% strain (Yun et al. 2018). (Reprinted under a Creative Commons license (CC BY-NC 3.0)) (**c**) Gas-sensing performance of PANI/CNT nanocomposites, PANI fiber aggregates, and CNT thin film devices for different concentrations of NH_3 ranging from 200 ppb–50 ppm and 200 ppb–1500 ppb. (Reprinted with permission from Xue et al. (2017))

The implemented material modifications substantially enhanced the device's operational characteristics, particularly regarding energy efficiency and ambient condition functionality. These advancements have facilitated the development of compact detection systems that are compatible with portable electronic platforms, making them particularly valuable for atmospheric quality assessment and environmental surveillance applications.

1.3.2.2 Graphene-Based Gas Sensor

As a two-dimensional carbon allotrope, graphene has become a fundamental material in gas detection technologies due to its unique combination of physicochemical properties. These include an extensive specific surface area, superior charge carrier mobility, exceptional electrical conduction, mechanical flexibility, and remarkable elastic modulus. The material's structure consists of a monoatomic layer of hexagonally arranged carbon atoms, with each sp^2-hybridized atom forming three covalent bonds within the plane. This atomic configuration confers extraordinary mechanical characteristics, including a tensile modulus of 1 TPa and fracture resistance of 130 GPa, while maintaining exceptional optical transparency exceeding 97% (Lee et al. 2012).

The gas detection mechanism in graphene-based systems involves subtle electronic interactions, where charge carrier exchange between adsorbed molecules and the graphene lattice induces Fermi level modulation (Tian et al. 2018). The versatility of graphene synthesis allows for the fabrication of diverse nanostructures, from two-dimensional monolayers to one-dimensional nanoribbons, enabling their application in photovoltaic systems, electrochemical energy storage, and electron emission technologies (Singh et al. 2017).

A significant advancement in flexible sensing technology was achieved through the development of a strain-tolerant nitrogen dioxide detector utilizing reduced graphene oxide (Yun et al. 2018). The researchers employed a prestrain-assisted coating technique to integrate the sensing material onto elastic fiber substrates. Comparative analysis between conventional (RGOEYs) and pre-strained (FRGOEYs) configurations revealed that the optimized sensor maintained 55% sensitivity at 5 ppm NO_2 concentration under 200% mechanical deformation. Furthermore, the device demonstrated exceptional durability, withstanding 5000 strain cycles at 400% elongation while maintaining detection performance, as detailed in Fig. 1.7b. The system exhibited remarkable responsiveness to trace nitrogen dioxide concentrations.

In another study, Abdali et al. synthesized cross-linked bacterial cellulose–amino graphene/polyaniline (CLBC-AmG/PANI) using amino-functionalized graphene (AmG) and bacterial cellulose (BC) as raw materials (Abdali et al. 2019). The synthesis involved a one-step esterification reaction, followed by in situ chemical polymerization of aniline monomer on its surface. The CLBC-AmG/PANI sensor displayed a strong resistance response to carbon dioxide (CO_2) at room temperature.

It exhibited high sensitivity and rapid response characteristics within a CO_2 concentration range of 50–2000 ppm, with the sensor response showing an almost linear dependence on CO_2 concentration.

1.3.2.3 Carbon Nanocomposite-Based Gas Sensor

Carbon nanocomposites exhibit poor sensitivity to certain gases and find it challenging to detect low concentrations of target gases effectively (Wang et al. 2015). The functionalization and recombination of carbon nanomaterials with other sensing materials—such as metal nanoparticles, conducting polymers, and metal oxides—have been widely employed to significantly enhance their gas-sensing performance (Jeong et al. 2010). To illustrate, Zhang et al. fabricated a stand-alone flexible gas-sensing film by combining graphene with ethyl cellulose (EC) nanomaterials and integrating it onto a polyethylene terephthalate (PET) flexible substrate (Zhang et al. 2018a).

The developed ambient-temperature detection system maintained exceptional structural integrity under mechanical stress, demonstrating minimal resistance variation (0.3%) after 400 deformation cycles at a curvature radius of 3.18 mm, with even lower variation (0.2%) at 5 mm radius. The device also showed efficient detection capabilities for volatile organic compounds at standard temperature conditions, characterized by substantial electrical resistance modulation upon VOC exposure, with sensitivity thresholds spanning 37–167 ppm.

In a parallel development, Xue et al. engineered a hybrid nanostructured network combining polyaniline-functionalized carbon nanotubes with polymeric nanofibers (Xue et al. 2017). This architecture was fabricated through ammonium sulfate-mediated polymerization of aniline in the presence of PANI nanoparticles, enabling simultaneous polymerization and thin film formation.

Due to the high surface area (54.187 m^2/g) resulting from the synergistic interaction between PANI and CNTs, as well as the effective sensing channels provided by the nanocomposite network, the room-temperature sensor exhibited remarkable sensitivity to NH_3 over a concentration range of 200 ppb to 50 ppm. It also demonstrated a fast response and recovery time (85 s/20 s) and high selectivity for NH_3 compared to other volatile organic compounds, as shown in Fig. 1.7c.

Carbon nanotubes have emerged as prominent candidates for atmospheric monitoring applications, leveraging their extensive surface-to-volume ratio and exceptional charge transport characteristics to achieve superior gas detection sensitivity. Nevertheless, unmodified CNT-based flexible detection systems frequently encounter limitations in detection thresholds and molecular discrimination capabilities (Jeong et al. 2010). These constraints can be effectively mitigated through the strategic integration of complementary sensing materials within the nanotube framework and precise concentration optimization. Furthermore, surface modification of CNTs with reactive moieties, including carboxyl, amino, and thiol groups,

Table 1.3 Related parameters of carbon nanomaterials-based gas sensors

Sensing material	Target gas	Conc. (ppm)	*T* (°C)	Res. (%)	Res./Rec. Time(s)	References
SWCNT	NH_3	100	25	51	–	Abdellah et al. (2013)
	CO	10	150	4	–	Dong et al. (2013)
Graphene	NO_2	200	22.76	75	135/477	Lee et al. (2012)
SWCNT-Fe_2O_3	H_2S	10	25	9.40	–	Hua et al. (2017)
CNT/rGO	NO_2	10	25	20	–	Jeong et al. (2010)

significantly improves detection performance by creating additional active sites for target molecule adsorption and surface reactions (Seekaew et al. 2014).

Parallel investigations in graphene-based systems have revealed that chemically modified graphene derivatives, particularly graphene oxide and its reduced form, demonstrate enhanced detection capabilities compared to pristine graphene. This performance enhancement stems from the presence of surface functional groups that promote stronger molecular interactions and accelerated adsorption kinetics (Lee et al. 2012). These chemical modifications also contribute to improved material stability when incorporated into flexible device architectures. A comprehensive comparison of performance metrics for carbon-based detection systems is presented in Table 1.3.

1.3.3 2D Transition Metal Dichalcogenide (TMD) Nanomaterials-Based Gas Sensor

Emerging two-dimensional layered compounds, particularly transition metal dichalcogenides (TMDs) such as $MoSe_2$, WSe_2, and SnS_2, have attracted considerable research interest for energy-efficient detection applications due to their distinctive electronic configurations and structural characteristics. These semiconducting materials follow a general MX_2 stoichiometry, where M denotes transition metals (e.g., Mo, W, V) and X represents chalcogen elements (S, Se, Te), forming a sandwich-like structure with a 6–7 Å thick metal layer between chalcogenide planes.

The exceptional attributes of TMDs, including tunable electronic properties, efficient charge transport, minimal energy requirements, structural flexibility, and environmental resilience, make them particularly suitable for atmospheric monitoring applications (Kumar et al. 2020). Their detection mechanism involves charge carrier exchange processes between the layered material and target analytes, with conductivity modulation patterns determined by the redox characteristics of the interacting gas molecules.

Capitalizing on the thickness-dependent properties of TMDs, recent investigations have focused on hybrid material systems incorporating metal oxides, polymeric matrices, and functionalized nanoparticles. These composite architectures, achieved through nanostructural engineering approaches, have demonstrated enhanced detection capabilities (Huang et al. 2014).

1.3.3.1 Molybdenum Diselenide ($MoSe_2$)-Based Gas Sensor

$MoSe_2$ represents a distinctive class of transition metal dichalcogenides characterized by a trilayer atomic configuration, where molybdenum atoms are covalently bonded between two selenium planes. The expanded interatomic spacing resulting from selenium's larger atomic radius confers unique electronic properties, including dual bandgap characteristics (1.1 eV indirect and 1.55 eV direct) and enhanced electrical conductivity, distinguishing it from other TMD materials (Shim et al. 2014).

This material exhibits several advantageous properties for sensing applications, including substantial molecular adsorption energy, narrow direct bandgap, efficient charge carrier transport, and rapid electrical signal transduction. These features have established $MoSe_2$ as a promising candidate for next-generation chemical detection and optical sensing platforms (Late et al. 2014). Consequently, $MoSe_2$ has been extensively implemented in various technological domains, including electrochemical systems, photonic devices, and photocatalytic applications.

Baek et al. successfully developed a high-sensitivity multilayer $MoSe_2$ field-effect transistor (FET) grown by chemical vapor deposition for use in NO_2 gas sensing (Baek et al. 2017). The sensor's ultra-high sensitivity ($S \approx 1907$ for NO_2 at 300 ppm), real-time response, and fast switching capabilities were attributed to the changes in the gap states near the valence band caused by NO_2 adsorption on $MoSe_2$. This interaction significantly increased the hole current in the non-equilibrium region.

The study also demonstrated a linear relationship between the sensitivity of the $MoSe_2$ FET and the concentration of NO_2 gas, ranging from 100 to 300 ppm. Analysis of device modeling and quantum transport simulations revealed that the impact of NO_2 concentration on the gap state changes is a crucial mechanism underlying the $MoSe_2$ FET NO_2 gas sensor's performance.

In pursuit of enhanced detection capabilities, Zhang et al. engineered an innovative self-sustaining ammonia detection system operational at ambient temperatures (Zhang et al. 2019a). This integrated device combines a flexible molybdenum disulfide (MoS_2) piezoelectric energy harvesting component with a gold-functionalized $MoSe_2$ detection layer. The researchers performed extensive material characterization on both the two-dimensional piezoelectric MoS_2 substrate and the gold-modified $MoSe_2$ sensing element.

The distinctive nanoarchitectured morphology of the gold–molybdenum diselenide (Au-$MoSe_2$) composite, characterized by its thin-film nanoflower configuration, significantly contributes to the device's enhanced functionality. The incorporation of gold nanoparticles establishes Schottky barrier junctions at the metal–semiconductor interface, while simultaneously increasing the density of catalytically active surface sites through nanoparticle functionalization, as illustrated in Fig. 1.8a.

As a result, the NH_3 sensor based on the Au-$MoSe_2$ composite and MoS_2 piezoelectric nanogenerator exhibited a higher response to NH_3 ($Va/Vg = 29$ for NH_3 at 100 ppm) compared to its $MoSe_2$-only counterpart ($Va/Vg = 25$ for NH_3 at 100 ppm). Additionally, the sensor demonstrated rapid response and recovery times of 18 s and 16 s, respectively, when exposed to 20 ppm NH_3.

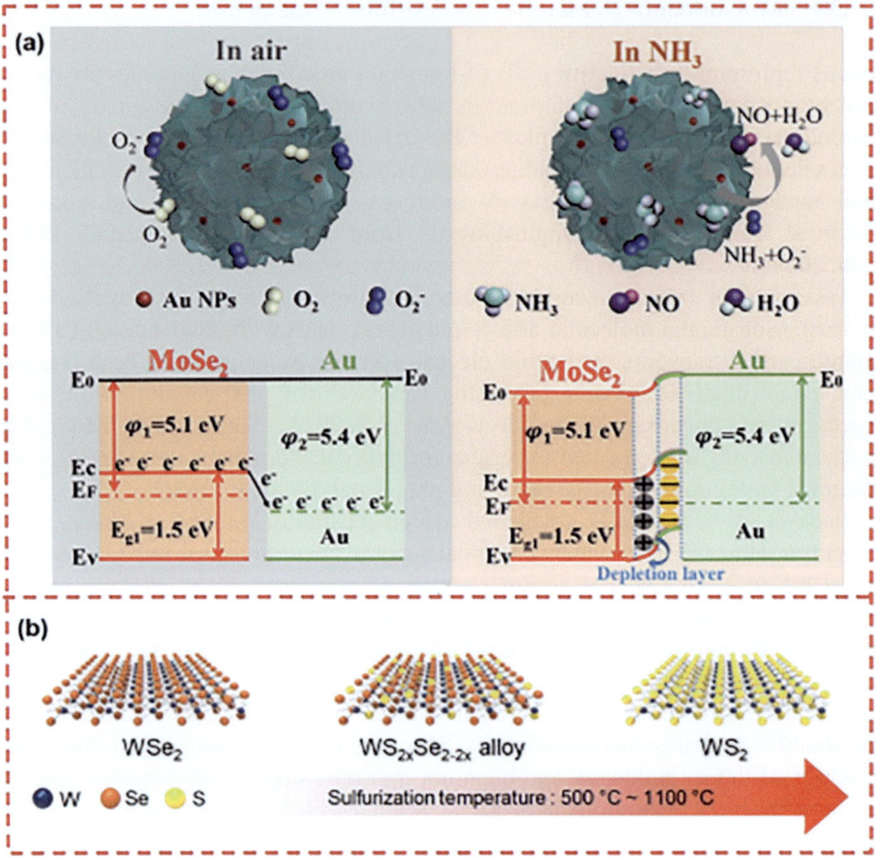

Fig. 1.8 (**a**) Schematic diagram of the gas-sensing mechanism and the energy band structure of Au-$MoSe_2$ composites in air and NH_3 gas. (Reprinted with permission from Zhang et al. (2019a). (**b**) Schematic images of the synthesis process of the $WS_{2x}Se_{2-2x}$ alloy. (Reprinted with permission from Ko et al. (2018))

1.3.3.2 Tungsten Diselenide (WSe_2)-Based Gas Sensor

As a representative p-type semiconductor within the transition metal dichalcogenide family, WSe_2 demonstrates exceptional charge transport characteristics, boasting the highest Hall mobility among TMD materials. Similar to its structural analogs, WSe_2 crystallizes in a layered configuration stabilized by van der Waals interactions between adjacent layers (Wang et al. 2012). The material exhibits dual bandgap characteristics, with an indirect gap of 1.2 eV and a direct gap of 1.65 eV in its bulk form. Notably, its charge carrier transport properties vary with layer thickness, achieving hole mobilities of 100 cm^2V^{-1} s^{-1} in monolayer configurations and up to 350 cm^2V^{-1} s^{-1} in few-layer structures, rendering it particularly advantageous for next-generation electronic and photonic devices (Zhou et al. 2015).

Lee and Choi developed a highly sensitive NO_2 sensor using two-dimensional WSe_2 nanolayers (Lee and Choi 2019). The WSe_2 nanolayers were integrated onto a paper substrate by exfoliating them from their bulk form. The sensor exhibited remarkable sensitivity to NO_2, reaching approximately 238.35% ppm^{-1} at low concentrations (0.1–0.5 ppm), facilitating early detection of NO_2.

Ko et al. employed chemical vapor deposition (CVD) and metal vulcanization techniques to fabricate a NO_2 gas sensor composed of large-area $WS_{2x}Se_{2-2x}$ alloy (Ko et al. 2018), as shown in Fig. 1.8b. The $WS_{2x}Se_{2-2x}$ alloy demonstrated precise compositional control due to its thermodynamically stable vulcanization process. Compared to a WSe_2 gas sensor, the $WS_{0.96}Se_{1.04}$ alloy sensor showed a 2.4-fold enhancement in response to 500 ppm NO_2, achieving a sensitivity of 2621%. This approach holds promise for improving 2D TMDC gas sensors and has potential applications in advanced gas-sensing devices.

1.3.3.3 Tin Disulfide (SnS_2)-Based Gas Sensor

Although SnS_2 does not belong to the TMD family due to the non-transition metal nature of tin, it exhibits structural and functional similarities to TMDs (Burton et al. 2014). This material demonstrates an n-type semiconducting behavior, characterized by a distinctive S-Sn-S layered configuration, where tin cations occupy both hexagonal and octahedral sites within the lattice (Huang et al. 2014). The interlayer interactions in SnS_2 are governed by van der Waals forces, while intralayer bonding is primarily covalent between sulfur and tin atoms (Whittles et al. 2016). Such structural features endow SnS_2 with exceptional properties, including a high surface-to-volume ratio, optimal band gap characteristics, and significant electronegativity, making it suitable for diverse technological applications.

SnS_2 has emerged as a promising material owing to its exceptional chemical and thermal stability, environmental friendliness, and cost-effectiveness. Its versatility has been demonstrated in various technological domains, ranging from photovoltaic devices (Chen et al. 2013) to biological detection systems (Yang et al. 2011) and gas detection platforms (Shi et al. 2006). Particularly in gas-sensing applications, SnS_2 exhibits notable advantages due to its abundant active sites, substantial surface area, tunable surface properties, and strong electronegativity (Burton and Walsh 2012).

Recent advancements in material synthesis have further enhanced the gas-sensing capabilities of SnS_2. Pyeon et al. developed a novel fabrication approach using atomic layer deposition (ALD) to create flexible SnS_2 sensors on Al_2O_3-coated polyimide substrates (Pyeon et al. 2020). This technique yielded single-phase SnS_2 with vertically oriented nanosheets, which demonstrated exceptional sensitivity to NO_2 at ambient temperature. The unique vertical alignment of the nanosheets promoted efficient charge transfer through physical adsorption, resulting in a significant response of 309 at 1 ppm NO_2. Furthermore, the sensor maintained consistent performance even at ultra-low NO_2 concentrations (<500 ppb), highlighting its potential for practical applications.

Moreover, the sensor exhibited excellent selectivity for NO_2, driven by SnS_2's high adsorption energy for NO_2 and the favorable Fermi energy associated with partial occupancy of NO_2 molecular orbitals.

Li et al. developed tubular SnO_2/SnS_2 composite materials through in situ hydrothermal vulcanization, using SnO_2 nanotubes synthesized by electrospinning as the structural framework (Li et al. 2017). They further fabricated flexible SnO_2/SnS_2 composite nanotube gas sensors on a polyethylene terephthalate (PET) substrate. Characterization revealed that the synthesized SnO_2/SnS_2 composite nanotubes had an average diameter of approximately 300 nm and were composed of numerous small nanoparticles.

The integration of a hollow architecture and the synergistic effects between SnO_2 and SnS_2 components substantially improved the gas-sensing capabilities. Notably, the SnO_2/SnS_2 nanotube-based sensor achieved a sensitivity of 2.48 upon exposure to 100 ppm NH_3 at ambient temperature, representing a twofold enhancement compared to pure SnO_2 nanotubes. Furthermore, the device demonstrated rapid response kinetics, consistent repeatability, long-term stability, and superior selectivity.

The versatility in synthesis approaches and deposition methods, combined with the morphological diversity of TMD, offers substantial surface area and abundant active sites for gas adsorption, thereby improving sensing efficiency (He et al. 2012). Studies on TMD-based sensors have revealed their pronounced sensitivity to NO_2, often achieving ultra-low detection thresholds. The efficient charge transport, distinctive layered architecture, and inherent chemical characteristics of TMD enable robust interactions with NO_2, resulting in measurable resistance variations during gas exposure (Cho et al. 2015). Nevertheless, limitations, including modest response magnitudes and prolonged reaction durations, persist. To overcome these constraints, researchers have integrated secondary components such as conjugated polymers (CPs), metallic nanoparticles, metal oxides, and additional TMD to augment the gas-sensing capabilities of pristine TMD (Ashraful Islam et al. 2018). The incorporation of these materials facilitates the creation of interfacial junctions, thereby elevating charge carrier density and improving sensor performance (Jung et al. 2018). A comprehensive summary of critical parameters for TMD-based nanomaterial gas sensors is presented in Table 1.4.

1.3.4 Metal Oxides-Based Gas Sensor

Metal oxides are widely recognized as effective gas-sensing materials owing to their precise signal transduction capabilities, rapid response characteristics, exceptional sensitivity and selectivity, excellent reversibility, cost-effective fabrication, and robust stability (Hahn et al. 2012). The fundamental sensing mechanism of these materials operates through two primary phenomena: charge carrier migration and resultant resistance modulation (Dey 2018). The extent of resistance variation, commonly referred to as sensor response, is influenced by the semiconductor type

Table 1.4 Related parameters of TMD nanomaterials-based gas sensors

Sensing material	Target gas	Conc. (ppm)	*T* (°C)	Res. (%)	Res./Rec. Time(s)	References
$MoSe_2$	NH_3	5	25	7.50	400/800	Guo et al. (2019c)
	NO_2	6	25	50	250/150	Guo et al. (2019d)
WSe_2	NO_2	10	25	1000	–	Lee and Choi (2019)
SnS_2	NH_3	1000	25	0.90	60/60	Ma et al. (2018)
	NO_2	1	25	309	–	Pyeon et al. (2020)

(p-type or n-type) and the chemical nature of the target gas (oxidizing or reducing). For instance, n-type metal oxides exhibit enhanced conductivity when interacting with reducing gases but demonstrate reduced conductivity upon exposure to oxidizing agents, while p-type counterparts display the inverse behavior.

Despite these advantages, a significant limitation of metal oxide-based sensors is their dependence on elevated operating temperatures, as their sensing functionality is contingent upon conductance alterations (Agarwal et al. 2021). To address this challenge, researchers have employed strategies such as doping or surface functionalization with complementary materials, including carbon-based nanostructures, metallic nanoparticles or CPs to optimize their gas-sensing performance (Mani and Rayappan 2013).

1.3.4.1 Zinc Oxide (ZnO)-Based Gas Sensor

ZnO has emerged as a particularly prominent material in gas-sensing research, distinguished by its superior semiconducting characteristics. These include a substantial direct band gap of 3.37 eV at ambient temperature, an exciton binding energy of 60 meV, and electron mobility reaching 200 $cm^2/V{\cdot}s$ (Rackauskas et al. 2017).

The combination of ZnO's n-type semiconducting behavior with its cost-effectiveness, straightforward synthesis and modification processes, chemical robustness, high surface-to-volume ratio, environmental compatibility, and responsiveness even under low-temperature conditions renders it exceptionally suitable for gas detection applications (Agarwal et al. 2019). These distinctive properties have established ZnO as one of the most comprehensively investigated materials in gas-sensing technology.

Ahn et al. successfully grew vertically aligned ZnO nanorods on polyimide films using the thermolysis-assisted chemical solution method to develop ethanol-sensing materials (Ahn et al. 2010). The ZnO nanorod sensor was tested by varying the operating temperature from 300 °C to 125 °C and by adjusting the ethanol concentration in synthetic air from 100 ppm to 10 ppm. The flexible ZnO nanorod sensor demonstrated the ability to detect ethanol gas at 100 ppm with a sensitivity of 3.11. The highest sensor response was achieved at a working temperature of 300 °C, attributed to the substantial adsorption of oxygen species (O^- or O_2^-) on the ZnO nanorod surface, which facilitates chemical reactions with ethanol gas molecules.

In a related development, researchers successfully engineered an economical and highly flexible hydrogen sensor utilizing ZnO nanorods, fabricated on a nylon substrate through hydrothermal synthesis combined with RF sputtering. The device demonstrated remarkable real-time sensitivity to varying hydrogen concentrations at 75 °C, with response values spanning 246–578%. At hydrogen levels between 750 ppm and 2000 ppm, the sensor exhibited response and recovery durations of 127 s and 103 s, respectively. This enhanced performance is primarily attributed to the nanorod architecture, which maximizes surface area for efficient hydrogen diffusion and adsorption.

Guan et al. developed a novel fabrication strategy for flexible ammonia sensors employing vapor phase polymerization (VPP) (Guan et al. 2020). This technique involved exposing ZnO sheets to polyaniline (PANI) vapor on nonwoven substrates, creating a composite material with superior ammonia detection capabilities. The research systematically evaluated the sensor's performance, demonstrating its enhanced sensitivity and flexibility in NH_3 detection applications.

At ambient temperature, the ZnO/PANI/nonwoven sensor demonstrated a remarkable increase in response value, escalating from 5% to 4300% across an NH_3 concentration range of 0.5–10 ppm. This composite outperformed conventional PANI-based sensors, exhibiting superior sensitivity alongside accelerated response (48 s) and recovery (185 s) durations. The improved performance stems from the synergistic interaction between the ZnO/PANI shell architecture and hydrogen bonding, which also enhances thermal stability (Fig. 1.9a). Additionally, the composite's morphology and the formation of p-n heterojunctions between p-type PANI and n-type ZnO significantly contribute to its enhanced gas-sensing capabilities.

1.3.4.2 Indium Oxide (In_2O_3)-Based Gas Sensor

In_2O_3, an n-type semiconductor with a broad band gap of around 3.7 eV, has attracted considerable research interest owing to its distinctive attributes. These include exceptional infrared reflectance, low electrical resistance, superior photoelectric efficiency, and negligible visible light absorption (Liu et al. 2010). Such properties position In_2O_3 as a highly suitable candidate for diverse technological applications, especially in the field of gas detection.

To enhance the performance of In_2O_3-based devices, researchers have synthesized various porous nanostructures, such as nanotubes, nanosheets, and nanowires, using multiple techniques (Ren et al. 2016). These nanostructures offer several advantages, including increased surface area and improved charge transport, which can significantly enhance gas-sensing capabilities. By utilizing these diverse forms, the optimal operating temperature for gas sensing can be reduced, while gas sensitivity parameters, such as response time, selectivity, and overall sensitivity, can be significantly improved.

You et al. developed flexible NO_2 sensor arrays that operate at room temperature by fabricating a composite film of In_2O_3 and graphene oxide (GO) using the direct laser writing (DLW) technique (You et al. 2018). The sensor, based on In_2O_3 and reduced graphene oxide (In_2O_3@RGO), demonstrated linear NO_2 sensing characteristics, good responsiveness and recovery behavior, and excellent selectivity for NO_2

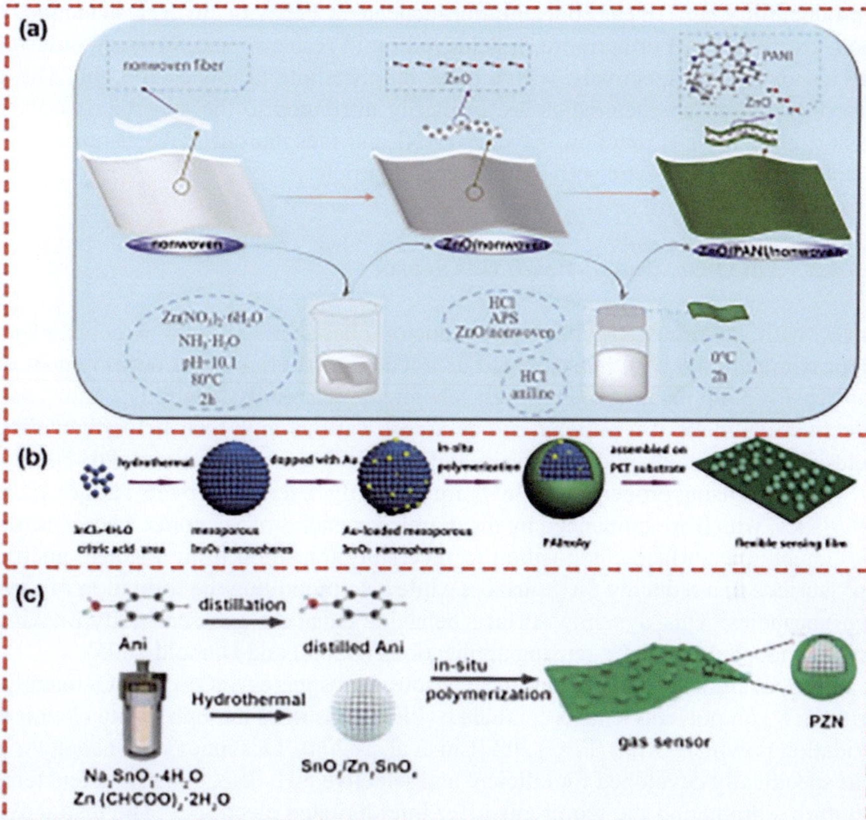

Fig. 1.9 (**a**) Schematic diagram for the preparation of the ZnO/PANI/non-woven nanocomposite (Guan et al. 2020). (Reprinted under a Creative Commons (CC BY 4.0)). (**b**) Schematic of the gold-loaded mesoporous In_2O_3 nanospheres @ PANI core–shell nanohybrids process. (Reprinted with permission from Li et al. (2018a). (**c**) PANI @ SnO_2/Zn_2SnO_4 nanohybrid (PZN) materials and preparation of gas sensors. (Reprinted with permission from Liu et al. (2020a))

within the concentration range of 50–1000 ppb at room temperature. Additionally, the sensor exhibited relatively fast response and recovery times of 4.2 min and 13.3 min, respectively.

Li et al. synthesized a series of sensing materials by combining hydrothermal methods with in situ chemical oxidation polymerization (Li et al. 2018a). These materials included mesoporous In_2O_3 nanospheres, Au-loaded mesoporous In_2O_3 nanospheres, polyaniline (PANI), and Au-loaded mesoporous In_2O_3 nanospheres@ PANI core–shell hybrids ($PAIn_xA_y$), as illustrated in Fig. 1.9b.

At ambient temperature, the mesoporous In_2O_3 nanospheres@PANI nanohybrid ($PAIn_{20}A_1$) containing 20% Au demonstrated a peak response value of ~46 to 100 ppm NH_3. This performance surpassed that of pure PANI and $PAIn_{10}$ by factors of 14 and 4, respectively. The sensor exhibited exceptional selectivity, reproducibility, and linear concentration-dependent behavior for NH_3 detection at room temperature. Furthermore, it achieved rapid response and recovery times of 118 s and 144 s, respectively, with a detection threshold as low as 500 ppb. These superior sensing characteristics are primarily attributed to the synergistic effects arising from the p-n junction between PANI and mesoporous In_2O_3 nanospheres, coupled with the catalytic activity of Au nanoparticles.

1.3.4.3 Tin Oxide (SnO_2)-Based Gas Sensor

SnO_2, with its robust n-type semiconductor characteristics and wide bandgap (approximately 3.6 eV), has emerged as a leading material in the development of gas-sensitive applications. Its high chemical sensitivity, stability, and cost-effectiveness make it a popular choice across various high-tech fields, including field-effect transistors, photocatalysis, and chemical sensors (Zou et al. 2017).

The gas-sensing properties of SnO_2 are primarily determined by its surface characteristics, which are influenced by the double vacancies of Sn atoms. These vacancies enable the surface composition to reversibly transition from a stoichiometric Sn^{4+} surface to a reducing Sn^{2+} surface, while also promoting the formation of oxygen vacancies. This dynamic surface behavior enhances gas sensitivity, making SnO_2 an ideal material for sensing applications (Batzill and Diebold 2005).

Liu et al. engineered a polyaniline@porous nanospheres SnO_2/Zn_2SnO_4 nanohybrid (PZN) on polyethylene terephthalate (PET) substrates through in situ chemical oxidation polymerization (Fig. 1.9c) (Liu et al. 2020a). This innovative nanohybrid was specifically developed for efficient and selective NH_3 detection at ambient temperature, eliminating the requirement for interdigitated electrodes. The distinctive morphology and the establishment of p-n heterojunctions between polyaniline (PANI) and the porous SnO_2/Zn_2SnO_4 nanospheres substantially improved the gas-sensing capabilities. Notably, when the SnO_2/Zn_2SnO_4 nanospheres (PZN10) were incorporated at 10 mol%, the sensor achieved a response of 20.4–100 ppm NH_3 at ambient temperature, representing a 2.6-fold enhancement compared to pure PANI (7.8). Furthermore, the PZN10 flexible sensor demonstrated a detection threshold as low as 500 ppb for NH_3, coupled with outstanding reproducibility, flexibility, and durability at ambient temperature. These features position it as a highly viable option for real-world NH_3 detection scenarios.

In a separate study, Kassem et al. fabricated a flexible SnO_2-based gas sensor through inkjet printing technology (Kassem et al. 2019). Utilizing an aqueous sol-gel approach, they synthesized SnO_2 sols and optimized them into printable inks with appropriate rheological characteristics. The SnO_2 crystallization temperature of 350 °C was compatible with polyimide substrates, rendering it ideal for flexible sensing applications. The inkjet-printed SnO_2 sensor exhibited remarkable capability in detecting trace ethanol concentrations, underscoring its potential for early-stage ethanol monitoring. The underlying sensing mechanism operates through two primary processes: initially, oxygen adsorption on the SnO_2 surface injects electrons into the conduction band; subsequently, the adsorbed oxygen interacts with ethanol molecules, altering the sensor's electrical characteristics. These combined processes drive the sensor's ethanol response.

Metal oxides are widely recognized for their exceptional performance in gas detection, primarily due to their distinctive attributes. These encompass diverse nanostructural configurations, straightforward synthesis methods, advantageous electrical properties, and robust chemical stability (Li et al. 2018b), Furthermore, they are characterized by swift response kinetics and excellent selectivity. Nevertheless, although capable of functioning at ambient temperature, metal oxide-based sensors frequently encounter issues such as elevated baseline resistance, which can diminish response magnitude and prolong detection times.

To mitigate these limitations, researchers have integrated room-temperature sensing materials such as CPs and carbon-based nanomaterials into the metal oxide framework. These additions create either homogeneous or heterogeneous junctions at the interface, enhancing charge carrier density and lowering baseline resistance through the composite's improved conductivity (Vallejos et al. 2015).

Moreover, to maximize surface area and active sites, metal oxides are designed into hierarchical architectures spanning zero to three dimensions. These configurations are frequently integrated with complementary sensing materials, including noble metal nanoparticles, to enhance the adsorption of target gases on the sensor surface (Asad and Sheikhi 2016). A comprehensive summary of critical parameters for metal oxide-based gas sensors is provided in Table 1.5.

The role of metallic nanoparticles in enhancing the sensing performance of diverse materials—such as carbon-based nanomaterials, CPs, and metal oxides—has been widely investigated. These metallic components can be integrated into metal oxides either as ions occupying distinct lattice positions or as nanoparticle clusters embedded within the oxide framework (Kohl 1990).

Jang et al. developed hollow Pd-Ag alloy nanowires (Pd@Ag HNWs) through a process involving the electrodeposition of lithographically patterned silver nanowires (Ag NWs), followed by the formation of palladium via galvanic replacement reaction (GRR) (Jang et al. 2017). The gas-sensing performance of these nanowires toward hydrogen was also examined (see Fig. 1.10a). By manipulating the duration of the GRR, where Ag NWs were immersed in an aqueous solution containing Pd^{2+} ions, a simultaneous compositional and morphological transformation from Ag

Table 1.5 Related parameters of metal oxides-based gas sensors

Sensing material	Target gas	Conc. (ppm)	*T* (°C)	Res. (%)	Res./Rec. Time(s)	References
ZnO	NO_2	50	270	2900	25/28.1	Kim et al. (2017)
	H_2S	20	25	2.68	404/275	Ugale et al. (2020)
In_2O_3	NO_2	5	150	575	105/785	Alvarado et al. (2018)
	NH_3	20	150	112	190/410	Alvarado et al. (2018)
SnO_2	CO	50	200	270	–	Rieu et al. (2016)
	NH_3	100	25	2040	46/54	Liu et al. (2020a)

NWs to Pd@Ag HNWs was achieved, with precise control over the atomic ratios of Pd and Ag. Specifically, a 17-h GRR resulted in Pd@Ag HNWs that exhibited a significantly improved hydrogen response and a 2.5-fold reduction in sensing time compared to those subjected to a 10-h GRR. The Pd-Ag 17h sample showed high sensitivity to varying hydrogen concentrations, with rapid response times of 120 s for low concentrations and 102 s for high concentrations.

Zhang et al. engineered a wireless ammonia (NH_3) detection system by combining silver nanoparticles (Ag NPs) with reduced graphene oxide (RGO) and integrating this composite with a commercial near-field communication (NFC) tag (Zhang et al. 2019b). The synergistic effects between Ag NPs and RGO significantly improved the sensor's NH_3 detection capabilities, achieving a swift response time of 7.5 s and a recovery duration of 20 s. Gas sensitivity tests revealed that the Ag-RGO composite showed a significantly higher response to NH_3 compared to pristine RGO, especially when the concentration of Ag nanoparticles was increased. In contrast, pristine RGO exhibited no response to NH_3. This suggests that Ag nanoparticles significantly enhance the sensing capability of RGO for ammonia, as illustrated in Fig. 1.10b. The wireless sensor demonstrated excellent detection performance and selectivity for NH_3 concentrations ranging from 5 to 100 ppm at room

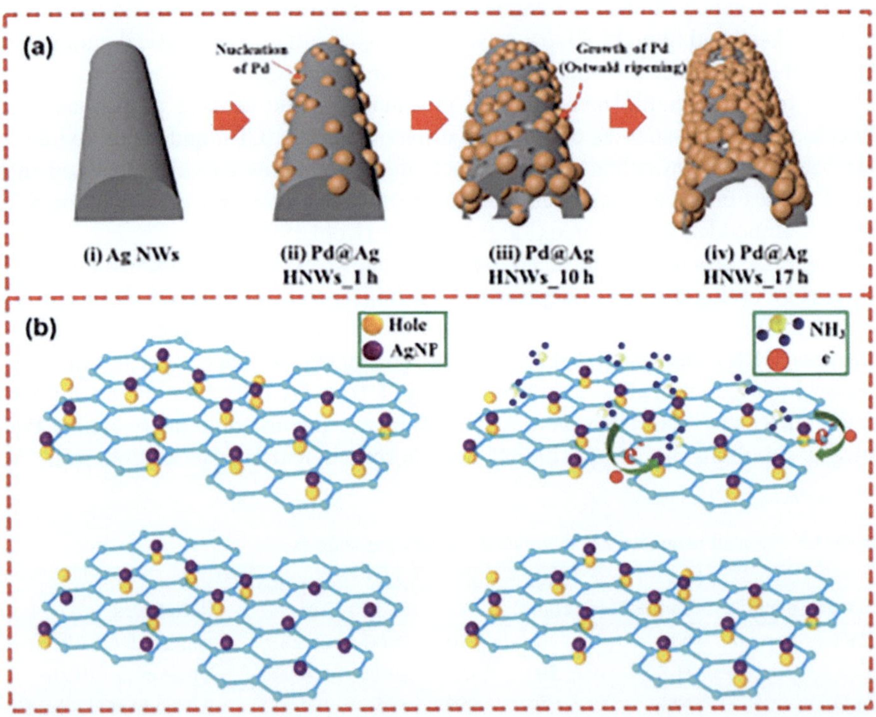

Fig. 1.10 (**a**) Schematic material synthesis from bare Ag NW to hollow Pd@Ag NW. (Reprinted with permission from Jang et al. (2017)). (**b**) Schematic diagram of the electron transfer principle of Ag-RGO in NH_3 gas and air (Zhang et al. 2019b). (Reprinted under the Creative Commons CC BY license)

Table 1.6 Related parameters of metal hybrid nanomaterials-based gas sensors

Sensing material	Target gas	Conc. (ppm)	*T* (°C)	Res. (%)	Res./Rec. Time(s)	References
Pd	H_2	100	25	20	–	Shin et al. (2015)
	H_2	1%	100	1.2	180/460	Öztürk and Kılınç (2016)
Ag/rGO	NO_2	5	25	200	220	Li et al. (2018c)
	NH_3	5	25	1.25	7.5/20	Zhang et al. (2019b)
Cu/SWCNT	H_2S	20	25	27	10/20	Asad et al. (2015)

temperature. A comprehensive overview of critical parameters for gas sensors utilizing metal hybrid nanomaterials is presented in Table 1.6.

1.4 Synthesis Method of Nanomaterials

Nanotechnology-derived nanomaterials hold significant promise for detecting air pollutants, spanning applications from nanosorbents used in gas sampling to nanostructured gas-sensitive materials employed in gas detection. Even materials with identical compositions can exhibit distinct properties based on their synthesis methods. Altering synthesis techniques may thus be a crucial approach to enhance the characteristics of nanomaterials. Broadly speaking, the synthesis of nanomaterials can be categorized into two primary methods: top-down and bottom-up approaches.

1.4.1 Top-Down Methods

1.4.1.1 Mechanical Milling

From the inception of nanoscience, synthesizing nanomaterials in an easy, cost-effective, and high-yield manner has been a significant challenge. To date, numerous top-down approaches have been developed to facilitate the commercial-scale production of nanomaterials. Among these, high-energy ball milling has emerged as a widely utilized method for creating a diverse range of nanomaterials, including nanoparticles, nanoalloys, nanocomposites, and nanoquasicrystals (Zhuang et al. 2016).

Mechanical alloying methods have been utilized to synthesize amorphous and nanocrystalline alloys, along with metal/nonmetal nanocomposites. This is accomplished through the milling and subsequent annealing of elemental or composite powders under inert conditions (Yadav et al. 2012). The technique, referred to as mechanical alloying, entails grinding diverse elemental powders in an inert environment to produce a homogenized mixed powder with a consistent composition.

Using mechanical milling, various Al/Ni/Mg/Cu-based nanoalloys, wear-resistant spray coatings, oxide- and carbide-reinforced aluminum alloys, and numerous other nanocomposites have been successfully synthesized (Lyu et al. 2017). Mechanical milling is applied either through direct milling and annealing or by mechanical activation followed by additional processing of the activated materials to produce nanomaterials.

Sun et al. investigated the microstructural transformation of an ordered $L1_2 + DO_{22}$ multiphase $Al_{67}Mn_8Ti_{24}Nb_1$ alloy during mechanical milling and annealing processes. Following 15 h of mechanical milling, the ordered $L1_2 + DO_{22}$ polyphase $Al_{67}Mn_8Ti_{24}Nb_1$ alloy transitioned into a disordered face-centered cubic (fcc) supersaturated solid solution (Sun et al. 2002). The complete amorphization of the $Al_{67}Mn_8Ti_{24}Nb_1$ alloy is attributed to the formation of high-density nanocrystalline grain boundaries, which elevate the alloy's free energy, coupled with the supersaturated dissolution of the Al_3 (TiNb) second phase in the fcc solid solution, inducing substantial internal lattice strain. During subsequent annealing, the amorphous fcc phase undergoes a morphological transition at 250 °C, followed by the transformation of the disordered fcc into an $L1_2$ ordered structure at approximately 300 °C. Between 500 and 650 °C, the DO_{22}-structured second phase precipitates within the $L1_2$ matrix, accompanied by grain growth to submicron dimensions.

Castro et al. examined the structural evolution of Mg, Nb, and a Mg-5 at% Nb mixture during reactive milling. Their results demonstrated that Nb rapidly incorporates hydrogen to form the $NbH_{0.9}$ phase (de Castro et al. 2004). Conversely, Mg undergoes partial transformation into β-MgH_2 and γ-MgH_2 phases. Importantly, the quantity of unreacted Mg remained constant when the milling duration was increased from 25 to 48 h, underscoring Nb′s catalytic role in enhancing Mg′s hydrogen absorption. Based on prolonged milling results, the study discussed the structural evolution of Mg-Nb composites. Both Mg and Mg-Nb samples exhibited lower dehydrogenation temperatures compared to polycrystalline Mg. The Mg-Nb nanocomposites demonstrated the lowest desorption temperature (270 °C) and the fastest hydrogen uptake kinetics.

Sheibani et al. synthesized copper nanoparticles with an average grain size of 27 nm using graphite as the starting material and employing high-energy planetary ball milling to reduce Cu_2O via a solid-phase reduction process (Sheibani et al. 2007). In the early stages of milling, graphite facilitated the mechanochemical reduction of Cu_2O. This process led to a continuous decrease in the average grain size of Cu_2O and an increase in internal strain. As milling time increased, finer copper powder was produced.

However, the grain size and internal strain of the copper particles fluctuated irregularly with extended milling time. This was attributed to the gradual formation of copper during the milling process. The copper particles underwent mechanical deformation and refinement until all copper oxide was fully converted to metallic copper. Prolonged milling eventually caused agglomeration of the copper powder. After 30 h of ball milling, the average grain size of the resulting copper was approximately 27 nm.

Additionally, Sheibani et al. prepared Cu-(2.5–10) vol.% Al_2O_3 nanocomposite powders using high-energy mechanical milling (HEMM).

1.4.1.2 Electrospinning

Electrospinning is a highly effective and straightforward technique for producing nanomaterials with diameters spanning from tens of nanometers to micrometers. First developed in 1934 for the synthesis of one-dimensional (1D) materials, electrospinning has proven capable of forming a wide range of materials into uniform 1D structures. These materials include polymers, inorganics, and hybrid (organic-inorganic) composites (Ostermann et al. 2011).

Ceramic nanofibers (NFs) can be produced by heating electrospun organometallic NFs in an oxidizing atmosphere. Metal nanotubes or nanowires can be created by heating NFs containing metal atoms in a reducing atmosphere. Additionally, hydrocarbon NFs, such as polyacrylonitrile NFs, can be transformed into carbon NFs through low-temperature oxidation followed by heating in an inert atmosphere (Suresh Kumar et al. 2014).

1D electrospun nanostructured materials have garnered significant attention for constructing active energy-harvesting devices due to their distinctive optical and electrical properties. These materials are particularly appealing for applications in photovoltaics, photocatalysis, hydrogen generation, and fuel cells (Du et al. 2012). Compared to bulk nanostructures, 1D nanostructures created through electrospinning exhibit superior characteristics, including high chemical reactivity, large surface area, low density, enhanced light absorption, and improved dye adsorption. As a result, the development of 1D nanostructured fibers using metal oxides, composites, dopants, and surface modifications has become a focal point of research (Song et al. 2004).

Anjusree et al. fabricated anisotropic TiO_2 nanostructures by controlling the chemical properties, particularly solubility, between TiO_2 and the polymer during electrospinning (Anjusree et al. 2013). In the electrospinning process, metal oxide nanofibers like TiO_2 can be generated by co-electrospinning a TiO_2 precursor with a polymer to produce composite nanofibers, followed by polymer removal through sintering. The resulting TiO_2 morphology is determined by the polymer's chemical characteristics and its interactions with TiO_2 during the sintering phase.

For instance, when the polymer is changed from polyvinyl pyrrolidone (PVP) to polyvinyl acetate (PVAc), or to a mixture of PVP and PVAc, the TiO_2 morphology evolves from continuous fibers to rice-like shapes and then to leaf-like shapes. In a device tested under 0.5 AM conditions, the current density reached 5.7 mA cm^{-2}, the open-circuit voltage was 0.65 V, the device efficiency was 4.5%, and the fill factor was 59.4%.

Li et al. successfully developed one-dimensional CuS/electrospun carbon nanofiber heterostructures (CuS/EC) with high catalytic activity by integrating electrospinning technology with a hydrothermal process (Li et al. 2014a). Through optimized reaction conditions, Cu nanoparticles were uniformly grown on electrospun carbon nanofibers as secondary nanostructures with excellent dispersibility (Fig. 1.11a). The study demonstrated that L-cysteine, acting as both a sulfur donor and chelator, facilitated the growth of CuS on carbon fibers.

The CuS/EC composite was applied to FTO glass and exhibited superior performance in quantum dot-sensitized solar cells (QDSCs), outperforming conventional Pt electrodes. The high catalytic activity of Cu nanoparticles played a crucial role in reducing oxidized polysulfides. Meanwhile, the 3D mat-like morphology of the carbon nanofibers served as a framework, bridging all CuS nanoparticles and enhancing charge transport during the catalytic process (Fig. 1.11b).

Kim et al. developed a versatile platinum nanofiber (PtNF) mesh capable of functioning both as a catalytic layer and a transparent counter electrode (CE) in dye-sensitized solar cells (DSSCs), thereby eliminating the reliance on conventional indium-doped tin oxide (ITO) or fluorine-doped tin oxide (FTO) glass (Kim et al. 2013). When employed as a counter electrode in DSSCs, the PtNF mesh demonstrated a power conversion efficiency of 6.0%, equivalent to 83% of the efficiency achieved by traditional DSSCs utilizing Pt-coated FTO glass (Fig. 1.11c).

The innovative DSSCs incorporating the PtNF mesh demonstrated highly stable photoelectric conversion efficiency, outstanding catalytic and conductive properties, transparency, and long-term stability. Additionally, the removal of the transparent conductive layer on the counter electrode significantly reduced manufacturing costs while maintaining DSSC functionality. The fabrication method based on PtNF

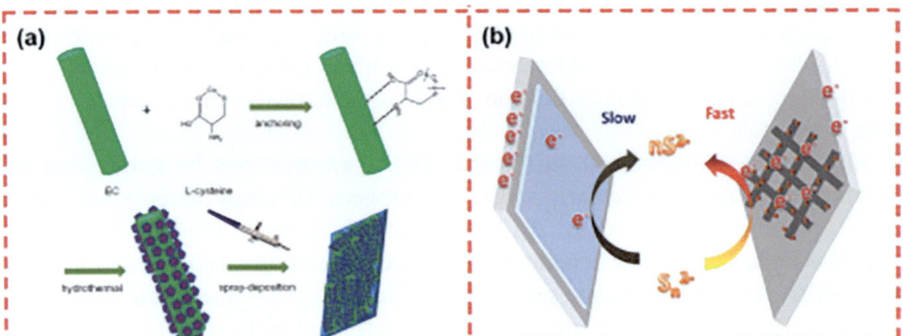

Fig. 1.11 (**a**) Schematic diagram of the preparation and precipitation formation process of CuS/electrospun carbon nanofiber heterostructures (CuS/EC). (Reprinted with permission from Li et al. (2014a)). Schematic representation of charge transport during catalysis. (Reprinted with permission from Li et al. (2014a)). (**c**) Schematic of a DSSC fabricated with a Pt nanofiber network as a counter electrode (DSSC-PtNF) and a Pt spin-coated FTO counter electrode (DSSC-FTO-Pt). (Reprinted with permission from Kim et al. (2013))

meshes (Fig. 1.11c) offers potential for broader application in other electrocatalytic photoelectronic devices, combining excellent catalytic activity, high conductivity, and transparency.

1.4.1.3 Lithography

Micro/nano lithography serves as a crucial enabling technology not only in the semiconductor and integrated circuit industries but also increasingly in the manufacturing of commercial microelectromechanical systems (MEMS) devices and in prototyping applications within emerging nanoscale science and engineering fields (Ko 2007). When combined with other manufacturing processes like deposition and etching, lithography can produce high-resolution structures, and this process can be repeated multiple times to create complex micro/nanoscale architectures. Lithography techniques are broadly categorized into two types based on the use of masks or templates: mask lithography and maskless lithography.

Mask lithography employs a mask or mold to transfer patterns over a large area simultaneously, enabling high-throughput production of up to tens of slices per hour. Common forms of mask lithography include photolithography (Sanders 2010), soft lithography (Xia and Whitesides 1998), and nanoimprint lithography (Chen et al. 2009).

In contrast, maskless lithography, such as electron beam lithography (Grigorescu and Hagen 2009), focused ion beam lithography (Tseng 2004), and scanning probe lithography (Shim et al. 2011), creates patterns by continuously writing without the need for a mask. These methods allow for ultra-high-resolution patterning of arbitrary shapes with minimum feature sizes as small as a few nanometers.

Roy et al. demonstrated the ability to create diverse nanostructures, including dots and lines, by etching quantum dots and magnetic nanoparticles onto gold and mica surfaces using dip-pen nanolithography (DPN) at room temperature (Roy et al. 2007). Their work showed that CdSe/ZnS particles retained their functionality after being written using DPN. Additionally, they found a linear correlation between writing speed and structure size over the studied time range. For the fabrication of thinner structures and single-particle devices, precise control over particle positioning is essential. This can be achieved by enhancing the stability of the atomic force microscope (AFM) scanner and employing a sharper tip. The application of DPN for imaging nanoparticles at the nanoscale could significantly broaden their potential applications.

Li et al. investigated the potential of DPN based on AFM and expanded its application to create protein nanopatterns along individual DNA molecules (Li et al. 2005). This method allows for real-time and in situ observation of the patterns using AFM, marking a significant advancement in nanopatterning on soft materials. Specifically, it enables the construction of nanopatterns on a single stretched DNA molecule. This innovative approach not only facilitates the study of site-specific interactions between DNA molecules and other molecules but also opens new possibilities in nanoscience and nanotechnology.

Chen et al. employed a combination of UV nanoimprint lithography (UV-NIL) and reactive ion etching (RIE) with Cr/Au lift-off to create sub-250 nm periodic gold nanopore arrays on glass substrates (Chen et al. 2009). They also proposed a soft UV nanoimprint and extraction-based replication technique for fabricating metal nanopore arrays. The plasma transport response of these nanopore arrays was found to be dependent on their diameter and periodicity. For sucrose solutions, an increase in transmission was observed, with higher sucrose concentrations resulting in greater transmission. This effect is consistent with changes in the dielectric index. Additionally, the sensitivity of sulfhydryl grafting at a thickness of 2 nm was investigated, confirming the high sensitivity of nanopore arrays for real-time chemical and biosensing applications using surface plasmon resonance (SPR).

1.4.1.4 Sputtering

In 1852, Grove first observed sputtering in a direct current gas discharge tube, noting that the cathode surface was eroded by high-energy ions formed in the gas phase, with the cathode material subsequently depositing on the inner wall of the discharge tube (Grove 1997). Since then, sputtering has been recognized as an effective technique for producing thin films of nanomaterials. During sputtering deposition, high-energy gas ions bombard the target surface, causing the physical ejection of small atomic clusters. The energy of the incident gas ions determines the size and nature of these ejected clusters (Chan and Teo 2005).

The sputtering process occurs in a vacuum chamber into which a sputtering gas is introduced. A high voltage is applied to the cathode target, causing free electrons to collide with the gas atoms and generate gas ions. These positively charged ions are then strongly accelerated toward the cathode target by the electric field. The continuous bombardment of the cathode target by these ions results in the ejection of atoms from the target surface. This mechanism is the basis for magnetron sputtering, a widely used technique in thin film deposition, etching, and analytical applications (Gao et al. 2000).

Wender et al. sputtered gold foil onto various ionic liquids (ILs), including 1-n-butyl-3-methylimidazolium tetrafluoroborate, hexafluorophosphorate, bis(trifluoromethylsul-fonyl)amide, and tris(fluoro)tris(perfluoroethane)phosphate, under conditions of 40 mA, 335 V, and 2 Pa Ar working pressure (Wender et al. 2010). This process yielded stable and well-dispersed gold nanoparticles (NPs) with sizes ranging from 3 to 5 nm. The size and distribution of these gold nanoparticles were found to depend on various experimental parameters, particularly the surface composition of the IL rather than its surface tension or viscosity. Nucleation and NP growth occurred simultaneously on the IL surface, with NP size varying according to the IL's surface composition. Specifically, increasing fluoride content influenced NP size, as illustrated in Fig. 1.12a. Additionally, NP size was independent of sputtering time but correlated with discharge current. Higher discharge currents resulted in more gold atoms reaching the IL surface per unit time, thereby altering the dynamics of particle growth on the IL surface.

Hyeon et al. utilized RF magnetron sputtering at ambient temperatures (<27 °C) to create an economical physical vapor deposition method for producing two-dimensional WSe_2 layered nanofilms, which served as electrochemical catalysts for the hydrogen evolution reaction (HER) (Nam et al. 2020). Through careful modulation of sputtering parameters, including RF power and deposition duration, the loading capacity and electrochemical surface area (ECSA) of the WSe_2 films on carbon paper could be finely tuned, as depicted in Fig. 1.12b. The surface of the sputtered WSe_2 film was partially oxidized, resulting in the formation of spherical particles. Regardless of the WSe_2 loading, the Tafel slope of the WSe_2 electrode during HER testing was consistently narrow, ranging from approximately 120 to 138 mV dec^{-1}. This indicated the good reproducibility and inherent catalytic activity of the WSe_2 films. When balancing loading capacity and electrochemical surface area (ECSA), the optimal hydrogen evolution reaction (HER) performance was achieved with the 200 W-15 min sample, which exhibited an overpotential of 220 mV and a current density of 10 mA cm^{-2}. This straightforward low-temperature sputtering technique can be readily adapted for other 2D transition metal dichalcogenide (TMD) electrochemical catalysts, holding promise for practical applications.

Nie et al. introduced an innovative approach for the sputtering deposition of metal nanoparticles within reactive ion etching cavities, characterized by extremely short distances between the target and substrate (Nie et al. 2009). The study revealed that sputtering distance, ion concentration, and sputtering time significantly influence the distribution and morphology of nanoparticles. Dense nanoparticles of various compositions were successfully deposited on substrates positioned 130 microns or less from the target. In contrast, when the distance was increased to 510μm, island-like structures emerged, suggesting a higher propensity for continuous film formation at greater distances. This observed trend in nanoparticle formation contrasts with previously reported mechanisms involving sputtering. It is hypothesized that the seeding effect and atomic neutrality on the substrate surface are crucial factors in facilitating nanoparticle formation. By leveraging these insights, a cost-effective nanofabrication method could be developed to produce nanoparticles and integrate them into functional devices.

1.4.1.5 Arc Discharge Method

The arc discharge method for producing nanoparticles has garnered significant attention due to its straightforward equipment setup and scalability in nanoparticle production rates. This method has been utilized to synthesize a variety of nanoparticles, including metals and refractory metals (Ashkarran et al. 2009). One of the most extensively reported applications of this method is the synthesis of carbon nanotubes (CNTs) using direct current (DC) arc discharge. While catalyst metals such as nickel (Ni), yttrium (Y), iron (Fe), cobalt (Co), or molybdenum (Mo) are necessary for the production of single-walled CNTs (SWNTs), they are not required for the synthesis of multi-walled CNTs (MWNTs) (Xu 2007).

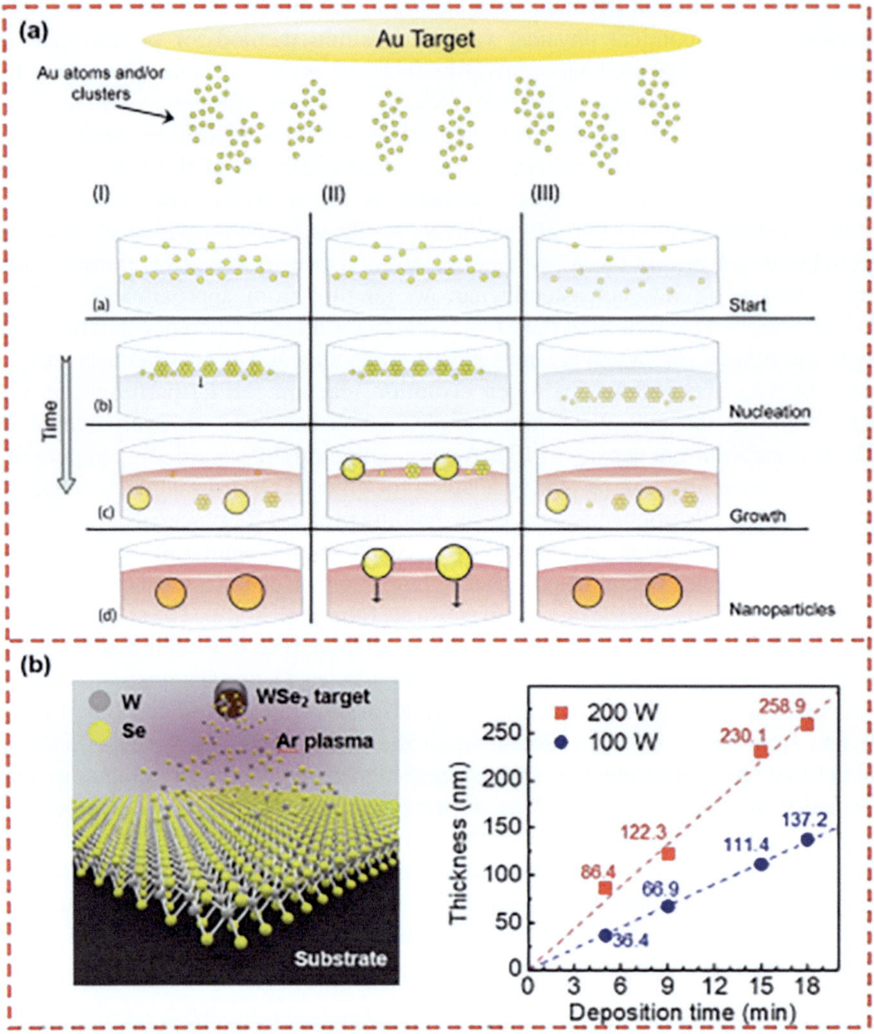

Fig. 1.12 (**a**) Schematic representation of the possible mechanisms for the nucleation and growth of sputtered gold nanoparticles on IL substrates. (Reprinted with permission from Wender et al. (2010)). (**b**) Schematic of the synthesis of a WSe_2 film deposited using RF magnetron sputtering and a plot of the thickness of this film as a function of deposition time for different RF powers. (Reprinted with permission from Nam et al. (2020))

Arc discharge is a straightforward process involving two electrodes that function as the cathode and anode. When a potential is applied across these electrodes, it induces electrical breakdown. This method typically operates at low voltages (a few to several tens of volts) and high current densities (ranging from 10^5 to 10^{11} A/m^2). Nanoparticle synthesis using arc discharge can be conducted in either pulsed or continuous modes (Zhao et al. 1997).

The synthesis of multi-walled carbon nanotubes (MWNTs) and single-walled carbon nanotubes (SWNTs) typically involves the use of high-purity graphite electrodes. These electrodes undergo arc discharge within an inert environment, such as helium or hydrogen, to enable the formation of the nanotubes.

Su et al. produced ZnO nanoparticles of varying sizes, with average rod-like diameters ranging from 10 to 23 nm, by adjusting process parameters such as electrode current, frequency, and spacing (Su et al. 2014). Elemental zinc was generated via arc discharge in ethanol, while ZnO nanoparticles were synthesized through arc discharge in deionized water.

Higher current and increased electrode movement frequency led to larger particle sizes and broader size distributions. This effect is likely due to the widening of the arc column and reduced cooling time, which promotes particle coalescence and growth. Conversely, increasing the electrode gap resulted in a slight decrease in the average particle size of nano-ZnO. The study indicated that nanoparticle quenching is driven by the combined effects of convection and surface vibration.

Every synthesized ZnO nanoparticle exhibited distinct ultraviolet (UV) and red visible emission bands. Significantly, the red visible emission intensity increased substantially with decreasing particle size. This phenomenon is attributed to a higher density of structural defects, such as nanopores and enhanced surface roughness.

Ashkarran et al. explored the synthesis and characterization of tungsten trioxide (WO_3) nanoparticles produced via arc discharge in deionized (DI) water (Ashkarran et al. 2008). The study examined the particle size and morphology of WO_3 nanoparticles synthesized at varying arc currents (25, 35, and 45 A). At an arc current of 25 A, the particle size was approximately 30 nm, increasing to 64 nm as the arc current rose. This increase in particle size led to a decrease in the optical band gap from 2.9 eV to 2.6 eV.

Photodegradation experiments using Rhodamine B demonstrated that samples prepared at the lowest arc current exhibited the highest photocatalytic activity. This was attributed to their smaller particle size and larger surface area. The findings indicate that the arc discharge method can effectively generate nano-WO_3 directly.

1.4.1.6 Laser Ablation

Laser ablation synthesis employs a high-energy laser beam to vaporize a target material, producing nanoparticles in the process. In this method, the precursor material is converted into a vapor phase by the laser's intense energy, resulting in nanoparticle formation. Since its introduction decades ago, laser ablation has been extensively utilized in diverse areas such as nanomaterial fabrication, surface cleaning, and nanopatterning (Zhang et al. 2017).

In nanoparticle synthesis via laser ablation, the source material is vaporized by laser irradiation, allowing for the production of a diverse range of materials at relatively low temperatures. This method is particularly advantageous for synthesizing high melting point elements or compounds, such as metals like silver, silicon, ceramics, oxide composites, and nitrides (Ismail et al. 2020).

The most commonly utilized laser ablation source is neodymium-doped yttrium aluminum garnet (Nd:YAG). In laser ablation techniques, the laser beam is focused onto the target material's surface at defined angles, usually 90° or 45°. The energy transfer to the target material is influenced by the laser's power, wavelength, and pulse duration. Photothermal ablation occurs when lower laser power (100–500 mW/cm^2) and longer pulse durations (>10 ns) are applied (Chrzanowska et al. 2015).

In photothermal ablation, the laser energy is absorbed by either free or bound electrons within the target material. These energized electrons subsequently transfer energy via collisions with other electrons and lattice phonons, resulting in heat diffusion throughout the material. This mechanism elevates the material's temperature, inducing melting (Duque et al. 2019).

Khang et al. synthesized nano-Si particles using pulsed nanosecond laser ablation (Khang et al. 2010). During the process, a positive voltage bias was applied during laser irradiation, which significantly narrowed the particle size distribution. The experimental setup used for this preparation is illustrated in Fig. 1.13a.

The study demonstrated that pulsed laser ablation is an effective technique for producing high-purity silicon nanoparticles. Applying a positive voltage bias was shown to be particularly effective in reducing the dispersion of nanoparticle sizes. The synthesized Si nanoparticles exhibited a density of $>1.0 \times 10^{12}$/cm^2 (within a 30-min sampling time), an average particle size of 4–5 nm, a geometric standard deviation of approximately 1.35, and a production rate of around 2×10^{-5} g/h.

Si nanoparticles produced through laser ablation possess desirable charge-trapping and -detrapping properties, making them suitable for applications in memory devices.

Duque et al. described the synthesis of metal oxide nanoparticles ($TiFe_2O_4$ and $ZnFe_2O_4$) in a liquid environment using pulsed laser ablation (Fig. 1.13b) (Duque et al. 2019). The study employed a nanosecond Nd:YAG laser system with wavelengths of 532 nm and 1064 nm, maintaining a constant laser beam energy of 80 mJ.

Changing the laser wavelength influenced the optical properties of the resulting nanoparticles. Specifically, altering the wavelength shifted the absorbance peak relative to the surface plasmon resonance peak in the UV-visible (UV-vis) spectrum, indicating that these properties are dependent on laser parameters. For nanoparticles grown at 532 nm, the absorbance peak near 150 nm exhibited a redshift with longer preparation times. In contrast, for those grown at 1064 nm, the absorption spectra remained unchanged, likely due to particle agglomeration and the formation of larger particles.

Characterization revealed that these nanoparticles possessed suitable plasmonic photocatalytic properties, making them promising for various applications.

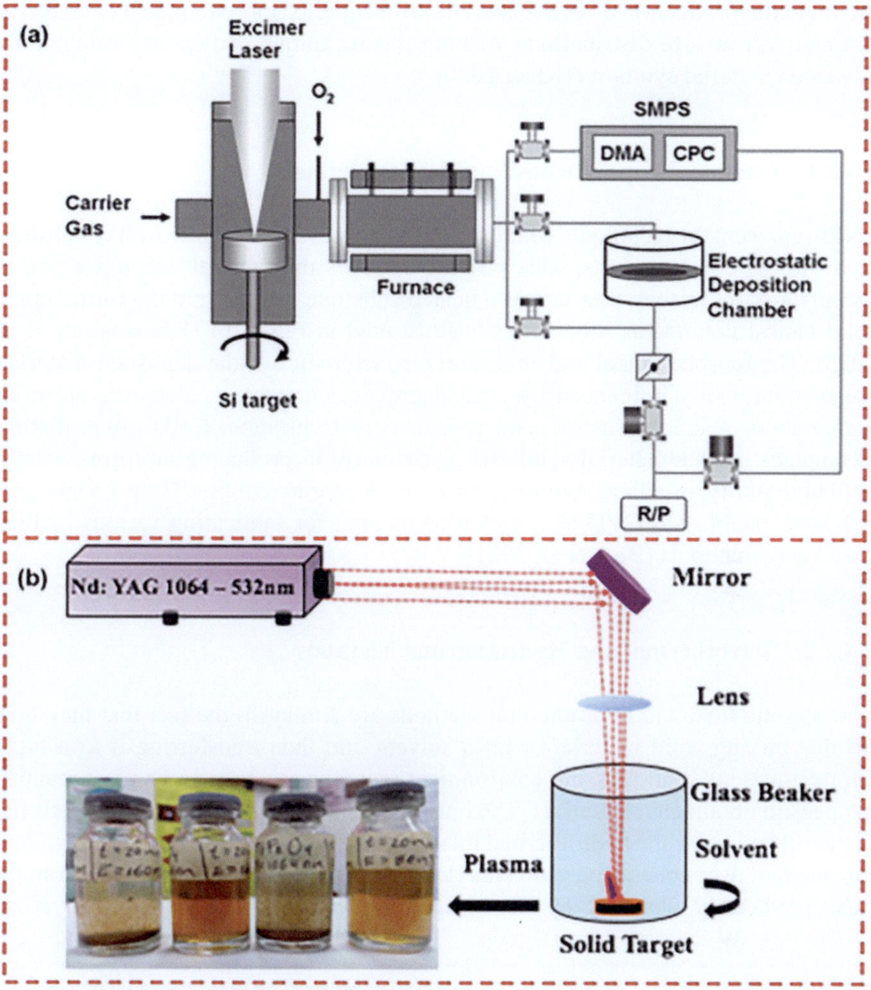

Fig. 1.13 (**a**) Schematic diagram of the laser ablation system and the reactor with the high-pressure bias electrode. (Reprinted with permission from Khang et al. (2010)). (**b**) Schematic of the experimental setup for the preparation of colloidal nanoparticles (NPs) by laser ablation in a liquid environment (Duque et al. 2019). (Reprinted under a Creative Commons license (CC BY 4.0))

1.4.2 Bottom-Up Methods

Bottom-up synthesis represents a fundamental approach in nanomaterial fabrication, involving the controlled assembly of atomic or molecular building blocks. This methodology enables precise regulation of chemical synthesis parameters, effectively suppressing the formation of undesired particle morphologies and size distributions. Through meticulous control of reaction conditions, bottom-up approaches facilitate the production of nanomaterials with superior morphological characteristics and narrow size distributions, making this technique particularly valuable for advanced material synthesis (Khan 2020).

1.4.2.1 Chemical Vapor Deposition (CVD) Method

CVD represents a technique wherein nanomaterials are deposited on a substrate from the gaseous reactants. This process involves the thermal decomposition of precursor gases as they flow across a heated substrate, resulting in the formation of solid nanostructures, as schematically illustrated in Fig. 1.14 (Esmaeilpour et al. 2023). The morphological and structural characteristics of the deposited materials are predominantly influenced by critical process parameters, including substrate temperature, reaction kinetics, and precursor concentration. CVD offers distinct advantages in nanomaterial synthesis, particularly in producing uniform coatings with high purity, excellent stability, and superior reproducibility. These characteristics have established CVD as a preferred method for fabricating various carbon-based nanomaterials (Baig et al. 2021).

1.4.2.2 Solvothermal and Hydrothermal Methods

The solvothermal and hydrothermal methods are similar in the fact that they both involve mixing solid materials with a solvent and then transferring it to a high-temperature and high-pressure environment (autoclave) where a chemical reaction happens to obtain nanomaterials. The only difference between the two methods lies in the solvent used: the hydrothermal method employs water, whereas the solvothermal method uses nonaqueous solvents (Denkbas et al. 2016). After the reaction, the final product is obtained by centrifugation, washing, and drying. Le et al.

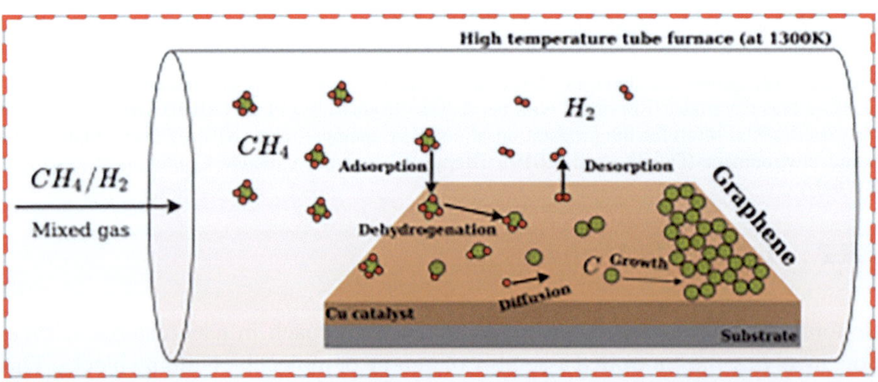

Fig. 1.14 Example of the chemical vapor deposition procedure used for graphene synthesis on Cu as the metal catalyst (Esmaeilpour et al. 2023). (Reprinted under a Creative Commons license (CC BY 4.0))

successfully synthesized high-quality porous CFO nanorods with hydrothermal synthesis method and achieved effective detection of VOCs (Thi Thanh Le et al. 2023). As illustrated in Fig. 1.15, these techniques enable precise control over nanocrystal characteristics, including crystallinity, morphology (e.g., nanorods, nanowires, nanosheets, nanospheres), and dimensional parameters through manipulation of reaction conditions (temperature, duration) and chemical parameters (solvent composition, surfactant selection, precursor choice) (Yan et al. 2008).

1.4.2.3 Sol–Gel Method

The sol–gel technique has gained widespread recognition as a versatile approach for synthesizing high-purity metal oxide nanoparticles and mixed oxide composites (Parashar et al. 2020). This method offers distinct advantages, including precise control over material texture and surface characteristics, coupled with energy-efficient operation due to its relatively low temperature requirements. Liu et al. successfully prepared CuO porous foam with abundant oxygen vacancies via the sol–gel method as depicted in Fig. 1.16 (Liu et al. 2023). The sol–gel process derives its name from its characteristic phase transitions and typically involves five sequential stages (Danks et al. 2016): (1) the precursor (metal alkoxides etc.) is hydrolyzed in the solvent (water as or organic solvents) to form "sol," with the process being classified as aqueous or nonaqueous based on solvent selection; (2) forms "gelby" polycondensation; (3) aging, which refers to continued condensation in a localized solution; (4) drying the gel to get xerogel, aerogel, or cryogel; and (5) thermal treatment for residue removal. Critical parameters influencing the final product characteristics include precursor selection, solution pH, drying methodology, and calcination temperature.

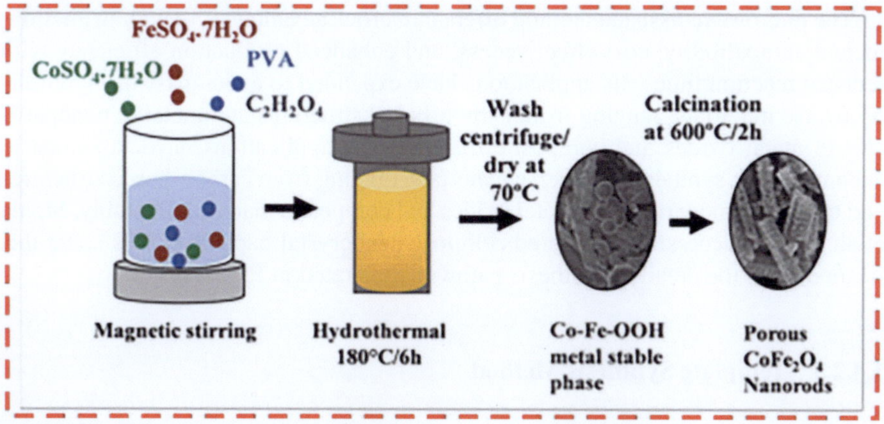

Fig. 1.15 Processes for the synthesis of $CoFe_2O_4$ materials by the hydrothermal synthesis. (Reprinted with permission from Thi Thanh Le et al. (2023))

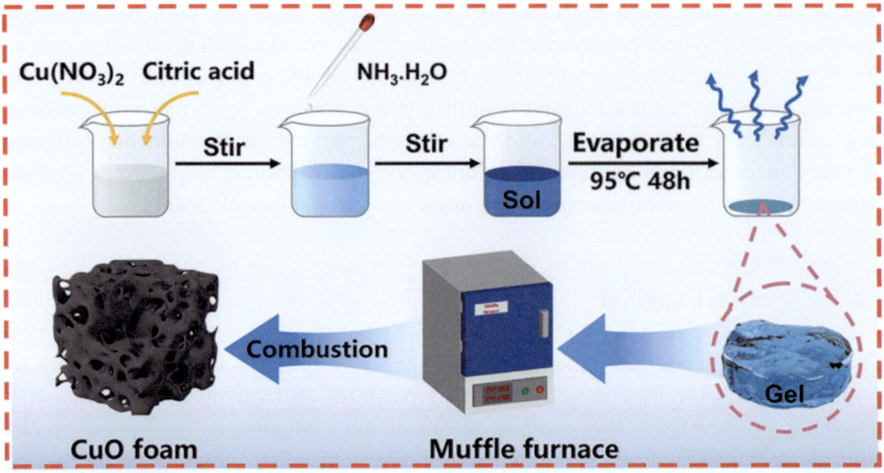

Fig. 1.16 Synthetic route of CuO foam through a sol–gel method. (Reprinted with permission from Liu et al. (2023))

1.4.2.4 Microwave-Assisted Method

Solution-based synthesis methods typically utilize thermal energy to drive chemical reactions. Among various heating techniques, microwave-assisted synthesis has emerged as a distinctive alternative to conventional heating methods. Unlike traditional surface heating that relies on convection and conduction for energy transfer (de la Hoz et al. 2016), microwave heating operates through dual mechanisms: dipole rotation and ionic conduction, resulting in more efficient, rapid, and uniform thermal distribution (Gao et al. 2023). This advanced heating approach enables precise control over nanomaterial size and morphology through manipulation of precursor concentration and irradiation duration (Rao et al. 2017).

The microwave-assisted method offers numerous advantages, including environmental compatibility, cost-effectiveness, and enhanced production efficiency with reduced reaction times. Its applications have expanded to encompass the synthesis of diverse materials, ranging from carbon-based structures and metallic nanoparticles to metal oxides and composite materials. Its applications have expanded to encompass the synthesis of diverse materials, ranging from carbon-based structures and metallic nanoparticles to metal oxides and composite materials. Notably, Mpala et al. (2023) successfully prepared cellulose nanocrystal-capped AgNPs using this method, with the detailed synthesis pathway illustrated in Fig. 1.17.

1.4.2.5 Template Synthesis Method

The template synthesis method has emerged as an innovative approach for fabricating high-quality nanomaterials through structural mediation using designed templates. This methodology typically involves three sequential stages: (1) template

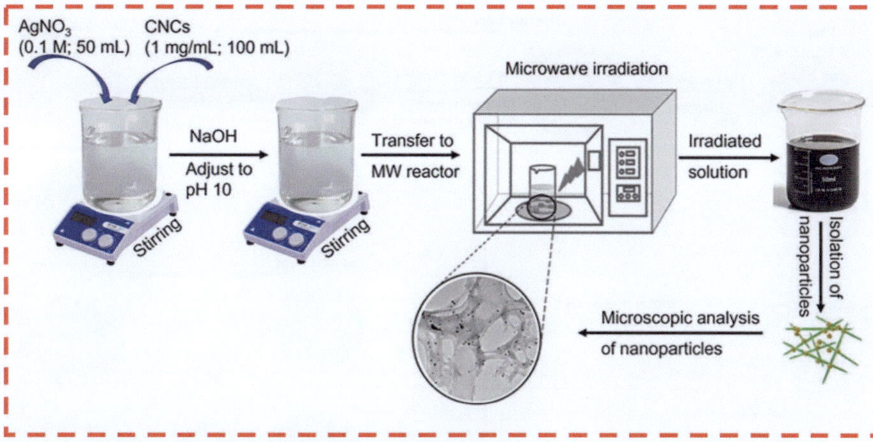

Fig. 1.17 Synthetic route of CuO foam through a sol–gel method. (Reprinted with permission from Mpala et al. (2023))

preparation, (2) nanomaterial fabrication within the template matrix, and (3) template removal through either physical dissolution or chemical calcination processes (Petkovich and Stein 2013). For this method, the template serves as a structural guide, dictating the morphological characteristics of the final product, while enabling precise control over critical material parameters including pore dimensions, grain morphology, phase composition, and particle size distribution. As shown in Fig. 1.18, by controlling the content of NaCl template, the morphology and structure of $ZnFe_2O_4$ can be changed (Lv et al. 2022). This synthesis approach offers multiple advantages, including mild reaction conditions, reduced processing times, minimal requirement for hazardous precursors and solvents, and elimination of toxic byproducts, while maintaining energy efficiency.

The versatility of template-directed synthesis is evidenced by its successful application in producing diverse nanostructures, including metal and metal oxide nanoparticles (Poolakkandy and Menamparambath 2020), nanoporous materials (Zhao et al. 2006), nanofibers (Bagherzadeh et al. 2017), MOFs (Doustkhah et al. 2021), etc.

Generally, the template-assisted synthesis can be systematically categorized into two distinct approaches based on template characteristics: hard-template and soft-template methodologies (Poolakkandy and Menamparambath 2020). The hard-template approach, alternatively termed nanocasting, employs rigid structural materials with micro- or nano-scale features as hard templates. These templates can include polymeric matrices, silica nanostructures, metal oxides, layered hydroxides, noble metals, graphene derivatives, zeolites, and MOFs (Petkovich and Stein 2013). In contrast, soft-template synthesis utilizes flexible nanostructures stabilized through intermolecular interactions, including surfactant assemblies, organic molecular aggregates, and block copolymer systems (Meng et al. 2005). These templates typically interact with precursor materials through weak noncovalent bonds, including electrostatic forces, van der Waals interactions, and hydrogen bonding networks (Li et al. 2014b).

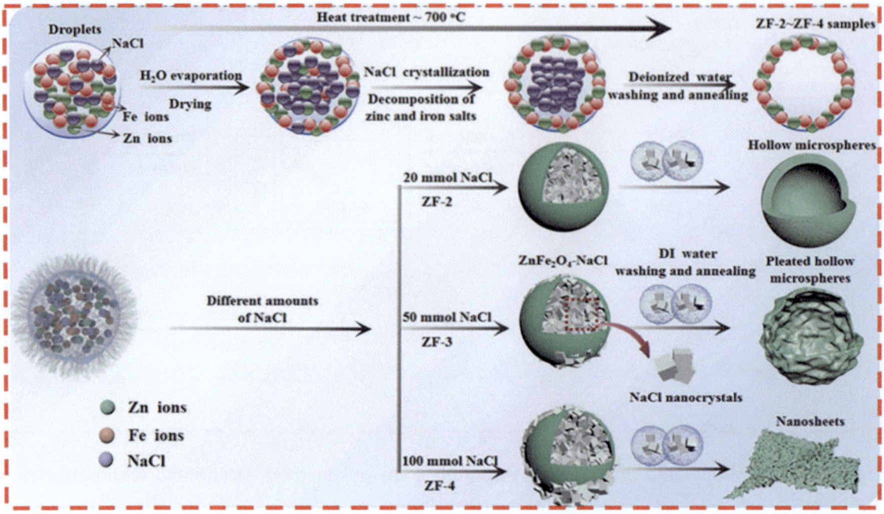

Fig. 1.18 Schematic diagram of the specific formation process of $ZnFe_2O_4$ (ZF-2–ZF-4) samples with hollow, pleated hollow, and nanosheet structures. (Reprinted with permission from Lv et al. (2022))

1.4.2.6 Reverse Micelle Method

The reverse micelle synthesis approach represents a specialized water-in-oil (W/O) microemulsion system comprising three essential components: an aqueous phase, a nonpolar organic solvent, and amphiphilic surfactant molecules (Malik et al. 2012). In this unique microemulsion structure, surfactant molecules organize themselves with hydrophilic head groups oriented toward aqueous cores, forming discrete nanoscale compartments termed "water pools." These aqueous domains function as nanoreactors, facilitating precursor solubilization and subsequent chemical reactions for nanomaterial synthesis. The scale of these nanoreactors can be precisely regulated by adjusting the water-to-surfactant ratio (*W*), with reduced W-values yielding smaller aqueous cores and consequently producing nanomaterials with decreased particle sizes (Soleimani Zohr Shiri et al. 2019). The reverse micelle method offers the following advantages: ease of preparation, high thermodynamic stability, precise control over particle size and composition, and the production of fine, monodisperse nanomaterials (Rane et al. 2018). Owing to these advantages, the reverse micelle technique has been successfully employed in synthesizing various nanomaterials, including metal nanoparticles, metal sulfide nanoparticles, metal salts, metal oxides, and their composite.

1.4.2.7 Co-precipitation Method

The co-precipitation method, also known as chemical precipitation (Su and Chang 2018), is a widely employed synthesis method for nanomaterials. This process initiates with the combination of aqueous metal salt solutions, followed by the addition of precipitation agents such as chlorides, hydroxides, oxalates, or carbonates. The subsequent precipitation reaction results in the simultaneous deposition of metal ions (Kolahalam et al. 2019). The mechanism of co-precipitation encompasses four primary stages: nucleation, crystal growth, coarsening (Ostwald ripening), and agglomeration. Nucleation, a crucial phase governed by supersaturation levels, leads to the formation of initial crystallites. Nucleation, a crucial phase governed by supersaturation levels, leads to the formation of initial crystallites. This versatile method facilitates the production of diverse nanostructures such as nanotubes, nanorods, and various nanoparticles (Rani et al. 2018; Martin Mark et al. 2021). Critical synthesis parameters influencing nanoparticle characteristics encompass zeta potential, reactant concentration, temperature, pH value, mixing efficiency, and precipitate recrystallization (Su and Chang 2018; Adam et al. 2018; Al-Abdallat et al. 2019). Precise control of these variables enables regulation of particle size distribution, morphological characteristics, and stoichiometric composition.

1.5 Conclusion and Outlook

The advent of nanotechnology, enabling precise manipulation of material properties at the atomic scale, has significantly advanced sustainable air quality management. Extensive research has been conducted on the application of nanomaterials in gas sampling and detection systems, representing crucial components in air quality monitoring and control. The remarkable potential of nanomaterials has simultaneously driven substantial progress in material synthesis methodologies. Despite considerable advancements in nanomaterial research, significant challenges persist in nanotechnology development, as illustrated in Fig. 1.19. Realizing the complete potential of nanomaterials necessitates the establishment of large-scale, controlled synthesis protocols while maintaining cost-effectiveness, which is essential for commercial viability. Advanced characterization techniques play a pivotal role in elucidating structure–property relationships in nanomaterials. As the demand for detailed nanostructural information increases, the development of sophisticated

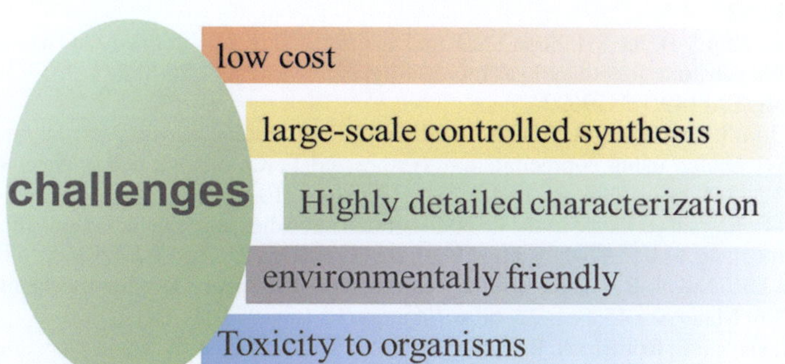

Fig. 1.19 Challenges of nanotechnology

characterization tools becomes imperative, facilitating the translation of nanotechnology into practical applications.

However, alongside these technical considerations, the environmental and ecological implications of nanomaterials warrant careful examination. The synthesis processes of nanomaterials present environmental concerns, primarily through solvent utilization and byproduct generation. Furthermore, chemically synthesized nanomaterials have demonstrated potential toxicity to biological systems, necessitating comprehensive toxicological assessments. These considerations have propelled environmental nanotechnology to the forefront of green chemistry research, emphasizing the development of sustainable and eco-friendly nanomaterial production and application strategies.

References

Abdali H, Heli B, Ajji A (2019) Cellulose nanopaper cross-linked amino graphene/polyaniline sensors to detect CO_2 gas at room temperature. Sensors 19(23):5215. https://doi.org/10.3390/s19235215

Abdellah A, Abdelhalim A, Loghin F, Köhler P, Ahmad Z, Scarpa G, Lugli P (2013) Flexible carbon nanotube based gas sensors fabricated by large-scale spray deposition. IEEE Sensors J 13(10):4014–4021. https://doi.org/10.1109/JSEN.2013.2265775

Adam RE, Pozina G, Willander M, Nur O (2018) Synthesis of ZnO nanoparticles by co-precipitation method for solar driven photodegradation of Congo red dye at different pH. Photonics Nanostruc Fundam Appl 32:11–18. https://doi.org/10.1016/j.photonics.2018.08.005

Agarwal S, Rai P, Gatell EN, Llobet E, Güell F, Kumar M, Awasthi K (2019) Gas sensing properties of ZnO nanostructures (flowers/rods) synthesized by hydrothermal method. Sens Actuators B Chem 292:24–31. https://doi.org/10.1016/j.snb.2019.04.083

Agarwal S, Kumar S, Agrawal H, Moinuddin MG, Kumar M, Sharma SK, Awasthi K (2021) An efficient hydrogen gas sensor based on hierarchical Ag/ZnO hollow microstructures. Sens Actuators B Chem 346:130510. https://doi.org/10.1016/j.snb.2021.130510

Ahmadi A, Abbaspour M, Arjmandi R, Abedi Z (2015) Air Quality Risk Index (AQRI) and its application for a megacity. Int J Environ Sci Technol 12(12):3773–3780. https://doi.org/10.1007/s13762-015-0837-7

Ahmadi M, Elmongy H, Madrakian T, Abdel-Rehim M (2017) Nanomaterials as sorbents for sample preparation in bioanalysis: a review. Anal Chim Acta 958:1–21. https://doi.org/10.1016/j.aca.2016.11.062

Ahn H, Park J-H, Kim S-B, Jee SH, Yoon YS, Kim D-J (2010) Vertically aligned ZnO nanorod sensor on flexible substrate for ethanol gas monitoring. Electrochem Solid-State Lett 13(11):J125. https://doi.org/10.1149/1.3479692

Al-Abdallat Y, Jum'h I, Al Bsoul A, Jumah R, Telfah A (2019) Photocatalytic degradation dynamics of methyl orange using coprecipitation synthesized Fe_3O_4 nanoparticles. Water Air Soil Pollut 230(12):277. https://doi.org/10.1007/s11270-019-4310-y

Aldy JE (2017) Policy surveillance in the G-20 fossil fuel subsidies agreement: lessons for climate policy. Clim Chang 144(1):97–110. https://doi.org/10.1007/s10584-015-1505-0

Ali FA, Mishra DK, Nayak R, Nanda B (2022) Solid-state gas sensors: sensing mechanisms and materials. Bull Mater Sci 45(1):1–22. https://doi.org/10.1007/s12034-021-02581-5

Alvarado M, Navarrete È, Romero A, Ramírez JL, Llobet E (2018) Flexible gas sensors employing octahedral indium oxide films. Sensors 18(4):999. https://doi.org/10.3390/s18040999

Anjusree GS, Bhupathi A, Balakrishnan A, Sree S, Subramanian KRV, Sivakumar N, Ramakrishna S, Nair S, Nair S (2013) Fabricating fiber, rice and leaf-shaped TiO2 by tuning the chemistry between TiO2 and the polymer during electrospinning. RSC Adv 3(37):16720–16727. https://doi.org/10.1039/C3RA42250J

Aqel A, El-Nour KMMA, Ammar RAA, Al-Warthan A (2012) Carbon nanotubes, science and technology part (I) structure, synthesis and characterisation. Arab J Chem 5(1):1–23. https://doi.org/10.1016/j.arabjc.2010.08.022

Arrhenius K, Brown AS, van der Veen AMH (2016) Suitability of different containers for the sampling and storage of biogas and biomethane for the determination of the trace-level impurities – a review. Anal Chim Acta 902:22–32. https://doi.org/10.1016/j.aca.2015.10.039

Arrigan D (2018) Sabine Szunerits, Rabah Boukherroub, Alison Downard, Jun-Jie Zhu (Eds.): nanocarbons for electroanalysis. Anal Bioanal Chem 410(13):3039–3040. https://doi.org/10.1007/s00216-018-0994-x

Arthur CL, Janusz P (1990) Solid phase microextraction with thermal desorption using fused silica optical fibers. Anal Chem 62(19):2145–2148. https://doi.org/10.1021/ac00218a019

Asad M, Sheikhi MH (2016) Highly sensitive wireless H_2S gas sensors at room temperature based on CuO-SWCNT hybrid nanomaterials. Sens Actuators B Chem 231:474–483. https://doi.org/10.1016/j.snb.2016.03.021

Asad M, Sheikhi MH, Pourfath M, Moradi M (2015) High sensitive and selective flexible H_2S gas sensors based on Cu nanoparticle decorated SWCNTs. Sens Actuators B Chem 210:1–8. https://doi.org/10.1016/j.snb.2014.12.086

Ashkarran AA, Zad AI, Ahadian MM, Ardakani SAM (2008) Synthesis and photocatalytic activity of WO_3 nanoparticles prepared by the arc discharge method in deionized water. Nanotechnology 19(19):195709. https://doi.org/10.1088/0957-4484/19/19/195709

Ashkarran AA, Irajizad A, Mahdavi SM, Ahadian MM (2009) ZnO nanoparticles prepared by electrical arc discharge method in water. Mater Chem Phys 118(1):6–8. https://doi.org/10.1016/j.matchemphys.2009.07.002

Ashraful Islam M, Han Kim J, Ko T-J, Noh C, Nehate S, Golam Kaium M, Ko M, Fox D, Zhai L, Cho C-H, Sundaram KB, Bae T-S, Jung Y, Chung H-S, Jung Y (2018) Three dimensionally-ordered 2D MoS 2 vertical layers integrated on flexible substrates with stretch-tunable functionality and improved sensing capability. Nanoscale 10(37):17525–17533. https://doi.org/10.1039/C8NR05362F

Atkinson R (2000) Atmospheric chemistry of VOCs and NO_x. Atmos Environ 34(12):2063–2101. https://doi.org/10.1016/S1352-2310(99)00460-4

Azizi A, Bottaro CS (2020) A critical review of molecularly imprinted polymers for the analysis of organic pollutants in environmental water samples. J Chromatogr A 1614:460603. https://doi.org/10.1016/j.chroma.2019.460603

Baek J, Yin D, Liu N, Omkaram I, Jung C, Im H, Hong S, Kim SM, Hong YK, Hur J, Yoon Y, Kim S (2017) A highly sensitive chemical gas detecting transistor based on highly crystalline CVD-grown $MoSe_2$ films. Nano Res 10(6):1861–1871. https://doi.org/10.1007/s12274-016-1291-7

Bagherzadeh R, Gorji M, Sorayani Bafgi MS, Saveh-Shemshaki N (2017) 18 – Electrospun conductive nanofibers for electronics. In: Afshari M (ed) Electrospun nanofibers. Woodhead Publishing, Duxford, pp 467–519

Baig N, Kammakakam I, Falath W (2021) Nanomaterials: a review of synthesis methods, properties, recent progress, and challenges. Mater Adv 2(6):1821–1871. https://doi.org/10.1039/D0MA00807A

Bandgar DK, Navale ST, Nalage SR, Mane RS, Stadler FJ, Aswal DK, Gupta SK, Patil VB (2015) Simple and low-temperature polyaniline-based flexible ammonia sensor: a step towards laboratory synthesis to economical device design. J Mater Chem C 3(36):9461–9468. https://doi.org/10.1039/C5TC01483B

Barnett HJ (1974) Economic growth and environmental quality are compatible. Policy Sci 5(2):137–147. https://doi.org/10.1007/BF00148037

Batzill M, Diebold U (2005) The surface and materials science of tin oxide. Prog Surf Sci 79(2):47–154. https://doi.org/10.1016/j.progsurf.2005.09.002

Bayda S, Adeel M, Tuccinardi T, Cordani M, Rizzolio F (2020) The history of nanoscience and nanotechnology: from chemical–physical applications to nanomedicine. Molecules 25(1):112. https://doi.org/10.3390/molecules25010112

Bezzon VDN, Montanheiro TLA, De Menezes BRC, Ribas RG, Righetti VAN, Rodrigues KF, Thim GP (2019) Carbon nanostructure-based sensors: a brief review on recent advances. Adv Mater Sci Eng 2019:1–21. https://doi.org/10.1155/2019/4293073

Bogue R (2015) Detecting gases with light: a review of optical gas sensor technologies. Sens Rev 35(2):133–140. https://doi.org/10.1108/SR-09-2014-696

Burlachenko J, Kruglenko I, Snopok B, Persaud K (2016) Sample handling for electronic nose technology: state of the art and future trends. TrAC Trends Anal Chem 82:222–236. https://doi.org/10.1016/j.trac.2016.06.007

Burton LA, Walsh A (2012) Phase stability of the earth-abundant tin sulfides SnS, SnS_2, and Sn_2S_3. J Phys Chem C 116(45):24262–24267. https://doi.org/10.1021/jp309154s

Burton LA, Colombara D, Abellon RD, Grozema FC, Peter LM, Savenije TJ, Dennler G, Walsh A (2014) ChemInform abstract: synthesis, characterization, and electronic structure of single-crystal SnS, Sn_2S_3, and SnS_2. ChemInform 45(10). https://doi.org/10.1002/chin.201410009

Cavicchi RE (2011) Calorimetric sensors. NIST. Momentum Press, Highland Park, pp 287–320

Chan K-Y, Teo B-S (2005) Sputtering power and deposition pressure effects on the electrical and structural properties of copper thin films. J Mater Sci 40(22):5971–5981. https://doi.org/10.1007/s10853-005-1362-8

Chang Q, Wang M, Zhang G, Zang X, Li H, Zhang S, Wang C, Wang Z (2020) Carbon nanospheres as solid-phase microextraction coating for the extraction of polycyclic aromatic hydrocarbons from water and soil samples. J Sep Sci 43(13):2594–2601. https://doi.org/10.1002/jssc.201901294

Chen J, Shi J, Decanini D, Cambril E, Chen Y, Haghiri-Gosnet A-M (2009) Gold nanohole arrays for biochemical sensing fabricated by soft UV nanoimprint lithography. Microelectron Eng 86(4):632–635. https://doi.org/10.1016/j.mee.2008.12.093

Chen X, Hou Y, Zhang B, Yang XH, Yang HG (2013) Low-cost SnSx counter electrodes for dye-sensitized solar cells. Chem Commun 49(51):5793–5795. https://doi.org/10.1039/C3CC42679C

Cho B, Yoon J, Lim SK, Kim AR, Choi S-Y, Kim D-H, Lee KH, Lee BH, Ko HC, Hahm MG (2015) Metal decoration effects on the gas-sensing properties of 2D hybrid-structures on flexible substrates. Sensors 15(10):24903–24913. https://doi.org/10.3390/s151024903

Chrzanowska J, Hoffman J, Małolepszy A, Mazurkiewicz M, Kowalewski TA, Szymanski Z, Stobinski L (2015) Synthesis of carbon nanotubes by the laser ablation method: effect of laser wavelength. Phys Status Solidi B 252(8):1860–1867. https://doi.org/10.1002/pssb.201451614

Danks AE, Hall SR, Schnepp Z (2016) The evolution of 'sol–gel' chemistry as a technique for materials synthesis. Mater Horiz 3(2):91–112. https://doi.org/10.1039/C5MH00260E

Das S, Chakraborty S, Parkash O, Kumar D, Bandyopadhyay S, Samudrala SK, Sen A, Maiti HS (2008) Vanadium doped tin dioxide as a novel sulfur dioxide sensor. Talanta 75(2):385–389. https://doi.org/10.1016/j.talanta.2007.11.010

Das S, Sen B, Debnath N (2015) Recent trends in nanomaterials applications in environmental monitoring and remediation. Environ Sci Pollut Res 22. https://doi.org/10.1007/s11356-015-5491-6

de Castro JFR, Santos SF, Costa ALM, Yavari AR, Botta FWJ, Ishikawa TT (2004) Structural characterization and dehydrogenation behavior of Mg–5 at.%Nb nano-composite processed by reactive milling. J Alloys Compd 376(1):251–256. https://doi.org/10.1016/j.jallcom.2004.01.021

de la Hoz A, Díaz-Ortiz A, Prieto P (2016) Microwave-assisted green organic synthesis. In: Alternative energy sources for green chemistry. https://doi.org/10.1039/9781782623632-00001

Denkbas EB, Celik E, Erdal E et al (2016) Magnetically based nanocarriers in drug delivery. In: Nanobiomaterials in drug delivery. William Andrew Publishing, Amsterdam, pp 285–331

Dey A (2018) Semiconductor metal oxide gas sensors: a review. Mater Sci Eng B 229:206–217. https://doi.org/10.1016/j.mseb.2017.12.036

Ding L, Yazaydin AO (2013) Hydrogen and methane storage in ultrahigh surface area metal–organic frameworks. Microporous Mesoporous Mater 182:185–190. https://doi.org/10.1016/j.micromeso.2013.08.048

Dong K-Y, Choi J, Lee YD, Kang BH, Yu Y-Y, Choi HH, Ju B-K (2013) Detection of a CO and NH_3 gas mixture using carboxylic acid-functionalized single-walled carbon nanotubes. Nanoscale Res Lett 8(1):1–6. https://doi.org/10.1186/1556-276X-8-12

Doustkhah E, Hassandoost R, Khataee A, Luque R, Assadi MHN (2021) Hard-templated metal–organic frameworks for advanced applications. Chem Soc Rev 50(5):2927–2953. https://doi.org/10.1039/C9CS00813F

Du P, Song L, Xiong J, Li N, Xi Z, Wang L, Jin D, Guo S, Yuan Y (2012) Coaxial electrospun TiO_2/ZnO core–sheath nanofibers film: novel structure for photoanode of dye-sensitized solar cells. Electrochim Acta 78:392–397. https://doi.org/10.1016/j.electacta.2012.06.034

Duque JS, Madrigal BM, Riascos H, Avila YP (2019) Colloidal metal oxide nanoparticles prepared by laser ablation technique and their antibacterial test. Colloids Interfaces 3(1):25. https://doi.org/10.3390/colloids3010025

Esmaeilpour M, Bügel P, Fink K, Studt F, Wenzel W, Kozlowska M (2023) Multiscale model of CVD growth of graphene on Cu(111) surface. Int J Mol Sci 24(10):8563. https://doi.org/10.3390/ijms24108563

Fauzi F, Rianjanu A, Santoso I, Triyana K (2021) Gas and humidity sensing with quartz crystal microbalance (QCM) coated with graphene-based materials – a mini review. Sens Actuators Phys 330:112837. https://doi.org/10.1016/j.sna.2021.112837

Feng J, Feng J, Ji X, Li C, Han S, Sun H, Sun M (2021) Recent advances of covalent organic frameworks for solid-phase microextraction. TrAC Trends Anal Chem 137:116208. https://doi.org/10.1016/j.trac.2021.116208

Feynman RP (2011) There's plenty of room at the bottom. Resonance 16:890

Gai S, Wang B, Wang X, Zhang R, Miao S, Wu Y (2022) Ultrafast NH_3 gas sensor based on phthalocyanine-optimized non-covalent hybrid of carbon nanotubes with pyrrole. Sens Actuators B Chem 357:131352. https://doi.org/10.1016/j.snb.2021.131352

Gangopadhyay RK, Das SK (2008) Ammonia leakage from refrigeration plant and the management practice. Process Saf Prog 27(1):15–20. https://doi.org/10.1002/prs.10208

Gao Y, Lu F (eds) (2023) Covalent organic frameworks. IntechOpen, London

Gao P, Meng LJ, dos Santos MP, Teixeira V, Andritschky M (2000) Influence of sputtering power and the substrate–target distance on the properties of ZrO_2 films prepared by RF reactive sputtering. Thin Solid Films 377–378:557–561. https://doi.org/10.1016/S0040-6090(00)01291-8

Gao Y, Liu Y, Zou D (2023) Microwave-assisted synthesis and environmental remediation: a review. Environ Chem Lett 21(4):2399–2416. https://doi.org/10.1007/s10311-023-01599-x

Ghorbani M, Kazemipour M, Ansari M, Mehdipour M, Mohamadi N (2023) Solid phase microextraction of some polycyclic aromatic hydrocarbons based on hydrogel containing nanohydroxyapatite and modified multi-walled carbon nanotubes composite coated on steel wire followed by GC–MS. Microchem J 193:109011. https://doi.org/10.1016/j.microc.2023.109011

Goel N, Kunal K, Kushwaha A, Kumar M (2023) Metal oxide semiconductors for gas sensing. Eng Rep 5(6):e12604. https://doi.org/10.1002/eng2.12604

Gong J, Chen Q, Lian M-R, Liu N-C, Stevenson RG, Adami F (2006) Micromachined nanocrystalline silver doped SnO_2 H_2S sensor. Sens Actuators B Chem 114(1):32–39. https://doi.org/10.1016/j.snb.2005.04.035

González-Hernández P, Pacheco-Fernández I, Bernardo F, Homem V, Pasán J, Ayala JH, Ratola N, Pino V (2021) Headspace solid-phase microextraction based on the metal-organic framework CIM-80(Al) coating to determine volatile methylsiloxanes and musk fragrances in water samples using gas chromatography and mass spectrometry. Talanta 232:122440. https://doi.org/10.1016/j.talanta.2021.122440

Grigorescu AE, Hagen CW (2009) Resists for sub-20-nm electron beam lithography with a focus on HSQ: state of the art. Nanotechnology 20(29):292001. https://doi.org/10.1088/0957-4484/20/29/292001

Grove WR (1997) VII. On the electro-chemical polarity of gases. Philos Trans R Soc Lond 142:87–101. https://doi.org/10.1098/rstl.1852.0008

Guan Y, Wang C, Yu H, Zou Z, Zhou Y, Cao G, Yao J (2020) Flexible ZnO/PANI/nonwoven nanocomposite based high-sensitive NH_3 gas sensor via vapor phase polymerization method. Mater Sci Energy Technol 3:862–867. https://doi.org/10.1016/j.mset.2020.10.011

Guerra FD, Attia MF, Whitehead DC, Alexis F (2018) Nanotechnology for environmental remediation: materials and applications. Molecules 23(7):1760. https://doi.org/10.3390/molecules23071760

Guo Y, Wang T, Chen F, Sun X, Li X, Yu Z, Wan P, Chen X (2016) Hierarchical graphene–polyaniline nanocomposite films for high-performance flexible electronic gas sensors. Nanoscale 8(23):12073–12080. https://doi.org/10.1039/C6NR02540D

Guo H, Chen G, Ma J, Jia Q (2019) A triazine based organic framework with micropores and mesopores for use in headspace solid phase microextraction of phthalate esters. Microchim Acta 186(1): https://doi.org/10.1007/s00604-018-3060-7

Guo H, Song N, Wang D, Ma J, Jia Q (2019a) A modulation approach for covalent organic frameworks: application to solid phase microextraction of phthalate esters. Talanta 198:277–283. https://doi.org/10.1016/j.talanta.2019.02.025

Guo J-X, Qian H-L, Zhao X, Yang C, Yan X-P (2019b) In situ room-temperature fabrication of a covalent organic framework and its bonded fiber for solid-phase microextraction of polychlorinated biphenyls in aquatic products. J Mater Chem A 7(21):13249–13255. https://doi.org/10.1039/C9TA02974E

Guo S, Yang D, Li B, Dong Q, Li Z, Zaghloul ME (2019c) An artificial intelligent flexible gas sensor based on ultra-large area $MoSe_2$ nanosheet. In: 2019 IEEE 62nd international midwest symposium on circuits and systems (MWSCAS), pp 884–887. https://doi.org/10.1109/MWSCAS.2019.8885301

Guo S, Yang D, Zhang S, Dong Q, Li B, Tran N, Li Z, Xiong Y, Zaghloul ME (2019d) Development of a cloud-based epidermal MoSe2 device for hazardous gas sensing. Adv Funct Mater 29(18):1900138. https://doi.org/10.1002/adfm.201900138

Guo S-Y, Hou P-X, Wang H-X, Shi C, Fang H-T, Liu C (2019e) Transparent and flexible hydrogen sensor based on semiconducting single-wall carbon nanotube networks. Carbon 151:156–159. https://doi.org/10.1016/j.carbon.2019.05.045

Gusain A, Joshi NJ, Varde PV, Aswal DK (2017) Flexible NO gas sensor based on conducting polymer poly[*N*-9′-heptadecanyl-2,7-carbazole-alt-5,5-(4′,7′-di-2-thienyl-2′,1′,3′-benzothiadiazole)] (PCDTBT). Sens Actuators B Chem 239:734–745. https://doi.org/10.1016/j.snb.2016.07.176

Gutiérrez-Serpa A, Pacheco-Fernández I, Pasán J, Pino V (2019) Metal–organic frameworks as key materials for solid-phase microextraction devices—a review. Separations 6(4):47. https://doi.org/10.3390/separations6040047

Haghi M, Danyali S, Ayasseh S, Wang J, Aazami R, Deserno TM (2021) Wearable devices in health monitoring from the environmental towards multiple domains: a survey. Sensors 21(6):2130. https://doi.org/10.3390/s21062130

Hahn Y-B, Ahmad R, Tripathy N (2012) Chemical and biological sensors based on metal oxide nanostructures. Chem Commun 48(84):10369–10385. https://doi.org/10.1039/C2CC34706G

Hashemi-Moghaddam H, Hosseni M, Mohammadhosseini M (2017) Preparation of molecularly imprinted polymers on the surface of optical fiber for HS-solid-phase microextraction of phenol. Sep Sci Technol 52(11). https://doi.org/10.1080/01496395.2017.1290110

Hebisch R, Breuer D, Krämer W, Maschmeier C-P, Tschickardt M (2009) Sampling and analysis of gases and vapours. https://doi.org/10.13140/2.1.1036.1925

He Q, Zeng Z, Yin Z, Li H, Wu S, Huang X, Zhang H (2012) Fabrication of flexible MoS2 thin-film transistor arrays for practical gas-sensing applications. Small Weinh Bergstr Ger 8:2994–2999. https://doi.org/10.1002/smll.201201224

Hu Q, Liu S, Chen X, Xu J, Zhu F, Ouyang G (2019) Enhancing enrichment ability of a nanoporous carbon based solid-phase microextraction device by a morphological modulation strategy. Anal Chim Acta 1047:1–8. https://doi.org/10.1016/j.aca.2018.10.063

Hua C, Shang Y, Wang Y, Xu J, Zhang Y, Li X, Cao A (2017) A flexible gas sensor based on single-walled carbon nanotube-Fe_2O_3 composite film. Appl Surf Sci 405:405–411. https://doi.org/10.1016/j.apsusc.2017.01.301

Huang Y, Sutter E, Sadowski JT, Cotlet M, Monti OLA, Racke DA, Neupane MR, Wickramaratne D, Lake RK, Parkinson BA, Sutter P (2014) Tin disulfide—an emerging layered metal dichalcogenide semiconductor: materials properties and device characteristics. ACS Nano 8(10):10743–10755. https://doi.org/10.1021/nn504481r

Ismail RA, Mohsin MH, Ali AK, Hassoon KI, Erten-Ela S (2020) Preparation and characterization of carbon nanotubes by pulsed laser ablation in water for optoelectronic application. Phys E Low-Dimens Syst Nanostruct 119:113997. https://doi.org/10.1016/j.physe.2020.113997

Jagirani MS, Soylak M (2020) A review: recent advances in solid phase microextraction of toxic pollutants using nanotechnology scenario. Microchem J 159:105436. https://doi.org/10.1016/j.microc.2020.105436

Jakubik WP (2011) Surface acoustic wave-based gas sensors. Thin Solid Films 520(3):986–993. https://doi.org/10.1016/j.tsf.2011.04.174

Jang J-S, Qiao S, Choi S-J, Jha G, Ogata AF, Koo W-T, Kim D-H, Kim I-D, Penner RM (2017) Hollow Pd–Ag composite nanowires for fast responding and transparent hydrogen sensors. ACS Appl Mater Interfaces 9(45):39464–39474. https://doi.org/10.1021/acsami.7b10908

Jeong HY, Lee D-S, Choi HK, Lee DH, Kim J-E, Lee JY, Lee WJ, Kim SO, Choi S-Y (2010) Flexible room-temperature NO_2 gas sensors based on carbon nanotubes/reduced graphene hybrid films. Appl Phys Lett 96(21):213105. https://doi.org/10.1063/1.3432446

Jiang X-X, Yu L-Q, Sun Y-N, Li Y, Li H-M, Lv Y-K (2021) Hollow zeolitic imidazolate framework-7 coated stainless steel fiber for solid phase microextraction of volatile biomarkers in headspace gas of breast cancer cell lines. Anal Chim Acta 1181:338901. https://doi.org/10.1016/j.aca.2021.338901

Jiang X-X, Sun Y-N, Zhang C, Yang Z, Yu L-Q, Lv Y-K (2022) Synthesis of SiO_2@MOF-199 as a fiber coating for headspace solid-phase microextraction of phthalates in plastic bottled milk. Chromatographia 85(9):851–863. https://doi.org/10.1007/s10337-022-04184-6

Jing W, Wang J, Kuipers B, Bi W, Chen DDY (2021) Recent applications of graphene and graphene-based materials as sorbents in trace analysis. TrAC Trends Anal Chem 137:116212. https://doi.org/10.1016/j.trac.2021.116212

John AT, Murugappan K, Nisbet DR, Tricoli A (2021) An outlook of recent advances in chemiresistive sensor-based electronic nose systems for food quality and environmental monitoring. Sensors 21(7):2271. https://doi.org/10.3390/s21072271

Jung MW, Kang SM, Nam K-H, An K-S, Ku B-C (2018) Highly transparent and flexible NO_2 gas sensor film based on MoS_2/rGO composites using soft lithographic patterning. Appl Surf Sci 456:7–12. https://doi.org/10.1016/j.apsusc.2018.06.086

Kassem O, Saadaoui M, Rieu M, Viricelle J-P (2019) A novel approach to a fully inkjet printed SnO_2-based gas sensor on a flexible foil. J Mater Chem C 7(39):12343–12353. https://doi.org/10.1039/C9TC04170B

Khan FA (2020) Synthesis of nanomaterials: methods & technology. In: Khan FA (ed) Applications of nanomaterials in human health. Springer, Singapore, pp 15–21

Khang Y, Lee J (2010) Synthesis of Si nanoparticles with narrow size distribution by pulsed laser ablation. J Nanopart Res 12(4):1349–1354. https://doi.org/10.1007/s11051-009-9669-z

Kim J, Kang J, Jeong U, Kim H, Lee H (2013) Catalytic, conductive, and transparent platinum nanofiber webs for FTO-free dye-sensitized solar cells. ACS Appl Mater Interfaces 5(8):3176–3181. https://doi.org/10.1021/am400179j

Kim J-W, Porte Y, Ko KY, Kim H, Myoung J-M (2017) Micropatternable double-faced ZnO nanoflowers for flexible gas sensor. ACS Appl Mater Interfaces 9(38):32876–32886. https://doi.org/10.1021/acsami.7b09251

Ko WH (2007) Trends and frontiers of MEMS. Sens Actuators Phys 136(1):62–67. https://doi.org/10.1016/j.sna.2007.02.001

Ko KY, Lee S, Park K, Kim Y, Woo WJ, Kim D, Song J-G, Park J, Kim JH, Lee Z, Kim H (2018) High-performance gas sensor using a large-area WS_xSe_{2-2x} alloy for low-power operation wearable applications. ACS Appl Mater Interfaces 10(40):34163–34171. https://doi.org/10.1021/acsami.8b10455

Kohl D (1990) The role of noble metals in the chemistry of solid-state gas sensors. Sens Actuators B Chem 1(1):158–165. https://doi.org/10.1016/0925-4005(90)80193-4

Kolahalam LA, Viswanath IVK, Diwakar BS, Govindh B, Reddy V, Murthy YLN (2019) Review on nanomaterials: synthesis and applications. Mater Today Proc 18:2182–2190. https://doi.org/10.1016/j.matpr.2019.07.371

Koonani S, Ghiasvand A (2024) A highly porous fiber coating based on a Zn-MOF/COF hybrid material for solid-phase microextraction of PAHs in soil. Talanta 267:125236. https://doi.org/10.1016/j.talanta.2023.125236

Kori AH, Jagirani MS, Soylak M (2023) Graphene-based nanomaterials: a sustainable material for solid-phase microextraction (SPME) for environmental applications. Anal Lett 56(15):2385–2400. https://doi.org/10.1080/00032719.2023.2173221

Kumar A, Prajesh R (2022) The potential of acoustic wave devices for gas sensing applications. Sens Actuators Phys 339:113498. https://doi.org/10.1016/j.sna.2022.113498

Kumar D, Sharma RC (1998) Advances in conductive polymers. Eur Polym J 34(8):1053–1060. https://doi.org/10.1016/S0014-3057(97)00204-8

Kumar R, Goel N, Hojamberdiev M, Kumar M (2020) Transition metal dichalcogenides-based flexible gas sensors. Sens Actuators Phys 303:111875. https://doi.org/10.1016/j.sna.2020.111875

Lancioni C, Castells C, Candal R, Tascon M (2022) Headspace solid-phase microextraction: fundamentals and recent advances. Adv Sample Prep 3:100035. https://doi.org/10.1016/j.sampre.2022.100035

Late DJ, Doneux T, Bougouma M (2014) Single-layer $MoSe_2$ based NH_3 gas sensor. Appl Phys Lett 105(23):233103. https://doi.org/10.1063/1.4903358

Lecharlier A, Carrier H, Le Hécho I (2022) Characterization of biogas and biomethane trace compounds: a critical review of advances in *in situ* sampling and preconcentration techniques. Anal Chim Acta 1229:340174. https://doi.org/10.1016/j.aca.2022.340174

Lee WS, Choi J (2019) Two-dimensional tungsten diselenides integrated on paper substrate for highly flexible and sensitive gas sensor. In: 2019 20th international conference on solid-state sensors, actuators and microsystems & eurosensors XXXIII (TRANSDUCERS & EUROSENSORS XXXIII), pp 681–684. Berlin, Germany. https://doi.org/10.1109/TRANSDUCERS.2019.8808482.

Lee SC, Hwang BW, Lee SJ, Choi HY, Kim SY, Jung SY, Ragupathy D, Lee DD, Kim JC (2011) A novel tin oxide-based recoverable thick film SO_2 gas sensor promoted with magnesium and vanadium oxides. Sens Actuators B Chem 160(1):1328–1334. https://doi.org/10.1016/j.snb.2011.09.070

Lee C, Ahn J, Lee KB, Kim D, Kim J (2012) Graphene-based flexible NO_2 chemical sensors. Thin Solid Films 520(16):5459–5462. https://doi.org/10.1016/j.tsf.2012.03.095

Lee K, Scardaci V, Kim H-Y, Hallam T, Nolan H, Bolf BE, Maltbie GS, Abbott JE, Duesberg GS (2013) Highly sensitive, transparent, and flexible gas sensors based on gold nanoparticle decorated carbon nanotubes. Sens Actuators B Chem 188:571–575. https://doi.org/10.1016/j.snb.2013.07.048

Li B, Zhang Y, Hu J, Li M (2005) Fabricating protein nanopatterns on a single DNA molecule with Dip-pen nanolithography. Ultramicroscopy 105(1):312–315. https://doi.org/10.1016/j.ultramic.2005.06.056

Li L, Zhu P, Peng S, Srinivasan M, Yan Q, Nair AS, Liu B, Samakrishna S (2014a) Controlled growth of CuS on electrospun carbon nanofibers as an efficient counter electrode for quantum dot-sensitized solar cells. J Phys Chem C 118(30):16526–16535. https://doi.org/10.1021/jp4117529

Li W, Wu Z, Wang J, Elzatahry AA, Zhao D (2014b) A perspective on mesoporous TiO_2 materials. Chem Mater 26(1):287–298. https://doi.org/10.1021/cm4014859

Li R, Jiang K, Chen S, Lou Z, Huang T, Chen D, Shen G (2017) SnO_2/SnS_2 nanotubes for flexible room-temperature NH_3 gas sensors. RSC Adv 7(83):52503–52509. https://doi.org/10.1039/C7RA10537A

Liu M, Liu J, Guo C, Li Y (2019) Metal azolate framework-66-coated fiber for headspace solid-phase microextraction of polycyclic aromatic hydrocarbons. J Chromatogr A 1584:57–63. https://doi.org/10.1016/j.chroma.2018.11.043

Li S, Diao Y, Yang Z, He J, Wang J, Liu C, Liu F, Lu H, Yan X, Sun P, Lu G (2018a) Enhanced room temperature gas sensor based on Au-loaded mesoporous In_2O_3 nanospheres@polyaniline core-shell nanohybrid assembled on flexible PET substrate for NH_3 detection. Sens Actuators B Chem 276:526–533. https://doi.org/10.1016/j.snb.2018.08.120

Li S, Lin P, Zhao L, Wang C, Liu D, Liu F, Sun P, Liang X, Liu F, Yan X, Gao Y, Lu G (2018b) The room temperature gas sensor based on Polyaniline@flower-like WO_3 nanocomposites and flexible PET substrate for NH_3 detection. Sens Actuators B Chem 259:505–513. https://doi.org/10.1016/j.snb.2017.11.081

Li W, Teng C, Sun Y, Cai L, Xu J-L, Sun M, Li X, Yang X, Xiang L, Xie D, Ren T (2018c) Sprayed, scalable, wearable, and portable NO_2 sensor array using fully flexible AgNPs-all-carbon nanostructures. ACS Appl Mater Interfaces 10(40):34485–34493. https://doi.org/10.1021/acsami.8b11254

Li H, Hou B, Wang L, Zang X, Wang C, Wang Z (2021) Boron nitride modified reduced graphene oxide as solid-phase microextraction coating material for the extraction of seven polycyclic aromatic hydrocarbons from water and soil samples. J Sep Sci 44(7):1521–1528. https://doi.org/10.1002/jssc.202001088

Li T, Liu G, Kong H, Yang G, Wei G, Zhou X (2023) Recent advances in photonic crystal-based sensors. Coord Chem Rev 475:214909. https://doi.org/10.1016/j.ccr.2022.214909

Liu J, Luo T, Meng F, Qian K, Wan Y, Liu J (2010) Porous hierarchical In_2O_3 micro-/nanostructures: preparation, formation mechanism, and their application in gas sensors for noxious volatile organic compound detection. J Phys Chem C 114(11):4887–4894. https://doi.org/10.1021/jp911768m

Liu Q, Shi J, Jiang G (2012) Application of graphene in analytical sample preparation. TrAC Trends Anal Chem 37:1–11. https://doi.org/10.1016/j.trac.2012.03.011

Liu A, Wang C, Yang X, Liu F, Li S, Wang J, You R, Yang Z, He J, Jiang L, Yan X, Sun P, Lu G (2020a) Polyaniline @ porous nanosphere SnO_2/Zn_2SnO_4 nanohybrid for selective room temperature flexible NH_3 sensor. Sens Actuators B Chem 317:128218. https://doi.org/10.1016/j.snb.2020.128218

Liu G, Liu H, Tong Y, Xu L, Ye Y-X, Wen C, Zhou N, Xu J, Ouyang G (2020b) Headspace solid-phase microextraction of semi-volatile ultraviolet filters based on a superhydrophobic metal-organic framework stable in high-temperature steam. Talanta 219:121175. https://doi.org/10.1016/j.talanta.2020.121175

Liu X, Yang D, Guo Y, Wang F, Han D, Kan Z, Li C (2023) Room temperature triethylamine gas sensor with excellent performances based on novel porous CuO foam via a simple and facile route. Sens Actuators B Chem 390:133934. https://doi.org/10.1016/j.snb.2023.133934

Liu Y, Zhou S, Kuang Y, Feng X, Wang Z, Shen Z, Zhou N, Zheng J, Ouyang G (2024) Nitrogen-rich covalent organic framework as a practical coating for effective determinations of polycyclic aromatic hydrocarbons. Talanta 271:125655. https://doi.org/10.1016/j.talanta.2024.125655

Lv L, Cheng P, Zhang Y, Zhang Y, Lei Z, Wang Y, Xu L, Weng Z, Li C (2022) Ultra-high response acetone gas sensor based on $ZnFe_2O_4$ pleated hollow microspheres prepared by green NaCl template. Sens Actuators B Chem 358:131490. https://doi.org/10.1016/j.snb.2022.131490

Lyu H, Gao B, He F, Ding C, Tang J, Crittenden JC (2017) Ball-milled carbon nanomaterials for energy and environmental applications. ACS Sustain Chem Eng 5(11):9568–9585. https://doi.org/10.1021/acssuschemeng.7b02170

Ma X, Xie Z, Yang Z, Zeng G, Xue M, Zhang X (2018) Inkjet printed 2D SnS_2 nanosheets for ammonia gas sensor. Mater Res Express 6(1):015025. https://doi.org/10.1088/2053-1591/aae5c4

Ma X, Huang P, Dang X, Ai Y, Zheng D, Chen H (2019) MWCNTs/MnO_2 nanocomposite-based polythiophene coating for solid-phase microextraction and determination of polycyclic aromatic hydrocarbons in soil. Microchem J 146:1026–1032. https://doi.org/10.1016/j.microc.2019.02.031

Malik MA, Wani MY, Hashim MA (2012) Microemulsion method: a novel route to synthesize organic and inorganic nanomaterials: 1st nano update. Arab J Chem 5(4):397–417. https://doi.org/10.1016/j.arabjc.2010.09.027

Mani GK, Rayappan JBB (2013) A highly selective room temperature ammonia sensor using spray deposited zinc oxide thin film. Sens Actuators B Chem 183:459–466. https://doi.org/10.1016/j.snb.2013.03.132

Martin Mark JA, Venkatachalam A, Pramothkumar, Senthilkumar, Jothivenkatachalam, Jesuraj JP (2021) Investigation on structural, optical and photocatalytic activity of $CoMn_2O_4$ nanoparticles prepared via simple co-precipitation method. Phys B Condens Matter 601:412349. https://doi.org/10.1016/j.physb.2020.412349

Maya F, Ghani M (2019) Ordered macro/micro-porous metal-organic framework of type ZIF-8 in a steel fiber as a sorbent for solid-phase microextraction of BTEX. Microchim Acta 186(7):425. https://doi.org/10.1007/s00604-019-3560-0

Meng Y, Gu D, Zhang F, Shi Y, Yang H, Li Z, Yu C, Tu B, Zhao D (2005) Ordered mesoporous polymers and homologous carbon frameworks: amphiphilic surfactant templating and direct transformation. Angew Chem Int Ed 44(43):7053–7059. https://doi.org/10.1002/anie.200501561

Mohammad Mahdi K, Yamini Y, Mostafa G (2020) Reduced graphene-decorated covalent organic framework as a novel coating for solid-phase microextraction of phthalate esters coupled to gas chromatography-mass spectrometry. Microchimica Acta 187(4):256. https://doi.org/10.1007/s00604-020-4224-9

Mokhtari N, Khataei MM, Dinari M, Hosseini Monjezi B, Yamini Y (2020) Imine-based covalent triazine framework: synthesis, characterization, and evaluation its adsorption. Mater Lett 263:127221. https://doi.org/10.1016/j.matlet.2019.127221

Mpala TJ, Serepa-Dlamini MH, Etale A, Richards H, Nthunya LN (2023) Cellulose nanocrystal-mediated synthesis of silver nanoparticles *via* microwave assisted method for biofouling control in membrane distillation. Mater Today Commun 34:105028. https://doi.org/10.1016/j.mtcomm.2022.105028

Murtada K (2020) Trends in nanomaterial-based solid-phase microextraction with a focus on environmental applications – a review. Trends Environ Anal Chem 25:e00077. https://doi.org/10.1016/j.teac.2019.e00077

Naganaboina VR, Singh SG (2023) Chemiresistive gas sensors: from novel gas-sensing materials to electrode structure. Chem Phys Rev 4(2). https://doi.org/10.1063/5.0151356

Nam JH, Jang MJ, Jang HY, Park W, Wang X, Choi SM, Cho B (2020) Room-temperature sputtered electrocatalyst WSe_2 nanomaterials for hydrogen evolution reaction. J Energy Chem 47:107–111. https://doi.org/10.1016/j.jechem.2019.11.027

Nie M, Sun K, Meng DD (2009) Formation of metal nanoparticles by short-distance sputter deposition in a reactive ion etching chamber. J Appl Phys 106(5):054314. https://doi.org/10.1063/1.3211326

Olyaie E, Banejad H (2012) Recent nanotechnology applied for soil and water remediation: a critical review. In: Conference: 8th international soil science congress on "land degradation and challenges in sustainable soil management", Izmir

Omarova A, Baizhan A, Baimatova N, Kenessov B, Kazemian H (2021) New *in situ* solvothermally synthesized metal-organic framework MOF-199 coating for solid-phase microextraction of volatile organic compounds from air samples. Microporous Mesoporous Mater 328:111493. https://doi.org/10.1016/j.micromeso.2021.111493

Ostermann R, Cravillon J, Weidmann C, Wiebcke M, Smarsly BM (2011) Metal–organic framework nanofibers via electrospinning. Chem Commun 47(1):442–444. https://doi.org/10.1039/C0CC02271C

Ouyang J, Rao S, Liu R, Wang L, Chen W, Tao W, Kong N (2022) 2D materials-based nanomedicine: from discovery to applications. Adv Drug Deliv Rev 185:114268. https://doi.org/10.1016/j.addr.2022.114268

Öztürk S, Kılınç N (2016) Pd thin films on flexible substrate for hydrogen sensor. J Alloys Compd 674:179–184. https://doi.org/10.1016/j.jallcom.2016.03.042

Parashar M, Shukla VK, Singh R (2020) Metal oxides nanoparticles via sol–gel method: a review on synthesis, characterization and applications. J Mater Sci Mater Electron 31(5):3729–3749. https://doi.org/10.1007/s10854-020-02994-8

Park S, Vosguerichian M, Bao Z (2013) A review of fabrication and applications of carbon nanotube film-based flexible electronics. Nanoscale 5(5):1727–1752. https://doi.org/10.1039/C3NR33560G

Peng L-M, Zhang Z, Wang S (2014) Carbon nanotube electronics: recent advances. Mater Today 17(9):433–442. https://doi.org/10.1016/j.mattod.2014.07.008

Petkovich ND, Stein A (2013) Controlling macro- and mesostructures with hierarchical porosity through combined hard and soft templating. Chem Soc Rev 42(9):3721–3739. https://doi.org/10.1039/C2CS35308C

Piryaei M, Amirifard H, Torabbeigi M (2023) Modified Graphenized Pencil Lead by $CoNi_2S_4$ Nanostructure as a SPME Fiber for Analysis of PAHs from Water Samples. Polycyclic Aromatic Compounds 43(10):8965–8977. https://doi.org/10.1080/10406638.2022.2157451

Poolakkandy RR, Menamparambath MM (2020) Soft-template-assisted synthesis: a promising approach for the fabrication of transition metal oxides. Nanoscale Adv 2(11):5015–5045. https://doi.org/10.1039/D0NA00599A

Popoola AO, Jimoda LA, Olu-Arotiowa OA, Ogunkunle O, Laseinde OT, Adebanjo SA, Raji WA (2023) Dispersion of PM and VOC pollutants from open burning of municipal solid wastes on host communities: emission inventory estimation and dispersion modelling study. Environ Sci Atmos 3(7):1090–1109. https://doi.org/10.1039/D3EA00041A

Portillo-Castillo OJ, Castro-Ríos R, Chávez-Montes A, González-Horta A, Cavazos-Rocha N, De Torres NHW, Garza-Tapia M (2018) Developments of solid-phase microextraction fiber coatings for environmental pharmaceutical and personal care products analysis. Rev Anal Chem 37(2):20170018. https://doi.org/10.1515/revac-2017-0018

Prize (2013) Scanning the past. Nat Nanotechnol 8(8):539–539. https://doi.org/10.1038/nnano.2013.167

Pyeon JJ, Baek I-H, Song YG, Kim GS, Cho A-J, Lee G-Y, Han JH, Chung T-M, Hwang CS, Kang C-Y, Kim SK (2020) Highly sensitive flexible NO_2 sensor composed of vertically aligned 2D SnS_2 operating at room temperature. J Mater Chem C 8(34):11874–11881. https://doi.org/10.1039/D0TC02242J

Rackauskas S, Barbero N, Barolo C, Viscardi G (2017) ZnO nanowire application in chemoresistive sensing: a review. Nanomaterials 7(11):381. https://doi.org/10.3390/nano7110381

Ramgir NS, Sharma PK, Datta N, Kaur M, Debnath AK, Aswal DK, Gupta SK (2013) Room temperature H_2S sensor based on Au modified ZnO nanowires. Sens Actuators B Chem 186:718–726. https://doi.org/10.1016/j.snb.2013.06.070

Rane AV, Kanny K, Abitha VK et al (2018) Methods for synthesis of nanoparticles and fabrication of nanocomposites. In: Synthesis of inorganic nanomaterials. Woodhead Publishing, Sawston, pp 121–139

Rani BJ, Ravina M, Ravi G, Ravichandran S, Ganesh V, Yuvakkumar R (2018) Synthesis and characterization of hausmannite (Mn_3O_4) nanostructures. Surf Interfaces 11:28–36. https://doi.org/10.1016/j.surfin.2018.02.007

Rao BG, Mukherjee D, Reddy BM (2017) Chapter 1 – novel approaches for preparation of nanoparticles. In: Ficai D, Grumezescu AM (eds) Nanostructures for novel therapy. Elsevier, Amsterdam, pp 1–36

Ren Y, Zhou X, Luo W, Xu P, Zhu Y, Li X, Cheng X, Deng Y, Zhao D (2016) Amphiphilic block copolymer templated synthesis of mesoporous indium oxides with nanosheet-assembled pore walls. ACS Publ. https://pubs.acs.org/doi/full/10.1021/acs.chemmater.6b03733. Accessed 6 Apr 2024

Rieu M, Camara M, Tournier G, Viricelle J-P, Pijolat C, de Rooij NF, Briand D (2016) Fully inkjet printed SnO_2 gas sensor on plastic substrate. Sens Actuators B Chem 236:1091–1097. https://doi.org/10.1016/j.snb.2016.06.042

Robbiani S, Lotesoriere BJ, Dellacà RL, Capelli L (2023) Physical confounding factors affecting gas sensors response: a review on effects and compensation strategies for electronic nose applications. Chemosensors 11(10):514. https://doi.org/10.3390/chemosensors11100514

Rowe MD, Novak KM, Moskowitz PD (1983) Health effects of oxidants. Environ Int 9(6):515–528. https://doi.org/10.1016/0160-4120(83)90008-9

Roy D, Munz M, Colombi P, Bhattacharyya S, Salvetat J-P, Cumpson PJ, Saboungi M-L (2007) Directly writing with nanoparticles at the nanoscale using dip-pen nanolithography. Appl Surf Sci 254(5):1394–1398. https://doi.org/10.1016/j.apsusc.2007.06.058

Sanders DP (2010) Advances in patterning materials for 193 nm immersion lithography. Chem Rev 110(1):321–360. https://doi.org/10.1021/cr900244n

Sandhu A (2006) Who invented nano? Nat Nanotechnol 1(2):87–87. https://doi.org/10.1038/nnano.2006.115

Saputra HA (2023) Electrochemical sensors: basic principles, engineering, and state of the art. Monatshefte Für Chem – Chem Mon 154(10):1083–1100. https://doi.org/10.1007/s00706-023-03113-z

Seekaew Y, Lokavee S, Phokharatkul D, Wisitsoraat A, Kerdcharoen T, Wongchoosuk C (2014) Low-cost and flexible printed graphene–PEDOT:PSS gas sensor for ammonia detection. Org Electron 15(11):2971–2981. https://doi.org/10.1016/j.orgel.2014.08.044

Sereshti H, Duman O, Tunç S, Nouri N, Khorram P (2020) Nanosorbent-based solid phase microextraction techniques for the monitoring of emerging organic contaminants in water and wastewater samples. Microchim Acta 187(9):541. https://doi.org/10.1007/s00604-020-04527-w

Sheibani S, Ataie A, Heshmati-Manesh S, Khayati GR (2007) Structural evolution in nanocrystalline Cu synthesized by high energy ball milling. Mater Lett 61(14):3204–3207. https://doi.org/10.1016/j.matlet.2006.11.027

Shi W, Huo L, Wang H, Zhang H, Yang J, Wei P (2006) Hydrothermal growth and gas sensing property of flower-shaped SnS_2 nanostructures. Nanotechnology 17(12):2918. https://doi.org/10.1088/0957-4484/17/12/016

Shim W, Braunschweig AB, Liao X, Chai J, Lim JK, Zheng G, Mirkin CA (2011) Hard-tip, soft-spring lithography. Nature 469(7331):516–520. https://doi.org/10.1038/nature09697

Shim GW, Yoo K, Seo S-B, Shin J, Jung DY, Kang I-S, Ahn CW, Cho BJ, Choi S-Y (2014) Large-area single-layer $MoSe_2$ and its van der Waals heterostructures. ACS Nano 8(7):6655–6662. https://doi.org/10.1021/nn405685j

Shin DH, Lee JS, Jun J, An JH, Kim SG, Cho KH, Jang J (2015) Flower-like palladium nanoclusters decorated graphene electrodes for ultrasensitive and flexible hydrogen gas sensing. Sci Rep 5(1):12294. https://doi.org/10.1038/srep12294

Singh E, Tiwari S, Agrawal M (2009) Effects of elevated ozone on photosynthesis and stomatal conductance of two soybean varieties: a case study to assess impacts of one component of predicted global climate change. Plant Biol 11(s1):101–108. https://doi.org/10.1111/j.1438-8677.2009.00263.x

Singh E, Meyyappan M, Nalwa HS (2017) Flexible graphene-based wearable gas and chemical sensors. ACS Appl Mater Interfaces 9(40):34544–34586. https://doi.org/10.1021/acsami.7b07063

Soleimani Zohr Shiri M, Henderson W, Mucalo MR (2019) A review of the lesser-studied microemulsion-based synthesis methodologies used for preparing nanoparticle systems of the noble metals, Os, Re, Ir and Rh. Materials 12(12):1896. https://doi.org/10.3390/ma12121896

Song MY, Kim DK, Ihn KJ, Jo SM, Kim DY (2004) Electrospun TiO_2 electrodes for dye-sensitized solar cells. Nanotechnology 15(12):1861. https://doi.org/10.1088/0957-4484/15/12/030

Su SS, Chang I (2018) Review of production routes of nanomaterials. In: Brabazon D, Pellicer E, Zivic F, Sort J, Dolors Baró M, Grujovic N, Choy K-L (eds) Commercialization of nanotechnologies–a case study approach. Springer International Publishing, Cham, pp 15–29

Su SS, Chang ITH, Kuo WCH, Price D, Pikramenou Z, Lead J (2014) Pulsed electrical discharge synthesis of red photoluminescence zinc oxide nanoparticles. J Nanopart Res 16(9):2611. https://doi.org/10.1007/s11051-014-2611-z

Su L, Zhang N, Tang J, Zhang L, Wu X (2021) In-situ fabrication of a chlorine-functionalized covalent organic framework coating for solid-Phase microextraction of polychlorinated biphenyls in surface water. Analytica Chimica Acta 1186:339120–339120.

Sun J, Zhang JX, Fu YY, Hu GX (2002) Microstructural evolution of an Al67Mn8Ti24Nb1 alloy during mechanical milling and subsequent annealing process. Mater Sci Eng A 329–331:703–707. https://doi.org/10.1016/S0921-5093(01)01629-X

Suresh Kumar P, Jayaraman S, Subramanian S, Veluru JB, Singh G, Allakhverdiev SI, Ramakrishna S (2014) Hierarchical electrospun nanofibers for energy harvesting, production and environmental remediation. Energy Environ Sci 7(10):3192–3222. https://doi.org/10.1039/C4EE00612G

Tang R, Shi Y, Hou Z, Wei L (2017) Carbon nanotube-based chemiresistive sensors. Sensors 17(4):882. https://doi.org/10.3390/s17040882

Taufik A, Asakura Y, Hasegawa T, Yin S (2021) $MoS_{2-x}Se_x$ nanoparticles for NO detection at room temperature. ACS Appl Nano Mater 4(7):6861–6871. https://doi.org/10.1021/acsanm.1c00926

Thi Thanh Le D, Long NDH, Thi Xuan C, Van Toan N, Hung CM, Van Duy N, Thi Theu L, Dinh VA, Hoa ND (2023) Porous $CoFe_2O_4$ nanorods: VOC gas-sensing characteristics and DFT calculation. Sens Actuators B Chem 379:133286. https://doi.org/10.1016/j.snb.2023.133286

Tian Y, Hou Y, Yu Q, Wang X, Tian M (2020) Layer-by-layer self-assembly of a novel covalent organic frameworks microextraction coating for analyzing polycyclic aromatic hydrocarbons from aqueous solutions via gas chromatography. J Sep Sci 43(5):896–904. https://doi.org/10.1002/jssc.v43.5. https://doi.org/10.1002/jssc.201900870

Tian W, Liu X, Yu W (2018) Research progress of gas sensor based on graphene and its derivatives: a review. Appl Sci 8(7):1118. https://doi.org/10.3390/app8071118

Tian X, Cui X, Lai T, Ren J, Yang Z, Xiao M, Wang B, Xiao X, Wang Y (2021) Gas sensors based on TiO_2 nanostructured materials for the detection of hazardous gases: a review. Nano Mater Sci 3(4):390–403. https://doi.org/10.1016/j.nanoms.2021.05.011

Tseng AA (2004) Recent developments in micromilling using focused ion beam technology. J Micromech Microeng 14(4):R15. https://doi.org/10.1088/0960-1317/14/4/R01

Ugale AD, Umarji GG, Jung SH, Deshpande NG, Lee W, Cho HK, Yoo JB (2020) ZnO decorated flexible and strong graphene fibers for sensing NO_2 and H_2S at room temperature. Sens Actuators B Chem 308:127690. https://doi.org/10.1016/j.snb.2020.127690

Vallejos S, Gràcia I, Figueras E, Cané C (2015) Nanoscale heterostructures based on $Fe_2O_3@WO_{3-x}$ nanoneedles and their direct integration into flexible transducing platforms for toluene sensing. ACS Appl Mater Interfaces 7(33):18638–18649. https://doi.org/10.1021/acsami.5b05081

Wang L (2020) Metal-organic frameworks for QCM-based gas sensors: a review. Sens Actuators Phys 307:111984. https://doi.org/10.1016/j.sna.2020.111984

Wang QH, Kalantar-Zadeh K, Kis A, Coleman JN, Strano MS (2012) Electronics and optoelectronics of two-dimensional transition metal dichalcogenides. Nat Nanotechnol 7(11):699–712. https://doi.org/10.1038/nnano.2012.193

Wang X, Xiong Z, Liu Z, Zhang T (2015) Exfoliation at the liquid/air interface to assemble reduced graphene oxide ultrathin films for a flexible noncontact sensing device. Adv Mater 27(8):1370–1375. https://doi.org/10.1002/adma.201404069

Wang X, Meng S, Tebyetekerwa M, Weng W, Pionteck J, Sun B, Qin Z, Zhu M (2017) Nanostructured polyaniline/poly(styrene-butadiene-styrene) composite fiber for use as highly sensitive and flexible ammonia sensor. Synth Met 233:86–93. https://doi.org/10.1016/j.synthmet.2017.09.012

Wang Y, Yao L, Xu L, Wu W, Lin W, Zheng C, Feng Y, Gao X (2021) Enhanced NO_2 gas sensing properties based on Rb-doped hierarchical flower-like In_2O_3 microspheres at low temperature. Sens Actuators B Chem 332:129497. https://doi.org/10.1016/j.snb.2021.129497

Wang X, Han Y, Cao J, Yan H (2022) Headspace solid-phase-microextraction using a graphene aerogel for gas chromatography–tandem mass spectrometry quantification of polychlorinated naphthalenes in shrimp. J Chromatogr A 1672:463012. https://doi.org/10.1016/j.chroma.2022.463012

Wang Y, Guo J, Xu D, Gu Z, Zhao Y (2023) Micro-/nano-structured flexible electronics for biomedical applications. Biomed Tech 2:1–14. https://doi.org/10.1016/j.bmt.2022.11.013

Wang W, Zhang S, Li Z, Li J, Yang X, Wang C, Wang Z (2020) Construction of covalent triazine-based frameworks and application to solid phase microextraction of polycyclic aromatic hydrocarbons from honey samples. Food Chem 322:126770–126770

Wender H, de Oliveira LF, Migowski P, Feil AF, Lissner E, Prechtl MHG, Teixeira SR, Dupont J (2010) Ionic liquid surface composition controls the size of gold nanoparticles prepared by sputtering deposition. J Phys Chem C 114(27):11764–11768. https://doi.org/10.1021/jp102231x

Wetchakun K, Samerjai T, Tamaekong N, Liewhiran C, Siriwong C, Kruefu V, Wisitsoraat A, Tuantranont A, Phanichphant S (2011) Semiconducting metal oxides as sensors for environmentally hazardous gases. Sens Actuators B Chem 160(1):580–591. https://doi.org/10.1016/j.snb.2011.08.032

Whittles TJ, Burton LA, Skelton JM, Walsh A, Veal TD, Dhanak VR (2016) Band alignments, valence bands, and core levels in the tin sulfides SnS, SnS_2, and Sn_2S_3: experiment and theory. Chem Mater 28(11):3718–3726. https://doi.org/10.1021/acs.chemmater.6b00397

Wu S, Liu P, Zhang Y, Zhang H, Qin X (2017) Flexible and conductive nanofiber-structured single yarn sensor for smart wearable devices. Sens Actuators B Chem 252:697–705. https://doi.org/10.1016/j.snb.2017.06.062

Xia Y, Whitesides GM (1998) Soft lithography. Annu Rev Mater Res 28:153–184. https://doi.org/10.1146/annurev.matsci.28.1.153

Xu Y (2007) Direct synthesis of multifunctional heterostructured magnetic nanoparticles in gas phase. University of Minnesota.

Xie W, Zhu X, Mei H, Guo H, Li H, Wang P, Li Y, Deng X, Zhu J, Hu C (2023) Metal-organic frameworks as solid-phase microextraction adsorbents for the determination of triacetone triperoxide by gas chromatography-mass spectrometry. Forensic Sci Int 352:111852. https://doi.org/10.1016/j.forsciint.2023.111852

Xue L, Wang W, Guo Y, Liu G, Wan P (2017) Flexible polyaniline/carbon nanotube nanocomposite film-based electronic gas sensors. Sens Actuators B Chem 244:47–53. https://doi.org/10.1016/j.snb.2016.12.064

Yadav TP, Yadav RM, Singh DP (2012) Mechanical Milling: a Top Down Approach for the Synthesis of Nanomaterials and Nanocomposites Nanoscience and Nanotechnology 2(3):22–48. https://doi.org/10.5923/j.nn.20120203.01

Yan Q, Huang L, Guo W, Ouyang L, Shuai Q (2021) Metal organic framework derived Zn/N co-doped hydrophilic porous carbon for efficient solid phase microextraction of polar phenols. Microchim Acta 188(11):400. https://doi.org/10.1007/s00604-021-05060-0

Yan L, Yu R, Chen J, Xing X (2008) Template-free hydrothermal synthesis of CeO_2 nano-octahedrons and nanorods: investigation of the morphology evolution. Cryst Growth Des 8(5):1474–1477. https://doi.org/10.1021/cg800117v

Yang Z, Ren Y, Zhang Y, Li J, Li H, Hu XHX, Xu Q (2011) Nanoflake-like SnS_2 matrix for glucose biosensing based on direct electrochemistry of glucose oxidase. Biosens Bioelectron 26(11):4337–4341. https://doi.org/10.1016/j.bios.2011.04.031

Yang R, Yan F, Zhao N (2017) Urban air quality based on Bayesian network. In: 2017 IEEE 9th international conference on communication software and networks (ICCSN). IEEE, Guangzhou, pp 1003–1006

Yang X, Wang J, Wang W, Zhang S, Wang C, Zhou J, Wang Z (2019) Solid phase microextraction of polycyclic aromatic hydrocarbons by using an etched stainless-steel fiber coated with a covalent organic framework. Microchim Acta 186(3):145. https://doi.org/10.1007/s00604-019-3258-3

Yang X, Zhang S, Wang J, Wang W, Li J, Chen J, Zhao Y, Wang C, Wang Z (2020) Modulated construction of imine-based covalent organic frameworks for efficient adsorption of polycyclic aromatic hydrocarbons from honey samples. Anal Chim Acta 1134:50–57. https://doi.org/10.1016/j.aca.2020.07.072

You R, Han D-D, Liu F, Zhang Y-L, Lu G (2018) Fabrication of flexible room-temperature NO_2 sensors by direct laser writing of In_2O_3 and graphene oxide composites. Sens Actuators B Chem 277:114–120. https://doi.org/10.1016/j.snb.2018.07.179

Yu C, Zhang J, Luo X, Zhang J (2023) Metal organic framework/covalent organic framework composite for solid-phase microextraction of polycyclic aromatic hydrocarbons in milk samples. Microchem J 187:108388. https://doi.org/10.1016/j.microc.2023.108388

Yuan Y, Lin X, Li T, Pang T, Dong Y, Zhuo R, Wang Q, Cao Y, Gan N (2019) A solid phase microextraction arrow with zirconium metal–organic framework/molybdenum disulfide coating coupled with gas chromatography–mass spectrometer for the determination of polycyclic aromatic hydrocarbons in fish samples. J Chromatogr A 1592:9–18. https://doi.org/10.1016/j.chroma.2019.01.066

Yue Q, Huang Y-Y, Shen X-F, Yang C, Pang Y-H (2020) *In situ* growth of covalent organic framework on titanium fiber for headspace solid-phase microextraction of 11 phthalate esters in vegetables. Food Chem 318:126507. https://doi.org/10.1016/j.foodchem.2020.126507

Yun YJ, Kim DY, Hong WG, Ha DH, Jun Y, Lee H-K (2018) Highly stretchable, mechanically stable, and weavable reduced graphene oxide yarn with high NO_2 sensitivity for wearable gas sensors. RSC Adv 8(14):7615–7621. https://doi.org/10.1039/C7RA12760J

Zhang T, Mubeen S, Myung NV, Deshusses MA (2008) Recent progress in carbon nanotube-based gas sensors. Nanotechnology 19(33):332001. https://doi.org/10.1088/0957-4484/19/33/332001

Zhang J, Chaker M, Ma D (2017) Pulsed laser ablation based synthesis of colloidal metal nanoparticles for catalytic applications. J Colloid Interface Sci 489:138–149. https://doi.org/10.1016/j.jcis.2016.07.050

Zhang Q, An C, Fan S, Shi S, Zhang R, Zhang J, Li Q, Zhang D, Hu X, Liu J (2018a) Flexible gas sensor based on graphene/ethyl cellulose nanocomposite with ultra-low strain response for volatile organic compounds rapid detection. Nanotechnology 29(28):285501. https://doi.org/10.1088/1361-6528/aabf2f

Zhang Y, Luo J, Flewitt AJ, Cai Z, Zhao X (2018b) Film bulk acoustic resonators (FBARs) as biosensors: a review. Biosens Bioelectron 116:1–15. https://doi.org/10.1016/j.bios.2018.05.028

Zhang D, Yang Z, Li P, Pang M, Xue Q (2019a) Flexible self-powered high-performance ammonia sensor based on Au-decorated $MoSe_2$ nanoflowers driven by single layer MoS_2-flake piezoelectric nanogenerator. Nano Energy 65:103974. https://doi.org/10.1016/j.nanoen.2019.103974

Zhang L, Tan Q, Kou H, Wu D, Zhang W, Xiong J (2019b) Highly sensitive NH_3 wireless sensor based on Ag-RGO composite operated at room-temperature. Sci Rep 9(1):9942. https://doi.org/10.1038/s41598-019-46213-9

Zhang J, Zhang B, Dang X, Song Z, Hu Y, Chen H (2020b) A polythiophene/UiO-66 composite coating for extraction of volatile organic compounds migrated from ion-exchange resins prior to their determination by gas chromatography. J Chromatogr A 1633:461627. https://doi.org/10.1016/j.chroma.2020.461627

Zhang L, Jiang R, Li W, Muir DCG, Zeng EY (2020) Development of a solid-phase microextraction method for fast analysis of cyclic volatile methylsiloxanes in water. Chemosphere 250:126304. https://doi.org/10.1016/j.chemosphere.2020.126304

Zhao X, Ohkohchi M, Wang M, Iijima S, Ichihashi T, Ando Y (1997) Preparation of high-grade carbon nanotubes by hydrogen arc discharge. Carbon 35(6):775–781. https://doi.org/10.1016/S0008-6223(97)00033-X

Zhao XS, Su F, Yan Q, Guo W, Bao XY, Lv L, Zhou Z (2006) Templating methods for preparation of porous structures. J Mater Chem 16(7):637–648. https://doi.org/10.1039/B513060C

Zheng W, Liu X, Xie J, Lu G, Zhang J (2021) Emerging van der Waals junctions based on TMDs materials for advanced gas sensors. Coord Chem Rev 447:214151. https://doi.org/10.1016/j.ccr.2021.214151

Zhou T, Zhang T (2021) Recent progress of nanostructured sensing materials from 0D to 3D: overview of structure–property-application relationship for gas sensors. Small Methods 5(9):2100515. https://doi.org/10.1002/smtd.202100515

Zhou H, Wang C, Shaw JC, Cheng R, Chen Y, Huang X, Liu Y, Weiss NO, Lin Z, Huang Y, Duan X (2015) Large area growth and electrical properties of p-type WSe_2 atomic layers. Nano Lett 15(1):709–713. https://doi.org/10.1021/nl504256y

Zhou X, Zhou X, Wang C, Zhou H (2023a) Environmental and human health impacts of volatile organic compounds: a perspective review. Chemosphere 313:137489. https://doi.org/10.1016/j.chemosphere.2022.137489

Zhou S, Kuang Y, Shi Y, Hu Y, Chen L, Zheng J, Ouyang G (2023) Modulated covalent organic frameworks with higher specific surface area for the ultrasensitive detection of polybrominated biphenyls. Chem Eng J 453:139743. https://doi.org/10.1016/j.cej.2022.139743

Ziegler D, Marchisio A, Montanaro L, Palmero P, Tulliani J-M (2018) Barium hexaferrite thick-films for ozone detection at low temperature. Solid State Ionics 320:24–32. https://doi.org/10.1016/j.ssi.2018.02.028

Zou Y, Chen S, Sun J, Liu J, Che Y, Liu X, Zhang J, Yang D (2017) Highly efficient gas sensor using a hollow SnO_2 microfiber for triethylamine detection. ACS Sens 2(7):897–902. https://doi.org/10.1021/acssensors.7b00276

Zhuang S, Lee ES, Lei L, Nunna BB, Kuang L, Zhang W (2016) Synthesis of nitrogen-doped graphene catalyst by high-energy wet ball milling for electrochemical systems. Int J Energy Res 40:2136–2149. https://doi.org/10.1002/er.3595

Chapter 2
Nanotechnology for Clean Air and Sustainable Futures

Tri Partono Adhi, Yohanes Andre Situmorang, Yano Surya Pradana, Wan Hanna Melini Wan Mohtar, and Antonius Indarto

Abstract Nanotechnology and nanomaterials present a transformative approach to air purification, offering innovative solutions that can significantly enhance the efficiency and effectiveness of pollutant removal. The unique properties of nanomaterials, such as increased surface area, enhanced reactivity, and improved catalytic performance enable the development of advanced filtration, adsorption, and catalytic systems that address various air quality challenges with better performance compared to conventional air purification technologies. This chapter provides a brief introduction to nanotechnology applications for improving air quality, such as nano-filtration, nano-adsorption, and nano-photocatalysis. The challenges of the application of these technologies for air pollutant removal in the real world and future perspectives are also discussed.

T. P. Adhi
Department of Chemical Engineering, Institut Teknologi Bandung, Bandung, Indonesia

Y. A. Situmorang
Department of Chemical Engineering, Parahyangan Catholic University, Bandung, Indonesia

Y. S. Pradana
Department of Chemical Engineering, Institut Teknologi Bandung, Bandung, Indonesia

Department of Chemical Engineering, Faculty of Engineering, Universitas Gadjah Mada, Yogyakarta, Indonesia

W. H. M. Wan Mohtar
Department of Civil Engineering, Faculty of Engineering and Built Environment, Universiti Kebangsaan Malaysia, Bangi, Selangor, Malaysia

Environmental Management Center, Institute of Climate Change, Universiti Kebangsaan Malaysia,
Bangi, Selangor, Malaysia

A. Indarto (✉)
Department of Chemical Engineering, Institut Teknologi Bandung, Bandung, Indonesia

Department of Bioenergy Engineering and Chemurgy, Institut Teknologi Bandung, Sumedang, Indonesia
e-mail: antonius.indarto@itb.ac.id

N. Ghosh et al. (eds.), *Nanotechnology in Air Quality Management*, Sustainable Development and Biodiversity 39, https://doi.org/10.1007/978-981-96-8339-0_2

Keywords Nanoadsorption · Nanofiltration · Nano-photocatalysis · Nanomaterials · Air quality

2.1 Introduction

Recently, awareness of public health and environmental issues has emerged. One of the critical subjects regarding those matters is air quality. While the rapid growth of urbanization and industrial activities increases air pollution, regions where low-tech manufacturing is still dominant contribute to worse air pollution (Cohen et al. 2017). Air pollution mainly results from a variety of activities, including industrial emissions, fossil fuel combustion, transportation, agricultural practices, and natural phenomena such as wildfires. The poor air quality caused by air pollution is linked to a range of serious health issues, including respiratory diseases, cardiovascular conditions, and even premature mortality. Moreover, it can adversely affect ecosystems, damaging vegetation, soil, and water resources, even changing climates, and causing global warming (Mohamed 2017; Ibrahim et al. 2015).

The air quality is considered poor when there is a change in harmful substances' composition in the atmosphere exceeding the threshold limit. Those harmful substances include H_2S, CO, CO_2, NOx, HONO, SOx, dioxins, volatile organic compounds (VOCs) and hydrocarbons, halogens and chlorofluorocarbons (CFCs), and particulate matters (PM). H_2S and SO_2 are hazardous gases produced from either the combustion of sulfur-containing fuels, the biological activity of anaerobic microbes in organic compounds such as organic waste, industrial sewage, manure disposal, sewer, and ports, etc., or often found in volcanic areas. Exposure to H_2S may cause skin burns, eye inflammation, and respiratory illness, and can be lethal at concentrations exceeding 100 ppm. SO_2 has the same toxicity level as H_2S. NO_2 is a gas pollutant emitted from the combustion of fossil fuels from stationary sources (heating, power generation) and automobile motor engines. The effects of NO_2 exposure are irritation of the respiratory system, pulmonary edema, and even chronic lung disease. Furthermore, CO is a particularly dangerous gas because, unlike H_2S and SO_2, it is odorless and can be deadly at levels above 35 ppm (Brunekreef and Holgate 2002; Yunus et al. 2012; Manisalidis et al. 2020; Kok et al. 2009).

Even though CO_2 is a common gas found in the atmosphere and produced daily by human activities, the emission of CO_2 is considered a primary contributor to air pollution, raising significant concerns because of its impact as a greenhouse gas (GHG). CO_2 is primarily released through the combustion of fossil fuels, mainly for energy production, transportation, and manufacturing, and also from wildfires. These emissions have contributed to a significant rise in atmospheric CO_2 levels, from about 280 parts per million (ppm) before the Industrial Revolution to over 410 ppm today. The impact of elevated CO_2 levels is not only affecting climate change but also contributing to respiratory problems in populations. Furthermore, higher CO_2 levels can alter the chemistry of oceans, leading to ocean acidification, which threatens marine life and ecosystems (Hadipoor et al. 2021).

The 2,3,7,8-tetrachlorodibenzo-p-dioxin (TCDD) more known as dioxin is a toxic gas produced from incineration of plastics or other chemical wastes and also be released during forest fires. Once released, dioxins can travel long distances through the air then accumulate in the fatty tissues of animals and contaminate the food chain. The effects of dioxin include cancer, immune system suppression, developmental and reproductive issues, and endocrine disruption (Hawkins et al. 2010; Matsumoto and Kondo 2008). Meanwhile, volatile organic compounds (VOCs) are a diverse group of organic chemicals characterized by their high vapor pressure at room temperature. Commonly, VOCs are hydrocarbons such as formaldehyde, toluene, benzene, ethylbenzene, and xylene that are formed in crude oil refineries, petrochemical industries, and vehicle emissions. Hydrocarbons are primarily categorized into two groups: aliphatic and aromatic hydrocarbons. Aliphatic hydrocarbons include straight-chain and branched forms, while aromatic hydrocarbons contain ring structures and often exhibit higher toxicity. In the atmosphere, VOCs can undergo complex chemical reactions, contributing to the formation of smog. These pollutants pose serious health risks, including respiratory problems and cancer. Certain hydrocarbons, such as methane, are potent greenhouse gases, contributing significantly to climate change (Guenther et al. 1995; Manisalidis et al. 2020; Seinfeld and Pandis 2016). Other compounds that have an impact on the climate and the environment are halogens and CFCs. The CFCs were once widely used as refrigerants, propellants in aerosol sprays, and solvents, containing chlorine, fluorine, and carbon. They were favored for their chemical stability and nontoxicity to humans. However, their release into the atmosphere has led to severe ozone depletion (Molina and Rowland 1974).

Besides gases, another harmful material affecting air quality is particulate matter. Particulate matter refers to a mixture of tiny solid particles and liquid droplets suspended in the air, which can vary in size, composition, and origin. PM is usually formed as a result of fossil fuel combustion, construction activities, natural events such as forest burning and volcano eruptions, and chemical reactions of other pollutants. The most commonly studied categories are PM10 (particles with a diameter of 10 micrometers or less) and PM 2.5 (particles with a diameter of 2.5 micrometers or less). Due to their small size, PM 2.5 particles can penetrate deep into the lungs and even enter the bloodstream, posing serious health risks. Categories of PM are particulate contaminants including smog, soot, tobacco smoke, oil smoke, and fly ash; biological contaminants including bacteria, viruses, fungi, mold, and pollen; and dust including suspended atmospheric dust, settling dust, and cement dust. The annual mean limits recommended by the World Health Organization (WHO) are for PM10 to be 15 $\mu g/m^3$ and PM 2.5 to be 5 $\mu g/m^3$ (Manisalidis et al. 2020; Heal et al. 2012; Mannucci et al. 2015).

There are many developed and developing technologies that have been applied to control air pollution. These technologies include incinerators; upgrading separation techniques such as settling chambers, electrostatic precipitators, cyclones, membranes, biofilters, adsorption columns, and scrubbers; and enhancement of processing techniques such as selective catalytic reduction systems, catalytic converters, photocatalytic remediation, and plasma combustion. Since the 1970s, developed countries have made significant strides in enhancing air quality, but there is still considerable work to be done in developing nations. The currently available technologies are still considered expensive. Thus, air pollution management strategies should include cost–benefit analysis, air quality standards, emission standards, and economic incentives. The selected control technology or process must be both technically and economically viable, as well as enforceable within the particular region or country to ensure compliance (He et al. 2024; Kalender 2019).

Lately, nanotechnologies have emerged as promisingly effective and efficient air pollution control technologies. Nanotechnologies use materials on the nanoscale that exhibit enhanced or unique properties not found in their normal size. The applications of nanotechnology for environmental purposes such as adsorption, absorption, chemical reaction, and filtration are found to be cost-effective with better performance compared to conventional technologies (Pramudita et al. 2018; Pramudita et al. 2022; Guerra et al. 2018). This chapter will discuss these available and potential nanotechnologies, the challenges, and the perspective on controlling air quality.

2.2 Nanotechnologies for Air Purification and Air Pollution Control

2.2.1 Nano-filtration

Among the technologies for controlling air pollution, filters have the highest application. Effective air filtration is essential for ensuring good air quality. So far, traditional air filtration methods primarily involve two primary categories: mechanical filtration for particulate air filters and chemical air filters. Particulate air filters are porous fibrous membranes that specifically capture solid particle pollutants. Conventional filters, such as HEPA (high-efficiency particulate air) filters, excel in capturing particulate matter. However, they often fall short in removing smaller particles, gases, and volatile organic compounds (VOCs). In contrast, chemical air filters are intended to eliminate airborne molecular contaminants like VOCs, odors, and ozone using an absorbent or catalyst. However, these types of filters are less effective at removing particulate matter. Those methods have been widely used in residential, commercial, and industrial settings (Garcıa-Mayagoitia et al. 2020).

Optimal air filtration materials should effectively capture aerosol particles while also allowing air to flow through them effortlessly. Lately, nano-fibers with fiber diameters ranging from 1 to 1000 nm have been developed for air filters. In comparison to standard air filters, they offer a greater specific surface area, higher porosity, and enhanced mechanical strength, along with improved airflow. Their filtration efficiency is significantly increased due to the higher number of active sites afforded by the large surface area, as well as improved selectivity (Pramudita et al. 2022; Chang et al. 2018).

In the last decade, several techniques have emerged for synthesizing nanofibers, including the stretching method, island spinning, and electrospinning. Among those techniques, electrospinning is considered the best. Electrospun nanofibers are appealing as air filters due to their remarkable characteristics, such as higher molecular orientation, greater tensile strength, higher efficiency, and lower pressure drop compared to conventional air filtration membranes. They also possess a large specific surface area, high porosity, small pore size, and a well-connected pore structure, which facilitate the capture of fine particles. Studies on electrospun nanofibers indicate that the filtration performance of a nonwoven fabric can be significantly enhanced by directly applying electrospun nanofibers to it. Others reported that small-size PM rapidly agglomerates and forms large-sized particles once they are contacted by electrospun nanofibers. This kind of nanofiber is considered energy-efficient, cost-effective, and easily scalable (Cao et al. 2021; Garcıa-Mayagoitia et al. 2020).

Some types of electrospun nanofibers have been developed for air filtration. Polyacrylonitrile (PAN) filters are nanofibers that operate on the principle of particle adhesion to reduce PM 2.5 pollutants. Ag/PAN nanofiber media, which consists of Ag nanoparticles on PAN nanofibers, can be widely used to eliminate PM 2.5 from the air and effectively deactivate bacteria. They are also beneficial for personal protection (such as masks), cleanroom environments (in biomedical, electronics, pharmaceuticals, and automotive industries), and indoor air purification. PAN is a transparent material that can be easily applied to house windows as a pollutant protector. PAN filters allow particulate matter to adhere to their surface while still permitting natural light to enter the room, offering added advantages such as excellent optical transparency, low air resistance, lightweight design, and high filtering efficiency with an efficiency of more than 95% for PM 2.5 filtration (Roy et al. 2019). PAN is not only effective in reducing PM 2.5 but can also be used to reduce VOCs. It is reported that PAN-activated carbon nanofiber with high microporosity and abundant nitrogen-containing functional groups can effectively adsorb formaldehyde even at low concentrations (Mohamed 2017).

The electrospinning technique enhances the polarity and surface electrostatic charges of the polymers used in nanofiber production to boost their filtration efficiency. Additionally, incorporating active components into the polymers can impart special functionalities to the nanofibers. For instance, since iron oxide constitutes a significant portion of airborne dust, adding magnetic metal oxide nanoparticles (Fe_3O_4) to electrospun nanofibers can help attract dust through magnetic forces (Cao et al. 2021). Another additional technique used to improve electrospun nanofiber is UV radiation to achieve the aim of antibacterial treatment. A bio-based chitosan/polyvinyl alcohol air filtration via green electrospinning and UV-cured was reported to have superior filtration performance as well as efficient antibacterial activities. The electrospinning technique to fabricate nanofiber is also proven to be effective in capturing coronavirus about 60–140 nm in diameter with an average size of 100 nm. It is reported that sodium chloride-based aerosols with a diameter of 50–500 nm and polyvinylidene fluoride nanofibers with dimensions of fibers of 84–349 nm have a coronavirus filtration efficiency of 90%. The addition of antimicrobial agents like silver to electrospun nanofibers is recognized for enhancing the antimicrobial properties of the filters. It is mentioned that electrospun nanofibers with Ag + and Ag particles demonstrated strong antimicrobial activity and suitable transport characteristics for air filtration applications. The antibacterial nylon nanofibers containing silver were highlighted for their effectiveness against *Escherichia coli* (Gram-negative) and *Staphylococcus aureus* (Mahmoudi et al. 2023; Zhang et al. 2011; Montazer and Malekzadeh 2012; Zhu et al. 2019; Sundarrajan et al. 2014).

2.2.2 Nano-adsorption

Adsorption is a nondestructive process that involves the accumulation of molecules from a gas, liquid, or dissolved solid onto a surface, forming a film of the adsorbate on the adsorbent. It has different purposes for filtration in improving air quality. Adsorption is divided into two categories depending on the interaction between the adsorbate and the adsorbent: physisorption (noncovalent) and chemisorption (covalent). The sorption capacity is influenced by several factors, such as (1) the physical structure and chemical makeup of the adsorbent materials, particularly their specific surface area; (2) the size and distribution of pores, as well as the presence of surface functional groups; (3) the molecular characteristics of the target pollutants, including their polarity, acidity, and wettability; and (4) the conditions under which adsorption occurs, mainly temperature and humidity. Traditional adsorbents, such as activated carbon, have been effective but often come with limitations in efficiency and specificity. Recent advancements in nanotechnology have led to the development of nanomaterials that significantly enhance adsorption capabilities due to their unique properties, such as increased surface area, tunable porosity, and high reactivity, as well as their ability to be regenerated. The adsorption process seems to be more likely to be used for water treatment. However, it is still efficient to be used for reducing air pollutants, such as H_2S, NOx, SOx, CO_2, dioxins, microbes, and VOCs (Cao et al. 2021; Zhang et al. 2024).

Adsorption of harmful gases is one of the important mechanisms of air purification. Among many nano-adsorbents that have been researched, carbon nanotubes (CNTs) are probably the most popular nano-adsorbents. CNTs are the graphene layer with a hexagonal arrangement of carbon atoms that surrounds the tube axis. They exhibit outstanding adsorption capabilities thanks to their high specific surface area, remarkable porosity, hollow and layered structures, abundant internal and external adsorption sites, π-conjugated framework, and the ability to undergo chemical activation and functionalization easily. Various mechanical, physical, and chemical methods have been employed to modify or functionalize carbon nanotubes, aiming to improve their overall adsorptive properties or to enhance their selectivity for specific types of pollutants (Sajid et al. 2022).

CNTs can be modified into three different groups depending on the number of inner walls. Single-walled carbon nanotubes (SWNTs) are a single layer of graphene sheet with a diameter of 1–2 nm. Double-walled carbon nanotubes (DWNTs) consist of two carbon nanotubes, with the outer tube surrounding the inner one. The outer tube has a diameter of 2–4 nm, while the inner tube measures 1–3 nm in diameter. Multiwalled carbon nanotubes (MWNTs) have multiple layers of graphene sheets rolled into cylinders, with diameters ranging from 2 to 50 nm. The inner layer radius of the tube is approximately 0.34 nm, while the outer layer can range from 2 nm to 20–30 nm. MWNTs can be classified into two structural models: the Russian doll model and the Parchment model. The Russian doll model features concentric cylinders of graphene sheets, while the Parchment model involves a single graphene sheet rolled up multiple times, resembling a scroll of paper. With these different types, CNTs have flexibility in adsorption applications for the removal of harmful gases like NOx and SOx, and VOCs (Rathinavel et al. 2021; Gulati et al. 2022).

The application of CNTs in the adsorption of gas molecules such as CO_2, NOx, H_2S, methane, dioxins, and VOCs has been extensively studied. Amine-functionalized CNTs have gained significant attention for CO_2 capture from flue gas, due to their exceptional physicochemical properties, including high thermal and chemical stability. Various studies have shown that both multiwalled and single-walled amine-functionalized CNTs possess excellent CO_2 adsorption capabilities (Liu et al. 2014; Wang et al. 2021).

Carbonaceous materials such as activated carbon fiber, graphene oxide, and CNTs are favorable for adsorbing VOC pollutants. The mechanisms of VOC adsorption on CNTs primarily involve physical adsorption, along with minor chemical reactions between the surface functional groups of CNTs and the VOC molecules. Due to the composition of graphene sheets, CNTs are highly hydrophobic. It enhances their ability to adsorb the aromatic ring of VOCs. In general, the adsorption of nonpolar VOCs on CNTs is mainly governed by physical processes, whereas the adsorption of polar VOCs involves chemical interactions (Zhang et al. 2017). CNTs are also proven to have superior performance in adsorbing dioxins. The findings indicated that CNTs are significantly more effective than activated carbon and Al_2O_3-γ in removing dioxins. It is shown that the interaction between dioxin and CNTs is approximately three times stronger than that between dioxin and activated carbon. The enhancement of CNTs is likely attributed to the curved surfaces of the nanotubes, which create stronger interaction forces with dioxins compared to the smooth surface of activated carbon (Beni and Jabbari 2022; Yunus et al. 2012).

Besides CNTs, metal-organic frameworks (MOFs) are promising materials for the selective adsorption of gaseous pollutants due to their numerous open metal sites and the potential for functionalization and modification of their frameworks. MOFs are crystalline materials formed through the self-assembly of organic ligands and metal ions or clusters into a periodic network structure. The metals used can be divalent (Cu, Zn, Mg), trivalent (Al, Cr, Fe), or tetravalent (V, Zr, Ti). MOFs have unique chemical structures, thanks to the integration of inorganic and organic components. Additionally, compared to traditional adsorbent materials, MOFs offer significant potential for the adsorption process due to their ultra-high specific surface area, adjustable pore sizes, ultra-high porosity, high crystallinity, and controllable structures (Wang et al. 2021).

Some MOFs are found to have excellent performance in adsorbing H_2S due to strong interaction with S atoms. It is found that metal cation-based MOFs, such as Cu, Zn, or Fe, can react with H_2S. The reaction creates a coordination bond between the open metal site and the functionality of MOFs and the redox-active H_2S to form metal sulfide. It is considered the primary reason for the extremely high adsorption capacity of MOFs. However, MOF applications for SO_2 adsorption are still minimal as the interactions of S atoms with metal clusters and organic linkers are relatively weak (Pramudita et al. 2022; Wen et al. 2018; Khan et al. 2013).

MOFs also showed outstanding adsorption capacity toward halogens because both the open metal site and the linker of MOFs are capable of interacting with halogens. However, exposing MOFs to high concentrations of light halogens can lead to the collapse of their framework due to the formation of corrosive acids that attack the nodes of the MOFs. In this context, developing redox-active MOFs to adsorb halogens through oxidation reactions between the halogens and the metal sites of the MOFs presents a promising strategy for preventing corrosive acid generation, thereby enabling reversible adsorption of halogens (Wen et al. 2018). Moreover, there are plenty of studies on MOF applications for the removal of VOCs, such as benzene, toluene, xylene, acetone, formaldehyde, and other hydrocarbons. Some MOFs are also extraordinary for CO_2 adsorption. MOFs featuring polar pores such as −OH, −N=N−, −NH2, and –N=C (R)− exhibit a greater CO_2 adsorption capacity because of the quadrupole moments of CO_2 molecules. Consequently, by structuring MOFs with these characteristics in mind, it is possible to significantly enhance their CO_2 adsorption capacity and selectivity as effective adsorbents (Ghanbari et al. 2020). The properties of MOFs can be easily modified and adjusted to the applications. Thus, MOF applications are flexible, selective, and adaptive depending on the pollutants that need to be removed. Table 2.1 shows several examples of nano-adsorption cases.

2.2.3 *Nano-photocatalysis*

The catalytic degradation process utilizing nano-catalyst materials can be employed to treat air pollutants. This method transforms harmful pollutants into less or non-toxic compounds. There are two main types of catalytic degradation: the common catalytic process and the photo-catalytic process. The common catalytic degradation process is useful for reducing the emission of CO, hydrocarbons, and NOx.

The nano-catalyst that is commonly used for NOx degradation is a vanadium-based catalyst. In a catalytic degradation system, a three-way catalyst consisting of platinum (Pt), palladium (Pd), and rhodium (Rd) is used to convert CO to CO_2 and to remove NOx. While Pd and Pt promoted the oxidation of CO, Rd was applied for NOx removal. Current catalytic degradation relies on expensive metal catalysts and requires a specific air-to-fuel ratio to operate effectively. This situation has led to a

Table 2.1 Examples of nano-adsorption cases for various air pollutants

Nano-adsorptive materials	Target pollutant gases	References
Aligned-CNT	Aerosols	Handojo et al. (2019) and Yildiz and Bradford (2013)
CNTs deposited on quartz filters	VOCs	Handojo et al. (2019) and Amade et al. (2014)
CNT-3-aminopropyltriethoxysilane (APTS)	CO_2	Su et al. (2009) and Mohamed (2017)
MNTs-(3-aminopropyl) triethoxysilane	CO_2	Saleem et al. (2022) and Gui et al. (2013)
SWCNTs doped with Si and B	CO and CH_3OH gases	Mohamed (2017) and Azam et al. (2017)
CNTs	PM	Attia et al. (2024)
MOF-177, MOF-170, MIL-101(Cr)-4F(1%)	SO_2	Martınez-Ahumada et al. (2021)
HKUST-21, MIL-47(V), MIL-53(Cr, Al)	H_2S	Wen et al. (2018)
MOF/graphene oxide composite	Acetone	Zhang et al. (2017)
MIL-100(Fe)	Toluene	Wang et al. (2023) and Ma et al. (2021)

strong interest in developing low-cost, high-efficiency catalysts. The Cu–CeO_2 and Au–CeO_2 catalysts show significantly higher activity and stability compared to copper oxide or sodium oxide catalysts for CO oxidation. Additionally, the catalytic performance of Cu–CeO_2 for CO oxidation even surpasses that of platinum-based catalysts (Beni and Jabbari 2022; Handojo et al. 2019).

Nano-photocatalysis is an effective method for biodegradation and pollution management. It utilizes oxidation and reduction reactions facilitated by photo-generated charge pairs when exposed to visible light. The process begins when a polluting compound is adsorbed on the surface of a photocatalyst. The utilization of nanomaterials as photocatalysts has the advantage of increasing reaction efficiency due to the large active surface. Photocatalysis is a promising technology as it can directly transform gaseous atmospheric pollutants, particularly VOCs, into less harmful CO_2 and H_2O under ambient conditions. The appealing ability of photocatalysts to purify air using UV-visible light has driven the development of innovative nano-catalytic materials for effective VOC removal (Jaison et al. 2023; Pramudita et al. 2022; Tahir et al. 2020).

Among various photocatalysts, titanium oxide (TiO_2) is the most popularly utilized due to its affordability, chemical stability, nontoxic properties, and strong oxidative capability. TiO_2 can reduce harmful compounds such as VOCs, NOx, and CO to become less toxic compounds. Unfortunately, TiO_2 has a large bandgap of 3.2 eV that can only absorb wavelengths shorter than 400 nm, which constitutes only a 5% fraction of solar light, making it less effective in indoor settings. Even so, research involving TiO_2 represents approximately two-thirds of the published articles on photocatalytic air purification. It is necessary to improve visible light absorption of

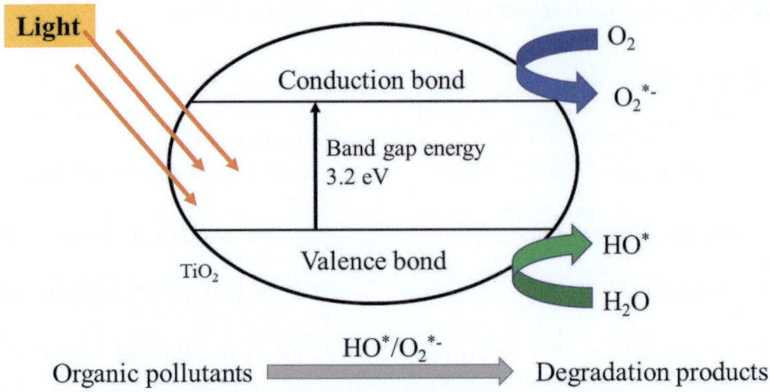

Fig. 2.1 TiO_2 photocatalytic mechanism. (Modified Sujatha et al. 2020)

TiO_2 by metal modification (doping and nanoparticle loading), non-metal doping, and heterojunction with other semiconductors (Weon et al. 2019). Figure 2.1 displays the TiO_2 photocatalytic mechanism.

In recent years, bismuth (Bi)-based materials have been second most researched nano-photocatalyst under visible light. It has a small bandgap and low toxicity. In particular, $BiVO_4$, Bi_2O_3, and $BiWO_6$ were mainly employed for CO_2 reduction and VOC degradation, such as hydrocarbons, alcohols, and aldehydes. It is reported that most Bi-based nano-photocatalysts have better performance than TiO_2 (Song et al. 2023). Zinc oxide (ZnO) is another photocatalyst material that attracted significant attention due to its excellent photocatalytic performance, affordability, and abundant availability. It has similar properties to TiO_2 but has a band gap of 3.37 eV, which is too wide to utilize visible light. It is required to modify the ZnO structure, so it can operate under visible light. ZnO has a 75% lower cost of production in comparison to TiO_2 and Al_2O_3, is nontoxic, and can absorb a larger fraction of the solar spectrum compared to TiO_2. In some cases, ZnO has better photocatalytic activity than TiO_2 when there are oxygen vacancies that make induced point defects increase the reaction rate and generate more hydroxyl ions. However, the ZnO nano-photocatalyst performs worse in aqueous solutions under UV light. The unique properties of ZnO and TiO_2 make them useful to be utilized as self-cleaning materials. The photocatalytic self-cleaning is a hydrophobic coating of metal oxide that can automatically decompose contaminant particles using solar light (Ong et al. 2018; Rafique et al. 2020). Table 2.2 shows several cases of nano-photocatalytic research in air pollutant degradation.

2.3 Challenges and Perspective

Nanotechnology emerges to improve the performance of conventional air purification as well as discover novel, environmentally friendly, cost-effective, and selective materials to be applied in air purification. Technologies explained in the previous

Table 2.2 Examples of nano-photocatalytic cases on air pollutants

Nano-photocatalytic materials	Target pollutant gases	References
Ag-loaded TiO_2	Acetaldehyde	Zeng et al. (2018) and Weon et al. (2019)
Fe-doped TiO_2	Formaldehyde	Weon et al. (2019) and Liu et al. (2011)
N-doped TiO_2	Benzene	Zeng et al. (2014) and Weon et al. (2019)
TiO_2 thin film	Acetone, toluene, p-xylene	Jaison et al. (2023) and Liang et al. (2010)
C-doped TiO_2	Toluene	Jaison et al. (2023) and Dong et al. (2011)
N-dopedTiO2	Ethylbenzene	Jaison et al. (2023) and Kamaei et al. (2018)
Bi co-catalyst/Bi_2MoO_6, Bi/$(BiO)_2CO_3$	NO	Rafique et al. (2020) and Huang et al. (2019)
N-doped ZnO	Formaldehyde	Ong et al. (2018) and Wu (2014)
CN-loaded WO_3	Acetaldehyde	Weon et al. (2019) and Jin et al. (2014)

Table 2.3 Summary of nanotechnology for air purification. (Modified from Rafique et al. 2020)

Technology	Main target pollutant	Advantages	Disadvantages
Filtration	PM 2.5, PM 10, some kind of VOCs	Based on the recovery of products, cost-effective, no extra treatment is required	Extremely costly, not easily available
Adsorption	Harmful gases (CO, CO_2, H_2S, SO_2), some kind of VOCs	Based on the recovery of products, cost-effective	Low efficiency due to moisture and organic compounds
Catalytic oxidation	CO, hydrocarbons, and NOx	Energy savings of about 70%	The efficiency of the process is changed by changing the operation conditions, chemicals affect the performance of the catalyst by deterioration, requiring special safety measures
Photocatalysis	VOCs	High efficiency, uses sunlight or artificial light as an energy source, low operating costs	Limited light activation, sensitive to humidity, low material durability

section introduce the possibility and current approach for applications of nanotechnology and the production of promising nanomaterials in reducing air pollutants to improve air quality sustainably. Table 2.3 summarizes different nanotechnologies for air pollutant reduction.

Nanotechnology is proven to have better performance and efficiency, offering innovative solutions for the removal and degradation of harmful airborne contaminants compared to conventional technologies. The unique properties of nanomaterials, such as increased surface area, enhanced reactivity, and improved catalytic performance, position them as effective agents for air pollutant reduction. Some nanotechnology applications for air purification have been commercialized and globally applied. However, despite their potential, several challenges impede the widespread application of nanotechnology in this area. First, one major challenge lies in the scalability of nanomaterial production. While laboratory-scale synthesis methods can yield highly effective nanomaterials with extraordinary performance, translating these processes to industrial levels poses significant technical and economic hurdles. Most of the nanomaterials for air purification are still in the research stage. Bringing them to the real application on a larger scale requires development and certain improvements. For example, the main serious problems of the most reported MOFs are still suffering from low stability and poor recyclability, preventing their practical application. Moreover, nano-filters are also facing the same challenge of the low mechanical strength of the fibrous layers (Pramudita et al. 2022; Wen et al. 2018). The challenge lies not only in technical issues but also in the fact that the production of nanomaterials remains largely expensive. Even though nanomaterials are selective, one process system that generates various pollutants requires many nanomaterials for specific purposes. Thus, the cost needed by one process system will be high to comply with all specifications required.

Second, the environmental and health implications of nanomaterials themselves are a growing concern; understanding their toxicity and behavior in the environment is critical for ensuring safety. For example, the current technique to produce nanofilters requires toxic solvents, which are harmful to the environment and humans. Green electrospinning technology has been introduced to address these issues, alleviating concerns related to safety, environmental impact, and toxicity. This process utilizes mild nontoxic solvents and biopolymers as raw materials. However, one challenge of the green electrospinning technique is the instability of the electrospun structures in humid conditions. Ongoing advancements are required to improve the stability of these nanofilters (Komaladewi et al. 2019). In accordance, the implementation of green chemistry principles in the production of nanomaterials needs significant development to create advanced materials that are stable, cost-effective, and environmentally friendly. Furthermore, integrating nanotechnology into existing air purification systems often requires overcoming engineering and design challenges. The effectiveness of nanomaterials can be influenced by factors such as humidity, temperature, and the presence of other pollutants, complicating their application in real-world scenarios.

The future of nanotechnology in air purification holds tremendous promise, driven by ongoing advancements in materials science and engineering. As air pollution levels continue to increase, the demand for innovative, efficient, and affordable air purification technology will grow. Here are several key perspectives for the future of this field:

- Enhanced Material Development: It is necessary for continuous research and development to discover nanomaterials with improved properties, such as higher reactivity, greater durability, and better selectivity for specific pollutants. Innovations in nanocomposites and hybrid materials could provide enhanced functionality for diverse air purification applications.
- Integration with Smart Technologies: The convergence of nanotechnology with smart technologies, such as sensors, will enable real-time monitoring and control of air quality. Smart air purification systems could automatically adjust their operations based on pollution levels, improving efficiency and effectiveness. The development of nano-sensors gives a quick and accurate sensing ability with better selectivity.
- Sustainable Production Processes: As the emphasis on sustainability increases, the development of eco-friendly synthesis methods for nanomaterials will become more important. Green chemistry approaches will help mitigate environmental impacts and enhance the overall sustainability of air purification technologies.
- Regulation Advances: Certain regulations for implementing nanotechnology to improve air quality are essential. Incentives support an economical mindset for the implementation. The regulations also have to focus on the lifecycle analysis of nanomaterials, addressing potential health risks and environmental impacts.
- Scalability and Commercialization: It is crucial to narrow the gap between laboratory research and commercial applications. Advances in manufacturing techniques will help scale up the production of nanomaterials while maintaining quality and performance, making them more accessible for widespread use.
- Multifunctional Applications: It is vital to enable nanomaterials to serve multiple functions in air purification systems, so they can remove PM, VOCs, and harmful gases simultaneously. This multifunctionality could lead to more comprehensive air quality solutions.

In summary, the future of nanotechnology in air purification is bright, with potential breakthroughs that could revolutionize the way we address air quality challenges. By focusing on material innovation, smart integrations, sustainability, and regulatory frameworks, we can harness the full potential of nanotechnology to create cleaner and healthier environments.

2.4 Conclusions

Nanotechnology is the future solution to better and sustainable air quality. Finding and producing the right nanomaterials are the keys to sustainable, selective, adaptive, flexible, and affordable processes to reduce air pollutants in the atmosphere. Continuous research and development are essential to narrow the gap between laboratory results and real-world implementations while producing nanomaterials that not only have outstanding performance and properties but are also affordable and environmentally safe. By fostering interdisciplinary collaboration and integrating sustainable practices, the future of nanotechnology in air purification holds great promise for creating cleaner, healthier environments for all.

References

Amade R, Hussain S, Ocaña IR, Bertran E (2014) Growth and functionalization of carbon nanotubes on quartz filter for environmental applications. J Environ Eng Ecol Sci 3(1):2

Attia YA, Ezet AE, Saeed S, Galmed AH (2024) Nano carbon-modified air purification filters for removal and detection of particulate matters from ambient air. Sci Rep 14:621. https://doi.org/10.1038/s41598-023-50902-x

Azam MA, Alias FM, Tack LW, Seman RNAR, Taib MFM (2017) Electronic properties and gas adsorption behavior of pristine, silicon-, and boron-doped (8, 0) single-walled carbon nanotube: a first principles study. J Mol Graph Model 75:85–93. https://doi.org/10.1016/j.jmgm.2017.05.003

Beni AA, Jabbari H (2022) Nanomaterials for environmental applications. Results Eng 15:100467. https://doi.org/10.1016/j.rineng.2022.100467

Brunekreef B, Holgate ST (2002) Air pollution and health. Lancet 360:1233. https://doi.org/10.1016/S0140-6736(02)11274-8

Cao J-j, Huang Y, Zhang Q (2021) Ambient air purification by nanotechnologies: from theory to application. Catalysts 11:1276. https://doi.org/10.3390/catal11111276

Chang J, Zhang L, Wang P (2018) Intelligent environmental nanomaterials. Environ Sci Nano 5:811–836. https://doi.org/10.1039/C7EN00760D

Cohen AJ, Brauer M, Burnett R, Anderson HR, Frostad J, Estep K (2017) Estimates and 25-year trends of the global burden of disease attributable to ambient air pollution: an analysis of data from the global burden of diseases study 2015. Lancet 389(10082):1907–1918. https://doi.org/10.1016/S0140-6736(17)30505-6

Dong F, Guo S, Wang H, Li X, Wu Z (2011) Enhancement of the visible light photocatalytic activity of C-doped TiO2 nanomaterials prepared by a green synthetic approach. J Phys Chem C 115(27):13285. https://doi.org/10.1021/jp111916q

Garcıa-Mayagoitia S, Perez-Hernandez H, Medina-Perez G, Campos-Montiel RG, Fernandez-Luqueno F (2020) Bio-nanomaterials in the air pollution treatment. In: Amrane Abdeltif AAA, Rtimi S (eds) Nanomaterials for air remediation. Elsevier, pp 227–248

Ghanbari T, Abnisa F, Daud WMAW (2020) A review on production of metal organic frameworks (MOF) for CO2 adsorption. Sci Total Environ 707:135090. https://doi.org/10.1016/j.scitotenv.2019.135090

Guenther A, Hewitt CN, Erickson D, Fall R, Geron C, Graedel T, Harley P, Klinger L, Lerdau M, Mckay WA, Pierce T, Scholes B, Steinbrecher R, Tallamraju R, Taylor J, Zimmerman P (1995) A global model of natural volatile organic compound emissions. J Geophys Res 100(D5):8873–8892. https://doi.org/10.1029/94JD02950

Guerra FD, Attia MF, Whitehead DC, Alexis F (2018) Nanotechnology for Environmental Remediation: Materials and Applications. Molecules 23(7):1760. https://doi.org/10.3390/molecules23071760

Gui MM, Yap YX, Chai S-P, Mohamed AR (2013) Multi-walled carbon nanotubes modified with (3-aminopropyl)triethoxysilane for effective carbon dioxide adsorption. Int J Greenhouse Gas Control 14:65–73. https://doi.org/10.1016/j.ijggc.2013.01.004

Gulati S, Lingam HN, Kumar S, Goyal K, Arora A, Varma RS (2022) Improving the air quality with functionalized carbon nanotubes: sensing and remediation applications in the real world. Chemosphere 299:134468. https://doi.org/10.1016/j.chemosphere.2022.134468

Hadipoor M, Keivanimehr F, Baghban A, Ganjali MR, Habibzadeh S (2021) Chapter 24 – Carbon dioxide as a main source of air pollution: prospective and current trends to control. In: Núñez-Delgado A (ed) Sorbents materials for controlling environmental pollution. Elsevier, pp 623–688

Handojo L, Nursanto EB, Indarto A (2019) Progress of nanomaterials application in environmental concerns. In: Sharma SK (ed) Nanohybrids in environmental & biomedical applications. Taylor & Francis

Hawkins B, Hess-Wilson J, Rice G, Swartout J, Teuschler L K (2010) Recommended Toxicity Equivalence Factors (TEFs) for Human Health Risk Assessments of 2,3,7,8-Tetrachlorodibenzo-p-dioxin and Dioxin-Li ke Compounds. https://semspub.epa.gov/work/HQ/190077.pdf

He J, Chen K, Xu J, Sun Y, Xu J, Sun Y (2024) Urban air pollution and control. In: Abraham MA (ed) Encyclopedia of sustainable technologies, vol 2, 2nd edn. Elsevier, pp 143–159

Heal MR, Kumar P, Harrison RM (2012) Particles, air quality, policy and health. Chem Soc Rev 41:6606–6630. https://doi.org/10.1039/C2CS35076A

Huang Y, Wang W, Zhang Y, Cao J, Huang R, Wang X (2019) Chapter 10 – Synthesis and applications of nanomaterials with high photocatalytic activity on air purification. In: Xiaoru Wang XC (ed) Novel nanomaterials for biomedical, environmental and energy applications. Elsevier, pp 299–325

Ibrahim RK, Hayyan M, AlSaadi MA, Hayyan A, Ibrahim S (2015) Environmental application of nanotechnology: air, soil, and water. Environ Sci Pollut Res 23:13754–13788. https://doi.org/10.1007/s11356-016-6457-z

Jaison A, Mohan A, Lee Y-C (2023) Recent developments in photocatalytic nanotechnology for purifying air polluted with volatile organic compounds: effect of operating parameters and catalyst deactivation. Catalyst 13(2):40713. https://doi.org/10.3390/catal13020407

Jin Z, Murakami N, Tsubota T, Ohno T (2014) Complete oxidation of acetaldehyde over a composite photocatalyst of graphitic carbon nitride and tungsten(VI) oxide under visible-light irradiation. Appl Catal B Environ 150–151:479–485. https://doi.org/10.1016/j.apcatb.2013.12.048

Kalender SS (2019) Air pollution prevention technologies. In: Hussain CM (ed) Handbook of environmental materials management. Springer Cham, pp 2871–2888

Kamaei M, Rashedi H, Dastgheib SMM, Tasharrofi S (2018) Comparing photocatalytic degradation of gaseous ethylbenzene using N-doped and pure TiO_2 Nano-catalysts coated on glass beads under both UV and visible light irradiation. Catalysts 8(10):466. https://doi.org/10.3390/catal8100466

Khan NA, Hasan Z, Jhung SH (2013) Adsorptive removal of hazardous materials using metal-organic frameworks (MOFs): a review. J Hazard Mater 244–245:444. https://doi.org/10.1016/j.jhazmat.2012.11.011

Kok LJD, Yang L, Stuiver CEE, Stulen I (2009) Chapter 5: negative vs. positive functional plant responses to air pollution: a study establishing cause–effect relationships of SO2 and H2S. Dev Environ Sci 9:121–135. https://doi.org/10.1016/S1474-8177(08)00205-2

Komaladewi AAIAS, Aryanti PTP, Subagia IDGA, Wenten IG (2019) Membrane technology in air pollution control: prospect and challenge. J Phys 1217(1):012046. https://doi.org/10.1088/1742-6596/1217/1/012046

Liang WJ, Li J, Jin YQ (2010) Photocatalytic degradation of gaseous acetone, toluene, and p-xylene using a TiO_2 thin film. J Environ Sci Health A 45(11):1384. https://doi.org/10.1080/10934529.2010.500925

Liu Q, Shi Y, Zheng S, Ning L, Ye Q, Tao M, He Y (2014) Amine-functionalized low-cost industrial grade multi-walled carbon nanotubes for the capture of carbon dioxide. J Energy Chem 23(1):111–118. https://doi.org/10.1016/S2095-4956(14)60124-8

Liu Y, Wei JH, Xiong R, Pan CX, Shi J (2011) Enhanced visible light photocatalytic properties of Fe-doped TiO2 nanorod clusters and monodispersed nanoparticles. Appl Surf Sci 257(18):8121–8126. https://doi.org/10.1016/j.apsusc.2011.04.121

Ma X, Wang W, Sun C, Li H, Sun J, Liu X (2021) Adsorption performance and kinetic study of hierarchical porous Fe-based MOFs for toluene removal. Sci Total Environ 793:148622. https://doi.org/10.1016/j.scitotenv.2021.148622

Mahmoudi A, Sany SBT, Salmasi MA, Bakhshi A, Bustan A, Heydari S, Rezayi M, Gheybi F (2023) Application of nanotechnology in air purifiers as a viable approach to protect against Corona virus. IET Nanobiotechnol 17(4):289–301. https://doi.org/10.1049/nbt2.12132

Manisalidis I, Stavropoulou E, Stavropoulos A, Bezirtzoglou E (2020) Environmental and health impacts of air pollution: a review. Public Health 8:14. https://doi.org/10.3389/fpubh.2020.00014

Mannucci PM, Harari S, Martinelli I, Franchini M (2015) Effects on health of air pollution: a narrative review. Intern Emerg Med 10(6):657–662. https://doi.org/10.1007/s11739-015-1276-7

Martınez-Ahumada E, Dıaz-Ramırez ML, Miriam JV-H, Jancik V, Ibarra IA (2021) Capture of toxic gases in MOFs: SO_2, H_2S, NH_3 and NOx. Chem Sci 12:6772. https://doi.org/10.1039/d1sc01609a

Matsumoto M, Kondo F (2008) Environmental behavior of dioxins and their health effects. Environ Toxicol Pharmacol 25(2):224–229

Mohamed EF (2017) Nanotechnology: future of environmental air pollution control. Environ Manag Sustain Dev 6(2):429. https://doi.org/10.5296/emsd.v6i2.12047

Molina MJ, Rowland FS (1974) Stratospheric sink for chlorofluoromethanes: chlorine atom-catalysed destruction of ozone. Nature 249:810–812

Montazer M, Malekzadeh SB (2012) Electrospun antibacterial nylon nanofibers through in situ synthesis of nanosilver: preparation and characteristics. J Polym Res 19:9980. https://doi.org/10.1007/s10965-012-9980-8

Ong CB, Ng LY, Mohammad AW (2018) A review of ZnO nanoparticles as solar photocatalysts: synthesis, mechanisms and applications. Renew Sustain Energy Rev 81:536–551. https://doi.org/10.1016/j.rser.2017.08.020

Pramudita D, Iskandar I, Indarto A (2018) Chapter 12 – Nano-enhanced materials for reclamation of mine spoils. In: de Campos Favas PJ, Prasad MNV, Maiti SK (eds) Bio-Geotechnologies for mine site rehabilitation. Elsevier, pp 201–214. https://doi.org/10.1016/B978-0-12-812986-9.00012-9

Pramudita D, Suendo V, Azis MM, Indarto A (2022) Nanomaterials for air purification advances and challenges. In: Barakat RKMAE-F (ed) *Nanomaterials for environmental applications*. Taylor & Francis

Rafique MS, Tahir MB, Rafique M, Shakil M (2020) Photocatalytic nanomaterials for air purification and self-cleaning. In: Bilal M, Rafique TM, Rafique MS (eds) Nanotechnology and Photocatalysis for environmental applications. Elsevier, pp 203–219

Rathinavel S, Priyadharshini K, Panda D (2021) A review on carbon nanotube: An overview of synthesis, properties, functionalization, characterization, and the application. Mater Sci Eng B 268:115095. https://doi.org/10.1016/j.mseb.2021.115095

Roy A, Mishra C, Jain S, Solanki N (2019) A review of general and modern methods of air purification. J Therm Eng 5(2):22–28

Sajid M, Asif M, Baig N, Kabeer M, Ihsanullah I, Mohammad AW (2022) Carbon nanotubes-based adsorbents: properties, functionalization, interaction mechanisms, and applications in water purification. J Water Process Eng 47:102815. https://doi.org/10.1016/j.jwpe.2022.102815

Saleem H, Zaidi SJ, Ismail AF, Goh PS (2022) Advances of nanomaterials for air pollution remediation and their impacts on the environment. Chemosphere 287:132083. https://doi.org/10.1016/j.chemosphere.2021.132083

Seinfeld JH, Pandis SN (2016) Atmospheric chemistry and physics: from air pollution to climate change, 3rd edn. Wiley

Song S, Xing Z, Zhao H, Li Z, Zhou W (2023) Recent advances in bismuth-based photocatalysts: environment and energy applications. Green Energy Environ 8(5):1232–1264. https://doi.org/10.1016/j.gee.2022.04.004

Su F, Lu C, Cnen W, Bai H, Hwang JF (2009) Capture of CO2 from flue gas via multi-walled carbon nanotubes. Sci Total Environ 407(8):3017–3023. https://doi.org/10.1016/j.scitotenv.2009.01.007

Sujatha G, Shanthakumar S, Chiampo F (2020) UV light-irradiated photocatalytic degradation of coffee processing wastewater using TiO2 as a catalyst. Environments 7(6):47. https://doi.org/10.3390/environments7060047

Sundarrajan S, Tan KL, Lim SH, Ramakrishna S (2014) Electrospun nanofibers for air filtration applications. Procedia Eng 75:159–163. https://doi.org/10.1016/j.proeng.2013.11.034

Tahir MB, Riaz KN, Rafique M, Rafique MS, Yousaf N, Sagir M (2020) Photocatalysis fundamentals. In: Tahir MB, Rafique MS (eds) Nanotechnology and photocatalysis for environmental applications. Elsevier

Wang K, Nie J, Huang H, He F (2023) Literature Review on the Indoor Air VOCs Purification Performance of Metal–Organic Frameworks. Sustain For 15(17):12923. https://doi.org/10.3390/su151712923

Wang X, He T, Hu J, Liu M (2021) The progress of nanomaterials for carbon dioxide capture via the adsorption process. Environ Sci Nano 8:890. https://doi.org/10.1039/d0en01140a

Wen M, Li G, Liu H, Chen J, An T, Yamashita H (2018) Metal–organic framework-based nanomaterials for adsorption and photocatalytic degradation of gaseous pollutants: recent progress and challenges. Environ Sci Nano 6(4):1006. https://doi.org/10.1039/c8en01167b

Weon S, He F, Choi W (2019) Status and challenges in photocatalytic nanotechnology for cleaning air polluted with volatile organic compounds: visible light utilization and catalyst deactivation. Environ Sci Nano 6:3185. https://doi.org/10.1039/c9en00891h

Wu C (2014) Facile one-step synthesis of N-doped ZnO micropolyhedrons for efficient photocatalytic degradation of formaldehyde under visible-light irradiation. Appl Surf Sci 319:237–243. https://doi.org/10.1016/j.apsusc.2014.04.217

Yildiz O, Bradford PD (2013) Aligned carbon nanotube sheet high efficiency particulate air filters. Carbon 64:295–304. https://doi.org/10.1016/j.carbon.2013.07.066

Yunus IS, Harwin AK, Adityawarman D, Indarto A (2012) Nanotechnologies in water and air pollution treatment. Environ Technol Rev 1(1):136–148. https://doi.org/10.1080/21622515.2012.733966

Zeng L, Song W, Li M, Jie X, Zeng D, Xie C (2014) Comparative study on the visible light driven photocatalytic activity between substitutional nitrogen doped and interstitial nitrogen doped TiO_2. Appl Catal A: Gen 488:239–247. https://doi.org/10.1016/j.apcata.2014.09.041

Zeng Q, Xie X, Wang X, Wang Y, Lu G, Pui DYH, Sun J (2018) Enhanced photocatalytic performance of Ag@TiO_2 for the gaseous acetaldehyde photodegradation under fluorescent lamp. Chem Eng J 341:83–92. https://doi.org/10.1016/j.cej.2018.02.015

Zhang L, Luo J, Menkhaus TJ, Varadaraju H, Sun Y, Fong H (2011) Antimicrobial nanofibrous membranes developed from electrospun polyacrylonitrile nanofibers. J Membr Sci 369(1–2):499–505. https://doi.org/10.1016/j.memsci.2010.12.032

Zhang X, Gao B, Creamer AE, Cao C, Li Y (2017) Adsorption of VOCs onto engineered carbon materials: a review. J Hazard Mater 338:102–123. https://doi.org/10.1016/j.jhazmat.2017.05.013

Zhang X, Zhang K, Shi Y, Xiang H, Yang W, Zhao F (2024) Surface engineering of multifunctional nanostructured adsorbents for enhanced wastewater treatment: a review. Sci Total Environ 920:170951. https://doi.org/10.1016/j.scitotenv.2024.170951

Zhu M, Xiong R, Huang C (2019) Bio-based and photocrosslinked electrospun antibacterial nanofibrous membranes for air filtration. Carbohydr Polym 205:55–62. https://doi.org/10.1016/j.carbpol.2018.09.075

Chapter 3
Promising Applications of Different Nanomaterials in Air Pollution Remediation

Ahmed M. El-Khawaga, Ibrahim W. Hasani, and Yosri A. Fahim

Abstract One of the major problems facing the globe today is environmental degradation, which is growing every year and has a detrimental impact on the planet. Many contaminants, including carbon monoxide, hydrocarbons, nitrogen oxides, volatile organic compounds, and chlorofluorocarbons, are currently prevalent in the air. Organic and inorganic substances can also contaminate water and soil; sewage water, industrial effluents, haphazard pesticide and fertilizer usage, and oil spills are the main sources of pollution. Simultaneously, the swift advancement of nanotechnology has sparked significant curiosity about the possible applications of nanomaterials in enhanced environmental monitoring and cleanup systems encompassing all three phases. It can advance the sensing and detection of contaminants and advance cutting-edge remediation solutions. We are focusing on the most recent advances in the application of nanotechnology to the many applications listed in this overview. We also discuss recent developments in nanomaterial engineering techniques and provide examples of how they have been applied to improve stability, catalytic efficiency, and the use of alternative oxidants. This chapter will provide an extensive comprehension of the definition of nanoparticles, the idea of nanotechnology, and the synthesis of nanomaterials for possibly useful applications. In this chapter, we offer an outlook on the future developments of nanotechnology and its potential uses in air pollution remediation. Owing to its many uses, academic researchers, environmental scientists, and healthcare professionals are anticipated to learn about the state-of-the-art and potential applications of nanomaterials in environmental sectors.

A. M. El-Khawaga (✉) · Y. A. Fahim
Department of Basic Medical Sciences, Faculty of Medicine, Galala University, Suez, Egypt
e-mail: Ahmed.Elkhawaga@gu.edu.eg

I. W. Hasani
Department of Pharmaceutics, Faculty of Pharmacy, S.P.U.,M.P.U and Idlib University, Idlib, Syria

N. Ghosh et al. (eds.), *Nanotechnology in Air Quality Management*, Sustainable Development and Biodiversity 39, https://doi.org/10.1007/978-981-96-8339-0_3

Keywords Environmental degradation · Nanomaterials · Contaminants · Air pollution

3.1 Introduction

Nanotechnology involves harnessing the distinctive characteristics of materials at the nanoscale. It entails meticulous observation, measurement, manipulation, assembly, control, and matter production at the nanoscale, achieving atomic-level accuracy (Mansoori 2017). Nanotechnology encompasses various technologies that rely on physical, chemical, and biological processes. It can be described as the capacity to translate theoretical concepts in nanoscience into practical and beneficial applications (Omran and Omran 2020). Various methods, including lithography, molecular self-assembly, and focused ion beam machining, are used to create materials and electronics at the nanoscale (Mahalik 2006). In nanotechnology, there are two primary methodologies: the "top-down" approach, which involves the creation of nano-objects by reducing larger entities, and the "bottom-up" approach, which utilizes molecular recognition principles to construct materials and gadgets chemically (Joachim 2009).

Nanotechnology has utility across diverse areas, including medicine, electronics, diagnostics, and the food business (Thiruvengadam et al. 2018). The industrialized world has witnessed tremendous technological change, particularly in the rapid growth of nanoscale markets during the past decade (Anton et al. 2001). Nanotechnology is no longer a novel notion, as it has evolved into a versatile technology with many applications. Four generations of nanomaterials have been identified on the surface and are being utilized in many scientific fields. These include active and passive nanoassemblies, general nanosystems, and small-scale molecular nanosystems (Klaine et al. 2012). The rapid progress in nanoscience has provided evidence that nanoscale production will soon be integrated into nearly every field of science and technology (Marcovich and Shinn 2014).

Nanotechnology is an interdisciplinary field that integrates the fundamental principles of biology, physics, and chemistry. It operates at the nanoscale—typically measured in nanometers—where unique physical, chemical, and biological phenomena occur. At this scale, reductions in physical size can lead to significant changes in chemical bonding, material properties, and biological interactions, enabling applications such as targeted drug delivery and controlled distribution within the body (El-Khawaga et al. 2023b). Nanotechnology connects classical and quantum physics within a region known as a mesoscopic system. This mesoscopic system produces nanoassemblies derived from natural sources, including agricultural goods, nanomedicine, and nanotools, which are utilized for medical treatment and diagnostic applications in the healthcare sector (Das 2010).

Nano-based drugs and diagnostic kits are already being utilized to mitigate diseases that were previously considered incurable (Chowdhury et al. 2017). This technology has also significantly impacted large-scale industrial manufacturing and production. Nanotechnology employs the principle of reverse engineering found in nature rather than relying on the conventional material production method, including significant material reduction. Nanotechnology enables the production of molecular-sized items, including individual atoms, and subsequently designs products to operate at a more microscopic level (Ramsden 2018).

Globally, in millions and billions, significant amounts of money are being invested in nanotechnology to harness its immense potential. This is particularly evident in industrialized regions such as Europe, China, and America. Nevertheless, emerging countries need to catch up, as they cannot keep up with the industrial advancement achieved in the previous decade (Malik et al. 2023b). The lag primarily stems from the ongoing economic struggles in these countries, necessitating a period for them to progress in the field of nanotechnology. Nevertheless, it is crucial to mention that scientific communities in the developed and developing world concur that nanotechnology will be the next advancement in technological innovation. Over the next few years, this will necessitate continuing the process of industrial upgrading and making investments in nanotechnology (Salamanca-Buentello et al. 2005).

Through scientific and technological advancements, the scientific community embraces cost-effective, secure, and more environmentally friendly technologies and products than their predecessors. Furthermore, they express apprehension regarding the financial viability of technology, given the significant depletion of natural resources worldwide (Shannon et al. 2008). Nanotechnology offers a solution to this challenge. This technology is characterized by its clarity, cleanliness, and affordability, in contrast to prior methods involving large-scale production and cumbersome apparatuses. Furthermore, nanotechnology can be applied in all facets of existence. Incorporating nanomaterial sciences, nanoelectronics, and nanomedicine will be pervasive throughout all chemistry and the physical and biological realms (Kumar et al. 2018). Thus, it is not absurd to speculate that nanotechnology will eventually become a subject that all students of the next generation must learn (Selin 2011).

Due to the potential harm to human health, researchers have extensively studied various locations affected by air pollutants (Lei et al. 2015). The World Health Organization (WHO) recently reported that air pollutants are accountable for a significant annual mortality rate of seven million people worldwide. Furthermore, nearly 90% of individuals inhale elevated concentrations of an airborne pollutant (Mammadova et al. 2022). Air pollution is the term used to describe changes in the natural composition of the atmosphere caused by various physicochemical and chemical substances originating from geogenic, biogenic, and anthropogenic sources. Air pollutants can be found outdoors and indoors in either gaseous or particulate form (Delay and Frimmel 2012). Fossil fuels are the predominant energy source globally and are crucial in powering daily activities and industrial production. However, fossil fuels are also the primary source of pollutants. Releasing pollutants, such as volatile organic compounds (VOCs), has a significant adverse ecological impact (Maroto-Valer et al. 2012; Rasheed et al. 2019). Nanotechnology has been utilized as a sophisticated therapeutic approach in this particular setting. Various nanomaterials have successfully addressed air pollution, including nanosensors, nanocatalysts, nano filters, and nano adsorbents (Asghar et al. 2024). This is crucial since chemical treatment options might lead to detrimental environmental consequences. This chapter explores the utilization of various nanomaterials (NMs) and the parameters that impact their efficacy in addressing air pollution through remediation methods.

3.2 Preparation Methods for the NPs

Different techniques are available for creating nanoparticles (NPs), but they may be broadly categorized into two primary classes: bottom-up approaches and top-down approaches, as illustrated in Scheme (1). These approaches can be further categorized into several subclasses based on their operation, reaction state, and accepted procedures.

3.2.1 *Top-Down Methods*

This method entails the transformation of large quantities of material into tiny particles, typically on the nanometer scale. The synthesis of nanoparticles decreases the size of the starting material through several physical and chemical techniques (Meyers et al. 2006). The methods include ball milling, mechanochemical synthesis, laser ablation, and ion sputtering (Zhao et al. 2022). Although top-down techniques are easy to carry out, a notable disadvantage of this strategy is the possibility of modifying the surface chemistry and physicochemical features of nanoparticles (Ghotekar et al. 2018).

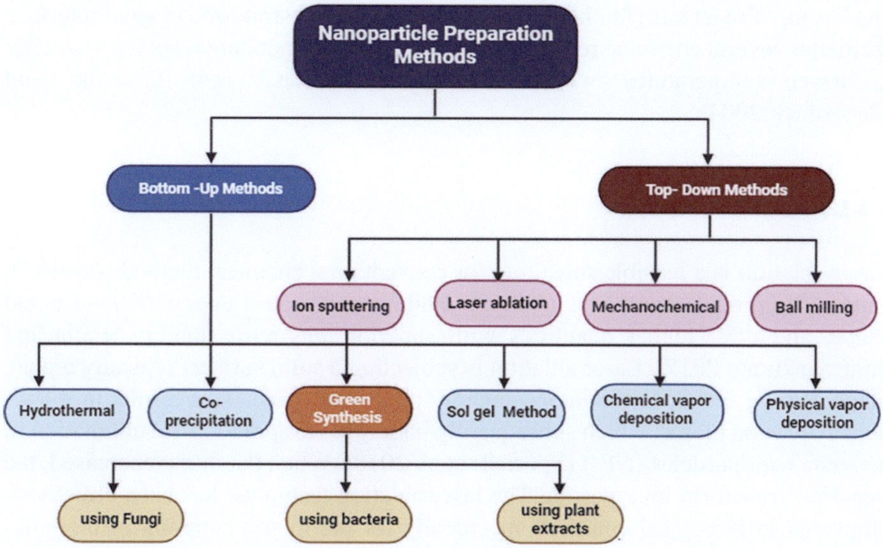

3.2.1.1 Ball Milling

This method entails adding a substantial amount of powdered material into a container and several heavy spheres. A significant amount of mechanical energy is applied to a considerable quantity of powdered material through a swiftly rotating ball (Jamkhande et al. 2019). Particle size reduction can be accomplished by employing different high-energy mills, such as attrition ball mills, planetary ball mills, vibrating ball mills, low-energy tumble mills, and high-energy ball mills (Rajput 2015). Among these different techniques, sizable, unobstructed, vigorously dynamic spheres can traverse the container's surface containing the pulverized substance in many parallel layers, or they can fall unrestrictedly and impact the powder (Yadav et al. 2012).

3.2.1.2 Mechanochemical Synthesis

The mechanochemical synthesis method utilizes an iterative process, including the deformation, welding, and fracturing of a combination of reactants (Tsuzuki 2021). Chemical modifications occur at the interface of nanoscale particles during the milling process (Fattahi et al. 2020). Usually, higher temperatures are required to start chemical reactions due to several factors, such as separating the reacting phases from the product phase. NPs can be produced by ball milling at low temperatures without external heating (Tsuzuki and Mccormick 2004). During the mechanochemical synthesis process, the initial components are precisely combined and then exposed to milling. The milling process involves the deformation, fragmentation, and fusing of reactants (Jamkhande et al. 2019). The substrate and reagent interface facilitates several chemical reactions, enabling high-temperature responses to occur at lower temperatures without requiring additional heat (Tsuzuki and Mccormick 2004).

3.2.1.3 Laser Ablation

Laser ablation is a feasible substitute for conventional chemical methods due to its capacity to process materials quickly, provide better control over particle size and shape, and offer higher quantities with superior long-term stability (Ealia and Saravanakumar 2017). Laser ablation is subjecting a solid surface, typically a plate made of pure metal, to the intense energy of a laser beam. This results in a low-density plasma plume, which subsequently undergoes evaporation or sublimation to generate nanoparticles (NPs) (Sportelli et al. 2018). When the flux is increased, the materials transform into plasma. The laser ablation technique has been effectively employed in biological contexts, specifically for the in situ coupling of biomolecules with metal nanoparticles (MNPs). This technique provides numerous benefits, such as producing magnetic nanoparticles (MNPs) in water-based and organic solvents and eliminating the requirement to eradicate excess reagents.

3.2.1.4 Ion Sputtering

Ion sputtering is a commonly used technique for producing nanoparticles. It involves depositing a thin layer of particles onto a substrate by causing ions to collide with it. Subsequently, annealing is carried out (Popok 2019). This process is alternatively referred to as the physical vapor deposition (PVD) method (Sharma et al. 2019). The efficacy of this procedure mostly depends on factors such as the thickness of the layers, the kind of substrate, and the time and temperature of annealing, which directly influence the dimensions and morphology of the nanoparticles (Hatakeyama et al. 2011).

3.2.2 Bottom-up Methods

The bottom-up approach for nanoparticle synthesis entails amalgamating smaller entities, such as atoms, molecules, or minuscule particles, to generate nanoparticles. This method involves the initial creation of nanostructured building blocks of nanoparticles, which are then assembled to form nanoparticles (Mukherjee et al. 2001).

3.2.2.1 Physical Vapor Deposition Method

Physical vapor deposition is a technique used to deposit material onto a surface, either as a thin film or as nanoparticles (Marichy et al. 2012). Applying tightly controlled vacuum processes, such as thermal evaporation and sputtered deposition, converts materials into vapor, which is then condensed onto a substrate (Macleod 2018).

3.2.2.2 Chemical Vapor Deposition Method

The chemical vapor deposition technique was first described and patented in the late nineteenth century. It was employed to manufacture carbon fibers as filaments and carbon powder as a color pigment in electric lamps (Collin 2021). This deposition method involves the production of a thin layer of target material on a surface using a chemical reaction with gaseous molecules that contain the essential atoms for film formation (Pedersen and Elliott 2014).

3.2.2.3 Sol–Gel Method

The sol–gel method for nanoparticle synthesis involves a) combining premade metal (oxide) colloids with a sol that contains the species responsible for producing the matrix, which then undergoes gel formation, or b) directly mixing metal and metal oxide or nanoparticles with a silica sol that has undergone hydrolysis. c) The process involves the creation of a compound between a metal and silane, which is then reduced before undergoing hydrolysis (Cushing et al. 2004). This approach entails the incorporation of a network framework by utilizing a colloidal suspension (sol) and gelatin to form a network within a continuous liquid phase (gel). Sol–gel production consists of four main stages: hydrolysis, condensation, particle development, and particle aggregation (Rajput 2015). In the direct precipitation method for producing metal or metal oxide particles, the formation of metal oxide particles occurs through the precipitation of silica sol, typically followed by low-temperature heat treatment. This technology is mostly utilized to fabricate thin films (Niederberger and Garnweitner 2006).

3.2.2.4 Coprecipitation Method

The coprecipitation method is a commonly used approach for producing nanoparticles. This process involves the application of precursor substances from a solution, followed by heat treatment to obtain the desired nanoparticles (El-Khawaga et al. 2023a). Various parameters, including reaction temperature, pH, mixing rate, and amounts of precursors and precipitating agents, influence the coprecipitation process. These parameters can significantly influence the phase transitions, particle shape, and properties of the generated nanoparticles (Roth et al. 2015).

3.2.2.5 Hydrothermal Methods

The hydrothermal process entails the reaction between vapor derived from an aqueous solution and a solid substance, carried out under elevated pressure and temperature conditions. This mechanism leads to the accumulation of minuscule particles (Kumar and Pandey 2018). This method involves the formation of cations as polymeric hydroxides, which subsequently undergo dehydration and facilitate the formation of metal oxide crystals (Jamkhande et al. 2019). The second metal cation serves a beneficial function in controlling the process of particle formation by impeding the creation of intricate hydroxides when a base is added to a solution of metal salt (Tavakoli et al. 2007).

3.2.2.6 Green Synthesis Method

The green synthesis of nanoparticles entails utilizing natural sources and environmentally acceptable ingredients as precursors for nanoparticle synthesis. This strategy aims to reduce the overall ecological footprint and promote the adoption of sustainable nanotechnology (Khan 2020). Plant extracts, microorganisms, and fungi are commonly employed for the eco-friendly production of nanoparticles through a process known as green synthesis. Natural substances serve as reducing agents, stabilizers, or templates during the synthesis process, producing nanoparticles (Khafaga et al. 2023). Microorganisms and plants possess enzymes and other chemical substances that can serve as reducing and inhibiting agents in biosynthesis. These methods can produce nanoparticles that exhibit stability and possess a restricted size distribution, leading to a high rate of synthesis (Osman et al. 2024).

3.3 Promising Properties of Nanomaterials

Nanomaterials exhibit distinctive characteristics because of their small dimensions and arrangement. Nanomaterials possess characteristics that render them well-suited for various applications, such as sensors, bioimaging, optoelectronic devices, the energy sector, textile creation, healthcare, the environment, and the medical industry (Singh et al. 2021a). Some of the unique nanomaterial properties are discussed below:

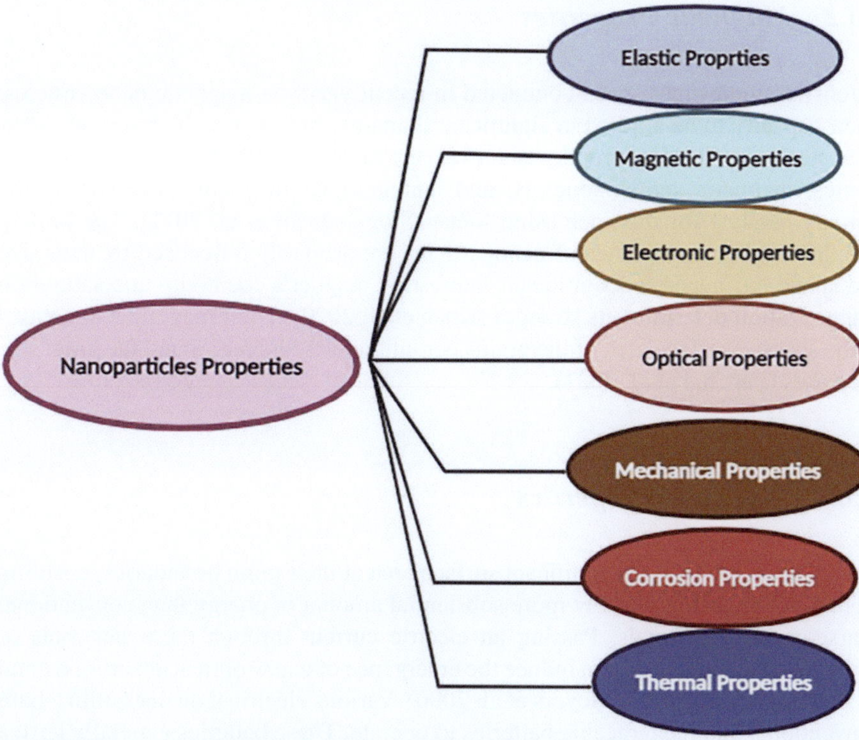

3.3.1 Elastic Properties

There is an inverse correlation between the bonding strength between atoms or molecules and the elastic modulus of a material. The melting temperature and elastic modulus exhibit a linear relationship with the bond strength, meaning that as the bond strength increases, both the melting temperature and elastic modulus increase proportionally (Tan and Cheetham 2011). The elastic properties of crystalline materials are commonly believed to be unaffected by their microstructure. As the temperature increases, the elastic modulus decreases due to the increase in the mean interatomic distance (Kachanov and Sevostianov 2005). Analogously, one can equate a substantial rise in vacancy and other defect concentrations to an increased temperature under consideration. Therefore, an increase in defect concentration will result in a decrease in the elastic modulus. Nevertheless, the influence of flaws on the elastic modulus becomes noticeable only when there are significantly more vacancies (Che et al. 2000). Many defects in nanomaterials can substantially reduce their elastic characteristics compared to those of bulk materials (Meyers et al. 2006).

3.3.2 Magnetic Properties

Extensive studies have been conducted in recent years on magnetic nanoparticles. Their capacity to be applied in significant domains, such as ultrahigh density magnetic storage systems, magnetic random access memory (MRAM), ferrofluids, spintronics, magnetic semiconductors, and nanogranular magnetic materials, is the primary catalyst for this increasing focus (Singamaneni et al. 2011). The finding that the magnetic properties of nanoparticles are primarily influenced by their size and shape has generated significant interest. A magnet's coercivity and saturation magnetization determine its strength (Khan et al. 2020). Their magnitude increases as the specific surface area of the grains per unit volume increases and the grain size decreases (Panchal et al. 2021).

3.3.3 Electrical Properties

Nanomaterials possess a significant surface area at their grain boundaries, enabling them to retain a considerably more substantial amount of energy than conventional coarse-grained materials. Passing an electric current through these materials or applying an electric field can induce the emergence of a new optical absorption band or alter an existing band (Meyers et al. 2006). Various electrical devices utilize both conventional and rechargeable batteries to operate. These batteries generally have a low energy density, meaning they have limited storage capacity and require frequent recharging (Dehghani-Sanij et al. 2019). Nanocrystalline materials show great potential as separator plates in batteries due to their superior energy storage capacity compared to conventional materials (Zhang et al. 2018). The dielectric properties substantially improve when these materials are transformed into nanocrystalline materials. Compared with bulk materials, nanoparticles exhibit lower thermal and electrical conductivity (Tjong 2013). The electrical conductivity of metallic solids is attributed to the unrestricted mobility of electrons, as explained by the classical free electron theory of metals (Barsan and Weimar 2001).

3.3.4 Optical Properties

The notable disparities in optical characteristics between nanocrystalline systems and bulk crystals have garnered considerable interest (Murray et al. 2000). The significant factors contributing to the behavior of nanoparticles include their ability to confine electrical carriers through quantum mechanics, the efficient passage of energy and charge over nanoscale distances, and the significant impact of surfaces (Minnich et al. 2009). By altering the dimensions and surface characteristics of the crystals, it is possible to precisely adjust the optical properties of these materials, both linearly and nonlinearly (Ganesh et al. 2017). These unusual optical features can be explained by the principles of quantum confinement and surface plasmon processes (Zhang and Zhang 2012).

3.3.5 Mechanical Properties

Extensive research has been conducted on the mechanical behavior of nanocrystalline materials due to their distinct and exceptional features (Meyers et al. 2006; Kumar et al. 2003). The intricate interplay of defect patterns gives rise to distinct tensile, work hardening, creep, fatigue, and deformation characteristics in nanocrystalline materials, setting them apart from their bulk counterparts. This is because nanocrystalline materials have significantly smaller dimensions (Tjong and Chen 2004).

3.3.6 Corrosion Properties

The significantly modified characteristics of nanocrystalline alloys can be attributed to the exceptionally high concentration of grain boundaries present in these substances, as indicated by several studies (Rupert et al. 2011; Lu 2016). Studies on the corrosion of nanocrystalline materials have shown a significantly greater level of active anodic dissolution than investigations on corrosion involving conventional materials (Liu et al. 2010; Mahesh and Raman 2014). The unique behavior observed can be ascribed to the material's decreased grain size, which results in a greater concentration of high-energy grain boundary defects compared to larger grain sizes. This type of grain boundary site with high energy levels is a preferred location for anodic dissolution (Raizada et al. 2021). As a result, it can lead to increased corrosion rates in nanocrystalline alloys. Nevertheless, higher fault density may only sometimes be considered harmful in all circumstances. Grain boundaries, which are routes that allow facile diffusion, can enhance surface diffusion coefficients (Zhang et al. 2017).

3.3.7 Thermal Properties

Generally, a greater quantity of grain boundaries will cause more photon scattering at the disordered borders, decreasing the material's thermal conductivity (Fox et al. 2019). Consequently, it is logical to expect nanocrystalline materials to exhibit lower thermal conductivity than conventional materials. Conversely, when the size of grains becomes very small, they resemble the average distance of phonons, which are responsible for transferring thermal energy (Toberer et al. 2012). Compared with coarse-grained materials, nanomaterials can have significantly distinct features due to photon transport's confinement and quantization effects (Haider et al. 2022). The variations can be ascribed to the dimensions of the particles. The thermal characteristics of nanomaterials are influenced not only by the particle size but also by the form of the grains (Andrievski 2014).

3.4 Different Applications of NPs

Nanotechnology is used for various applications in different sectors (Fig. 3.1). Some of the applications of nanotechnology are discussed below:

3.4.1 Agriculture

Agrochemicals are frequently used to protect plants from illnesses, increase crop output, and enhance the quality and quantity of harvests. Agrochemicals offer plant protection by mitigating damage and preventing infections. Nevertheless, plants can only assimilate some agrochemicals, leaving the remaining portion unused and ultimately discarded (Himani et al. 2022). Using nano-enabled agrochemicals results in controlled effectiveness and reduced toxicity of these substances. Nano-enhanced agrochemicals improve the uptake of agricultural nutrients, increase nutrient

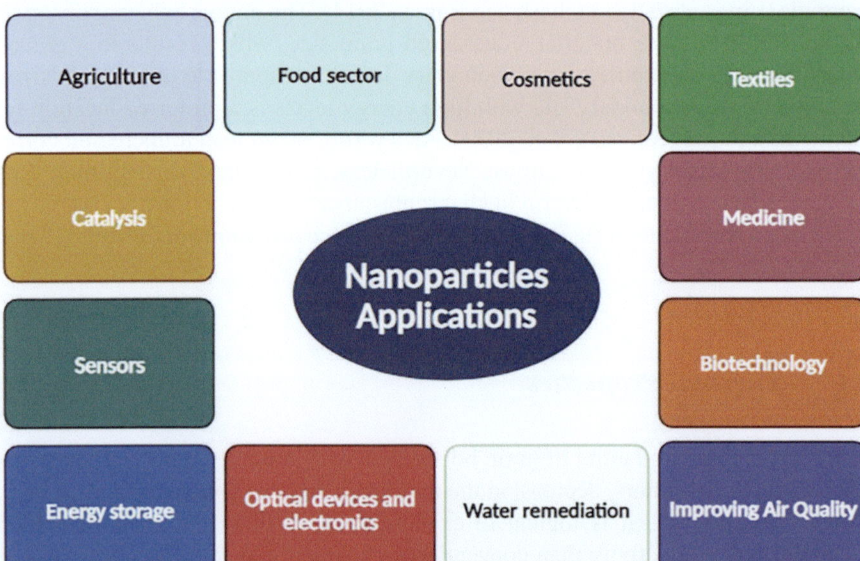

Fig. 3.1 Applications of nanotechnology across various sectors

stability and solubility, and regulate crop diseases (Kaushani and Priyadarshana 2023). Nanomaterials enhance plant development, improve seed and blossom quality, and enhance soil quality, and the utilization of nanomaterials and their functions is essential in various aspects of agriculture (Singh et al. 2021b).

3.4.2 Food Sector

Nanotechnology is thought to offer significant benefits to the food industry in the twenty-first century (Thakur et al. 2018). Nanotechnology is applied across various sectors, including agriculture, food processing, packaging, safety, and nutrition. The application of nanotechnology in food systems has resulted in the development of innovative products that exhibit enhanced food quality attributes, including flavor, texture, sensory qualities, and stability (Nile et al. 2020). Food processing involves a series of steps, including obtaining appropriate raw materials, improving taste, eliminating toxins, deactivating enzymes, reducing spoiling organisms, minimizing pathogens, adding micronutrients, packaging, storing, and transporting (Fellows 2022). Due to the complexity of the process, there is a significant likelihood of nanotechnology-based applications being involved. Nanomaterials are commonly employed as fillers in food and additives to enhance the packaging materials' durability and mechanical strength. The use of various nanomaterials in food packaging fulfils its intended objective (Bumbudsanpharoke et al. 2015).

3.4.3 Cosmetics

In technical terms, cosmetics encompass goods that enhance the skin's visual aspect and cleansing properties and promote its beauty. The utilization of nanoparticles commenced in the early 1960s with the application of liposome technology to moisturizers and skin creams (Gupta et al. 2022). By employing liposomes, the optical characteristics of the materials were enhanced, leading to increased solubility and absorption. The cosmetic industry has been utilizing nanoparticles in various goods, such as lotions, sunscreens, hair care products, and anti-aging creams, for a considerable period (Nafisi and Maibach 2017). Nanocosmetics are classified based on the system's properties, the preparation technique, and the components involved. Regulations for toxicity should be established alongside the development of nanotechnology-based cosmetics (Santos et al. 2019).

3.4.4 Textiles

Nanotechnology has been utilized to create intelligent fabrics in recent years. Nanomaterials employed in the textile industry produce fabrics that possess features such as ultraviolet resistance, antibacterial activity, optical characteristics, electrical conductivity, hydrophobicity, and flame retardancy (Shah et al. 2022). When smart devices based on nanomaterials are combined with textiles, they can perform several activities, such as sensing, drug release, energy harvesting and storage, and optics (Yetisen et al. 2016). This technology is currently being utilized in the fashion industry, in healthcare, and in on-body energy harnessing and military applications. Integrating nanotechnology with textiles has led to pioneering applications in various domains (Syduzzaman et al. 2023).

3.4.5 Medicine

Patients undergoing treatment for chronic conditions such as cancer (Diller et al. 2009) and brain tumor (Daly and Schiff 2007) may experience significant adverse effects from the prescribed medications and therapies. Nanoparticles deliver the drug to sick cells rather than distribute it across the entire body. In addition, a diverse array of nanomaterials is being employed to optimize the performance of imaging devices (Villa et al. 2016). Nanotechnology has diverse applications in gene therapy, wound treatment, and other fields beyond those mentioned (Malik et al. 2023a).

3.4.6 Biotechnology

The costly, antiquated production process will be replaced with more affordable, eco-friendly products with adaptable, accurate, and durable designs. This technology enables the production of robust yet lightweight materials suitable for applications such as sensors, microrobots, surgical instruments, electronic devices, and circuits (Berger 2016). Another example in the field of nanobiotechnology research involves the use of nanospheres that are coated with luminescent polymers (Sobha et al. 2010). Scientists are currently attempting to create polymer designs that can reduce the intensity of fluorescent light when it comes into contact with a specific group of molecules (Bosch et al. 2005). DNA nanotechnology is a prominent example of bionanotechnology. The promising field of this modern technology involves using the inherent characteristics of nucleic acids, such as DNA, to develop synthetic membranes and self-assembling proteins (Yadav et al. 2020).

3.4.7 Improving Air Quality

Nanomaterials have shown significant potential in reducing air pollution. Studies have demonstrated that compared with pure nanoparticles, modified nanomaterials have greater efficiency, cost-effectiveness, and robustness. Metal nanoparticles can mitigate nanomaterials' aggregation, increase metals' bioavailability, and accelerate the decomposition of noxious compounds (Ahmad et al. 2023). Furthermore, they exhibit enhanced responsiveness, flexibility, and accurate discharge of active constituents, making them highly effective at eradicating diverse chemical pollutants from complex environmental matrices (Bilal et al. 2020). Nanomaterials with dimensions less than 100 nm display remarkable physicochemical properties, such as a high specific surface area and rapid reactivity with pollutants. This makes them very ideal for addressing environmental contamination (Mehndiratta et al. 2013). Nanomaterials can be customized for particular uses, enhancing their efficacy in reducing air pollution (Zhang and Fang 2010). Carbon-based and metal-based nanoparticles are widely used to remove air pollutants (Afreen et al. 2018). Concurrently, research is underway to enhance the efficacy of catalysts, thereby mitigating the impact of air pollutants emitted by industrial facilities, automobiles, air conditioners, and similar sources. These catalysts, composed of nanoparticles, offer a substantial surface area for chemical reactions to occur (Li et al. 2021).

3.4.8 Water Remediation

Nanotechnology has great potential in developing systems that can obtain purified water from different sources. This approach offers a distinct opportunity to enhance existing advancements in wastewater treatment (Qu et al. 2013). Most organic contaminants exhibit chemical stability, which poses challenges for effectively removing these pollutants from water using conventional physicochemical methods, often resulting in various limitations. Recent research suggests that nanotechnology involving photocatalysis has the potential to efficiently decompose multiple organic pollutants, such as dyes, into harmless substances (Tahir et al. 2020).

3.4.9 Optical Devices and Electronics

Nanomaterials have a diverse array of uses in the fields of optical devices and electronics. Nanophotonics, a subdivision of nanotechnology, specifically investigates the characteristics of light at tiny scales and the interplay between materials of nanoscale dimensions and light (Minoli 2005). Nanomaterials with dimensions smaller than the wavelength of light can improve the optical properties of materials and allow for manipulation of the spatial distribution, phase, polarization, and spectral distribution of light (Wang et al. 2020). Graphene has been extensively studied in electronics due to its distinctive electrical properties and prospective uses in transistors, phototransistors, diodes, and logic gates (Sun et al. 2018). In addition, advanced nanofabrication methods such as directed assembly and additive manufacturing present novel opportunities for creating functional nanostructures and incorporating nanomaterials into electronic devices (Roco et al. 2011). The progress made in nanomaterials and fabrication processes can fundamentally transform the advancement of optical devices and electronics.

3.4.10 Energy Storage

The applications of nanomaterials in energy storage devices have been thoroughly investigated. These materials possess distinctive characteristics, such as a large surface area, an enhanced ability to transport charges, and improved electrochemical activity. These attributes can significantly boost the performance of energy storage devices in terms of their energy density, power density, cycling stability, and safety (Pomerantseva et al. 2019). Metal–organic framework (MOF) templates are receiving increasing interest for fabricating porous metal oxides and nanocomposites. This is due to the limitations of metal oxides, such as reduced capacitive performance caused by constant faradic interactions between the electrolyte and the electrode (Wu et al. 2019). MOFs can serve as electrode materials, separators, and catalysts, resulting in increased electrochemical performance, enhanced charge/discharge rates, and extended cycling stability (Zheng et al. 2021). To improve their electrochemical storage capacity, they can augment metal–organic framework-based oxides with carbon-based materials such as graphene, carbon nanotubes, and reduced graphene oxide (Wu and Lou 2017). Nanomaterial-based devices offer energy conversion and storage advantages due to their high specific surface area, high surface-to-volume ratio, and high ion mobility during intercalation (Zhu et al. 2018). Nanomaterials have become favorable contenders for advanced energy storage technologies, presenting possibilities for enhanced performance, longevity, and environmental friendliness (Abdalla et al. 2020).

3.4.11 Sensors

Sensors benefit from the diverse applications of nanomaterials. They have demonstrated substantial promise in biological sensing, enabling the acquisition of real-time data on analyte composition's quantitative and qualitative properties (Thakur and Kumar 2022). Nanomaterials, including gold nanoparticles, quantum dots, polymeric nanoparticles, carbon nanotubes, nanodiamonds, and graphene nanostructured materials, are currently under investigation and application as biosensors (Holzinger et al. 2014). Carbon-based nanomaterials, such as carbon nanotubes and graphene, possess exceptional physicochemical characteristics that render them highly desirable for chemical sensing and biosensing applications (Karimi et al. 2018). Nanomaterial sensors possess significant potential for use in medicine, healthcare, the environment, and biology. They can produce flexible, bendable, combined materials for various sensing purposes (Nyabadza et al. 2021). Nanomaterials are utilized as electrode materials in chemical sensors, providing selectivity, sensitivity, and fast sensing capacities (Abdel-Karim et al. 2020). In addition, nanoparticles, independently or enclosed within biomolecules, are being employed as sensors in biosensing and bioapplications. These materials exhibit biocompatibility, selectivity, and nontoxicity (Kumar et al. 2023).

3.4.12 Catalysis

Nanomaterials possess several uses in catalysis, including heterogeneous catalysis, photocatalysis, and organic synthesis. Nanotechnology is vital to advancing science and technology, particularly in catalysis (Hodges et al. 2018). The rapid progress in nanomaterials research and the utilization of sophisticated analysis techniques have resulted in the expansion of heterogeneous catalysis and its transformation into what is now known as "nanocatalysis" (Wei et al. 2022). Extensive research has been conducted on intelligent nanomaterials, namely photocatalysts, due to their potential for harnessing solar energy for many purposes, such as eliminating environmental contaminants, splitting water to produce hydrogen, and reducing CO_2 (Tatarchuk et al. 2018). There is increasing interest in multifunctional nanomaterials due to their exceptional catalytic performance in hydrogen production, environmental clean-up, and CO_2 reduction (Ezendam et al. 2022). Nanoparticles, by their diminutive dimensions and significant surface-to-volume ratio, have been shown to augment the efficacy and specificity of catalytic processes, rendering them highly advantageous for diverse organic conversions (Gao et al. 2020).

3.5 Nanotechnology for Air Pollution Control

Environmental pollution is a pressing global issue progressively deteriorating, resulting in substantial harm to the earth. Air pollutants encompass carbon monoxide (CO), chlorofluorocarbons (CFCs), volatile organic compounds (VOCs), nitrogen oxides, and other similar substances (Gheorghe and Ion 2011). Nanomaterials possess a significant ratio of surface area to volume, enabling enhanced interaction with airborne pollutants and heightened sensitivity and detection capabilities (Bhardwaj et al. 2021). Nanostructured materials can be intentionally engineered for specific characteristics, such as selectively targeting particular contaminants and enabling targeted surveillance of specific air pollutants. The distinctive characteristics of nanomaterials, such as their diminutive size and elevated reactivity, allow them to effectively trap and absorb dangerous gases and pollutants, improving the precision of air pollution monitoring (Das et al. 2015). Nanotechnology provides diverse materials to clean up, such as carbon nanotubes, nanoscale zeolites, dendrimers, enzymes, bimetallic particles, and metal oxides (Bhandari 2018). Nanotechnology techniques, such as nanocatalysts, are employed to mitigate air pollution. Carbon nanotubes (CNTs) and gold particle adsorption are the predominant nanotechnologies that remove harmful gases responsible for air pollution. They have proven efficient absorbents for eliminating organic and inorganic air pollutants, including nitrous oxide, ammonia gases, carbon dioxide, and dioxins (Abdillah et al. 2018). In addition, manganese oxide nanofiber catalysts are employed to purify industrial smokestack emissions by removing volatile organic contaminants (Prasanna 2017). Carbon nanotubes (CNTs) are graphite-based tubular structures composed of carbon (Fahim et al. 2024). They are utilized to mitigate the release of greenhouse gases resulting from coal mining and energy generation. Unlike traditional membranes, which are limited to performing either one function or the other, this groundbreaking technology efficiently handles and segregates large quantities of gas (Mittal and Kumar 2014). Manganese oxide and gold nanoparticles have been employed to combat pollutants due to the significant surface area of manganese oxide-based nanoparticles, which enables more efficient adsorption of volatile compounds and hence enhances the degradation rate (Dessie et al. 2019).

The emission of isopropyl alcohol (IPA) in the semiconductor manufacturing industries has become a significant concern due to the uncontrolled discharge of IPA vapors into the environment. Nanoparticles containing an adsorbent of nitric acid and sodium hypochlorite were used to capture and effectively eliminate IPA vapors from the air (Bansal et al. 2023). Nanomaterial-based sensors possess exceptional sensitivity and responsiveness, enabling the instantaneous monitoring of air pollution levels. This capability is crucial for timely interventions and mitigation endeavors (Ali et al. 2024). Nanomaterials can be utilized in the development of affordable sensors that can accurately measure air pollution levels in indoor and outdoor environments. These sensors use nanostructured materials as receptors in air/gas sensors, enabling them to detect and monitor a wide range of air pollutants (Kumar et al. 2016). Nanostructured materials have distinctive attributes, including elevated surface area, enhanced reactivity, and heightened sensitivity, rendering them valuable for absorbing and adsorbing harmful gases and pollutants (Guerra et al. 2018). The optimization of the production and characteristics of nanostructured materials enhances their sensing capacities, enabling accurate monitoring of air pollution levels. By employing nanomaterial-based sensors, it is possible to continuously monitor air pollution in real time, providing valuable data for managing and mitigating air quality (Das et al. 2015).

Nano-based technologies and materials aimed at mitigating air pollution and enhancing air quality can be categorized into two groups based on their specific applications: those designed to address outdoor air pollution and quality concerns and those designed to tackle indoor air quality issues. These can be generically categorized into adsorption, degradation, and filtering (Yadav et al. 2021).

3.5.1 Adsorption

Adsorption is a surface phenomenon in which gas or liquid molecules accumulate on the surface of a solid. This technique consists of two components: the adsorbent, which refers to the solid substance, and the adsorbent (Rathi and Kumar 2021). Nanotechnology focuses on developing systematic methodologies for the efficient design and production of different types of adsorbents based on nanoparticles to suit market requirements. It also emphasizes the significance of surface area in the process of adsorption. Nanoparticles can be used as adsorbents to remediate and treat air contaminants through several methods (Agboola et al. 2020).

Adsorption is a nondestructive technique that can eliminate gaseous contaminants from the air at average temperatures. Two distinct mechanisms, physisorption (non-covalent) and chemisorption (covalent), can be employed. The following factors primarily determine the sorption capacity: (1) the physical structure and chemical composition of the sorption materials, which includes their specific surface area; (2) the size and distribution of the pores and the functional groups on the surface; (3) the molecular structure, polarity, acidity, and wettability of the pollutants being targeted; and (4) the conditions under which adsorption occurs, primarily temperature and humidity (Nayak and Pathan 2023). Nanotechnology has progressed to the point where 1D carbon nanotubes and 2D graphene nanosheets (in oxidized and reduced forms) are now being utilized in various fields such as electronics, catalysis, photonics, and environmental remediation (Khan et al. 2022).

Graphene oxide nanoparticles have the benefit of having many surface oxygen functional groups, making it easier for hydrophilic VOC molecules to be adsorbed. Nevertheless, the clustering of graphene oxide has significantly hindered its practical application as an adsorbent in the commercial sector. It can be processed into different self-supporting structures or used as a support material for dispersed catalysts that are designed to adsorb and degrade VOCs simultaneously (Li et al. 2020). Additional carbon-based compounds, such as biochar and ordered mesoporous carbon, have been examined for their potential as adsorbents for VOCs (Shen 2023).

3.5.2 Degradation

In addition to adsorption, catalytic degradation is commonly used to reduce air pollution. The field of nanomaterial engineering has led to the creation of a wide variety of nanocatalysts that exhibit exceptional catalytic efficiency (Mitchell et al. 2021). Nanocatalysts exhibit the potential to enhance air quality and reduce air pollutants to a limited degree. The nanocatalysis system allows for fast and specific chemical transformations with a high yield of products and the capability to retrieve the catalyst, in contrast to other traditional catalysts (Gebre 2023). Photocatalysts may consist of nanomaterials such as titanium dioxide (TiO_2), zinc oxide (ZnO), and tungsten oxide (WO_3) (Gunti et al. 2018). Photocatalysts can convert organic pollutants into innocuous molecules by oxidation, making them valuable for environmental remediation. Light can catalyze the oxidation of organic molecules by utilizing the principles of semiconductors. Under adequate illumination, the charge transfer mechanism occurs from the valence band to the conduction band, leading to the oxidation of the surrounding material (Gusain et al. 2019). The reactivity and selectivity of semiconductor photocatalysts have been modified by nanotechnology (Xu et al. 2019). Recently, researchers conducted studies to determine the effectiveness of incorporating nanocatalysts into air purifiers to remove airborne germs in indoor spaces (Pramudita et al. 2022; Vivek et al. 2021; Zhong and Haghighat 2015).

3.5.3 Filtration by Nanofilters

Nanofibers are crucial nanomaterials used in various environmental applications. Nanotechnology's ability to manipulate matter at the molecular level, along with the small size and large surface-to-volume ratio of nanoparticles employed in manufacturing, makes it highly promising in this industry (Pramudita et al. 2022). A group of researchers conducted a study on integrating nanoparticles into nanofibers for air filtration (Zhu et al. 2017; Li et al. 2019; Zhou et al. 2022). Silver nanoparticle filters have recently gained significant popularity in air filtering systems. These air filtration systems can be effectively used in hospitals to purify the indoor air (Zhao et al. 2010).

In their study, Le et al. (2015) discovered that an air purifier with a pre-filter coated with nano silver, four photocatalytic filters, and four UV-A lamps could eliminate 69% of bacteria and 63% of fungi for 6 h. Additionally, it was revealed that 91.6% of butanol was eradicated within just 55 min of the air passing through the equipment (Le et al. 2015). In their study, Zhang et al. (2016) discovered that nanofibrous filters made of metal–organic framework (MOF) have high efficacy in removing PM2.5, PM10, and harmful SO_2 pollutants (Zhang et al. 2016).

Several specific capabilities can be introduced by incorporating active components into the polymers used to produce nanofibers (Lim 2017). Iron oxide, a significant component of airborne dust, can be incorporated into electrospun nanofibers as magnetic metal oxide nanoparticles (Fe_3O_4) to attract dust through magnetic attraction. Due to their low weight and ability to withstand high temperatures, polymer-derived nonwoven nanofiber mats are highly suitable for environmental applications (Kim et al. 2017). γ-Al_2O_3 nanofiber membranes, specifically ceramic nanofibers, exhibit excellent thermal and chemical resilience, rendering them well-suited for the treatment of automotive exhaust and industrial filtration purposes (Brayer 2022) (Table 3.1).

Table 3.1 Some nanoparticle applications in air remediation

Nanoparticles	Target pollutants	References
Titanium dioxide nanoparticles (TNPs)	VOCs (ethylene)	Hussain et al. (2010)
Silica nanoparticle (SiNPs)	Atmospheric lead (Pb)	Yang et al. (2013)
Ag/SBA-15 nanocomposites	Carbon monoxide (CO)	Zhang et al. (2011)
Aligned carbon nanotube	Aerosols	Yildiz and Bradford (2013)
Nanosilver-TiO_2 nanofibers	NOx, VOCs	Srisitthiratkul et al. (2011)
Titanium dioxide nanoparticles (TNPs)	Sulfur dioxide (SO_2)	Baltrusaitis et al. (2011)
Ruthenium nanoparticle catalysts (Ru/Al_2O_3)	N_2, N_2O	Komvokis et al. (2011)
Calcium-based nanoparticles	Carbon dioxide	Abanades and Alvarez (2003)
Titanium oxide-coated calcium carbonate NPs	Carbon dioxide	Wang et al. (2013)
Carbon nanotubes	Carbon dioxide	Wang et al. (2013)
Nickel nanoparticles	Methane	Wang and Lua (2012)
Titanate nanotubes	Nitrous oxide and nitric oxide	Nguyen and Bai (2015)
WO_3 (tungsten trioxide) nanoparticles	Airborne contaminants such as ground-level smoke, ozone, smog, viruses, and bacteria	Haliq et al. (2022)

3.6 Risks Associated with the Use of Nanoparticles

There is a significant rise in the use of nanoparticles (NPs) in both industrial and domestic settings, leading to the discharge of NPs into the environment. The over-utilization of nanomaterials leads to a higher concentration of these substances in groundwater and soil, resulting in environmental toxicity (Saeed et al. 2023). Nanoparticles' distribution is influenced by their ability to move, their survival, and the amount released into the environment. This phase split assesses the extent to which NPs are exposed to the environment. It is essential to examine the levels of NP concentrations in the atmosphere to evaluate the potential dangers posed by nanoparticles (NPs) in the environment (Martínez et al. 2020). Nanoparticles (NPs) possess a significant ratio of surface area to mass, resulting in the absorption of contaminants being distributed throughout the solid and aqueous phases. Pollutants were assimilated by nanoparticles (NPs) either by coprecipitation during NP production or through the aggregation of NPs with pollutants that were adsorbed onto their surface (Chauhan et al. 2023). The adsorption of contaminants on nanoparticles' surfaces is influenced by their size, composition, shape, permeability, aggregation/disaggregation, and aggregate architecture. Nevertheless, luminophores are not ecologically sustainable and are shielded from oxygen when incorporated into a silica matrix (Bansal et al. 2023).

3.7 Conclusion

The relevance of the field is evident as environmental degradation continues to worsen each year, posing a significant threat to our planet. The air is currently contaminated with various contaminants such as carbon monoxide, hydrocarbons, nitrogen oxides, volatile organic compounds, and chlorofluorocarbons. Water and soil contamination can also occur due to the presence of organic and inorganic substances. Main sources of pollution include sewage water, industrial effluents, improper usage of fertilizers and pesticides, and oil spills. The rapid progress of nanotechnology has generated considerable interest in exploring the potential uses of nanomaterials in improving environmental monitoring and cleanup systems across all three phases. This technology has the potential to greatly improve the sensing and detection of contaminants, leading to significant advancements in cutting-edge remediation solutions. In this overview, we discuss the latest advancements in the application of nanotechnology to various fields. We also cover recent advancements in nanomaterial engineering techniques and demonstrate their practical applications in enhancing stability, catalytic efficiency, and the utilization of alternative oxidants. This review will provide a comprehensive understanding of the definition of nanoparticles, the concept of nanotechnology, and the development of nanomaterials for potential practical applications. Finally, we provide an overview of the future of nanotechnology and its possible applications in addressing air pollution.

References

Abanades JC, Alvarez D (2003) Conversion limits in the reaction of CO_2 with lime. Energy Fuel 17:308–315

Abdalla AM, Elnaghi BE, Hossain S, Dawood M, Abdelrehim O, Azad AK (2020) Nanotechnology utilization in energy conversion, storage and efficiency: a perspective review. Adv Energy Convers Mater 1:30–54

Abdel-Karim R, Reda Y, Abdel-Fattah A (2020) Nanostructured materials-based nanosensors. J Electrochem Soc 167:037554

Abdillah F, Pramudita D, Indarto A, Handojo LA (2018) Nanotechnology: an emerging technology for bioremediation of environmental pollutants. In: Recent advances in environmental management. CRC Press, pp 109–143

Afreen S, Omar RA, Talreja N, Chauhan D, Ashfaq M (2018) Carbon-based nanostructured materials for energy and environmental remediation applications. In: Approaches in bioremediation: the new era of environmental microbiology and nanobiotechnology, pp 369–392

Agboola O, Popoola P, Sadiku R, Sanni SE, Fayomi SO, Fatoba OS (2020) Nanotechnology in wastewater and the capacity of nanotechnology for sustainability. Environ Nanotechnol 3:1–45

Ahmad MA, Adeel M, Shakoor N, Javed R, Ishfaq M, Peng Y, Zain M, Azeem I, Ali I, Usman M (2023) Modifying engineered nanomaterials to produce next generation agents for environmental remediation. Sci Total Environ 894:164861

Ali K, Nishtar Z, Sabir RM, Safdar M (2024) Biosensors for environmental monitoring in the smart agriculture sector. In: Agriculture and aquaculture applications of biosensors and bioelectronics. IGI Global

Andrievski RA (2014) Review of thermal stability of nanomaterials. J Mater Sci 49:1449–1460

Anton PS, Silberglitt R, Schneider J (2001) The global technology revolution: bio/nano/materials trends and their synergies with information technology by 2015. Rand Corporation

Asghar N, Hussain A, Nguyen DA, Ali S, Hussain I, Junejo A, Ali A (2024) Advancement in nanomaterials for environmental pollutants remediation: a systematic review on bibliometrics analysis, material types, synthesis pathways, and related mechanisms. J Nanobiotechnol 22:26

Baltrusaitis J, Jayaweera PM, Grassian VH (2011) Sulfur dioxide adsorption on TiO2 nanoparticles: influence of particle size, coadsorbates, sample pretreatment, and light on surface speciation and surface coverage. J Phys Chem C 115:492–500

Bansal A, Sachan RSK, Devgon J, Devgon I, Karnwal A (2023) Nanotechnology in environmental clean-up. Mater Res Proc 145:281–310

Barsan N, Weimar U (2001) Conduction model of metal oxide gas sensors. J Electroceram 7:143–167

Berger M (2016) Nanotechnology: the future is Tiny. Royal Society of Chemistry

Bhandari G (2018) Environmental nanotechnology: applications of nanoparticles for bioremediation.In. In: Prasad R, Aranda E (eds) Approaches in Bioremediation. Nanotechnology in the Life Sciences. Springer, Cham

Bhardwaj SK, Bhardwaj N, Kumar V, Bhatt D, Azzouz A, Bhaumik J, Kim K-H, Deep A (2021) Recent progress in nanomaterial-based sensing of airborne viral and bacterial pathogens. Environ Int 146:106183

Bilal M, Rasheed T, Mehmood S, Tang H, Ferreira LFR, Bharagava RN, Iqbal HMN (2020) Mitigation of environmentally-related hazardous pollutants from water matrices using nanostructured materials–a review. Chemosphere 253:126770

Bosch P, Catalina F, Corrales T, Peinado C (2005) Fluorescent probes for sensing processes in polymers. Chem–A Eur J 11:4314–4325

Brayer WA (2022) Alternating field electrospinning for fabrication of inorganic nanofibers in the AL2O3-SIO2-MGO system All ETDs from UAB. 505

Bumbudsanpharoke N, Choi J, Ko S (2015) Applications of nanomaterials in food packaging. J Nanosci Nanotechnol 15:6357–6372

Chauhan P, Imam A, Kanaujia PK, Suman SK (2023) Nano-bioremediation: an eco-friendly and effective step towards petroleum hydrocarbon removal from environment. Environ Res 231:116224

Che J, Cagin T, Goddard WA III (2000) Thermal conductivity of carbon nanotubes. Nanotechnology 11:65

Chowdhury A, Kunjiappan S, Panneerselvam T, Somasundaram B, Bhattacharjee C (2017) Nanotechnology and nanocarrier-based approaches on treatment of degenerative diseases. Int Nano Lett 7:91–122

Collin G (2021) History of carbon materials. In: Industrial carbon and graphite materials, volume I: raw materials, production and applications, vol 1, pp 33–43

Cushing BL, Kolesnichenko VL, O'connor CJJCR (2004) Recent advances in the liquid-phase syntheses of inorganic nanoparticles. Chem Rev 104:3893–3946

Daly FN, Schiff D (2007) Supportive management of patients with brain tumors. Expert Rev Neurother 7:1327–1336

Das MP (2010) Mesoscopic systems in the quantum realm: fundamental science and applications. Adv Nat Sci Nanosci Nanotechnol 1:043001

Das S, Sen B, Debnath N (2015) Recent trends in nanomaterials applications in environmental monitoring and remediation. Environ Sci Pollut Res 22:18333–18344

Dehghani-Sanij AR, Tharumalingam E, Dusseault MB, Fraser R (2019) Study of energy storage systems and environmental challenges of batteries. Renew Sust Energ Rev 104:192–208
Delay M, Frimmel FH (2012) Nanoparticles in aquatic systems. Anal Bioanal Chem 402:583–592
Dessie Y, Tadesse S, Eswaramoorthy R, Abebe B (2019) Recent developments in manganese oxide based nanomaterials with oxygen reduction reaction functionalities for energy conversion and storage applications: a review. J Sci: Adv Mater Devices 4:353–369
Diller L, Chow EJ, Gurney JG, Hudson MM, Kadin-Lottick NS, Kawashima TI, Leisenring WM, Meacham LR, Mertens AC, Mulrooney DA (2009) Chronic disease in the childhood cancer survivor study cohort: a review of published findings. J Clin Oncol 27:2339
Ealia SAM, Saravanakumar M (2017) A review on the classification, characterisation, synthesis of nanoparticles and their application. IOP Conf. Ser.: Mater. Sci. Eng 263:032019
El-Khawaga AM, Elsayed MA, Fahim YA, Shalaby RE (2023a) Promising photocatalytic and antimicrobial activity of novel capsaicin coated cobalt ferrite nanocatalyst. Sci Rep 13:5353
El-Khawaga AM, Zidan A, El-Mageed AIAA (2023b) Preparation methods of different nanomaterials for various potential applications: a review. J Mol Struct 1281:135148
Ezendam S, Herran M, Nan L, Gruber C, Kang Y, Gröbmeyer F, Lin R, Gargiulo J, Sousa-Castillo A, Cortés E (2022) Hybrid plasmonic nanomaterials for hydrogen generation and carbon dioxide reduction. ACS Energy Letters 7:778–815
Fahim YA, El-Khawaga AM, Sallam RM, Elsayed MA, Assar MFA (2024) A review on lipases: sources, assays, immobilization techniques on nanomaterials and applications. BioNanoScience 14:1–18
Fattahi Z, Sajjadi SA, Babakhani A, Saba F (2020) Ni–Cr matrix composites reinforced with nano- and micron-sized surface-modified zirconia: synthesis, microstructure and mechanical properties. J Alloys Compd 817:152755
Fellows PJ (2022) Food processing technology: principles and practice. Woodhead publishing
Fox A, Ray U, Li T (2019) Thermal conductivity of graphene grain boundaries along arbitrary in-plane directions: a comprehensive molecular dynamics study. J Appl Phys 125:015101
Ganesh V, Yahia IS, Alfaify S, Shkir M (2017) Sn-doped ZnO nanocrystalline thin films with enhanced linear and nonlinear optical properties for optoelectronic applications. J Phys Chem Solids 100:115–125
Gao C, Lyu F, Yin Y (2020) Encapsulated metal nanoparticles for catalysis. Chem Rev 121:834–881
Gebre SH (2023) Recent developments of supported and magnetic nanocatalysts for organic transformations: an up-to-date review. Appl Nanosci 13:15–63
Gheorghe IF, Ion B (2011) The effects of air pollutants on vegetation and the role of vegetation in reducing atmospheric pollution. In: The impact of air pollution on health, economy, environment and agricultural sources, vol 29, pp 241–280
Ghotekar S, Savale A, Pansambal SJJOW, Nanotechnology E (2018) Phytofabrication of fluorescent silver nanoparticles from Leucaena leucocephala L. leaves and their biological activities. J Water Environ Nanotechnol 3:95–105
Guerra FD, Attia MF, Whitehead DC, Alexis F (2018) Nanotechnology for environmental remediation: materials and applications. Molecules 23:1760
Gunti S, Kumar A, Ram MK (2018) Nanostructured photocatalysis in the visible spectrum for the decontamination of air and water. Int Mater Rev 63:257–282
Gupta V, Mohapatra S, Mishra H, Farooq U, Kumar K, Ansari MJ, Aldawsari MF, Alalaiwe AS, Mirza MA, Iqbal Z (2022) Nanotechnology in cosmetics and cosmeceuticals—a review of latest advancements. Gels 8:173
Gusain R, Gupta K, Joshi P, Khatri OP (2019) Adsorptive removal and photocatalytic degradation of organic pollutants using metal oxides and their composites: a comprehensive review. Adv Colloid Interf Sci 272:102009
Haider A, Ikram M, Rafiq A (2022) Properties of nanomaterials. Green nanomaterials as potential antimicrobials. Springer

Haliq R, Tarigan OC, Aprilio MR, Wijaya FTD, Aziz GDA, Catherine S, Abdullah H (2022) Syntheses and applications of nanomaterials-based Photocatalysts for air purification. In: Photocatalytic activities for environmental remediation and energy conversion. Springer

Hatakeyama Y, Onishi K, Nishikawa KJRA (2011) Effects of sputtering conditions on formation of gold nanoparticles in sputter deposition technique. RSC Adv 1:1815–1821

Himani PU, Mahawer SK, Kumar R, Prakash O (2022) Plant protection through agrochemicals and its consequences. In: Plant protection*:* from chemicals to biologicals. de Gruyter, p 25

Hodges BC, Cates EL, Kim J-H (2018) Challenges and prospects of advanced oxidation water treatment processes using catalytic nanomaterials. Nat Nanotechnol 13:642–650

Holzinger M, LE Goff A, Cosnier S (2014) Nanomaterials for biosensing applications: a review. Front Chem 2:63

Hussain M, Ceccarelli R, Marchisio DL, Fino D, Russo N, Geobaldo F (2010) Synthesis, characterization, and photocatalytic application of novel TiO_2 nanoparticles. Chem Eng J 157:45–51

Jamkhande PG, Ghule NW, Bamer AH, Kalaskar MG (2019) Metal nanoparticles synthesis: an overview on methods of preparation, advantages and disadvantages, and applications. J Drug Deliv Sci Technol 53:101174

Joachim C (2009) Nanosciences: the invisible revolution. World Scientific

Kachanov M, Sevostianov I (2005) On quantitative characterization of microstructures and effective properties. Int J Solids Struct 42:309–336

Karimi M, Mansouri MR, Rabiee N, Hamblin MR (2018) Carbon-based nanomaterials. In: Advances in nanomaterials for drug delivery: polymeric, nanocarbon, and bio-inspired

Kaushani KG, Priyadarshana G (2023) Use of nanoformulations and nano-enabled products in mitigating the risk associated with the current use of agrochemicals. In: Nanoformulations for Sustainable Agriculture and Environmental Risk Mitigation. CABI GB

Khafaga DSR, El-Khawaga AM, Elfattah Mohammed RA, Abdelhakim HK (2023) Green synthesis of nano-based drug delivery systems developed for hepatocellular carcinoma treatment: a review. Mol Biol Rep 50:10351–10364

Khan SH (2020) Green nanotechnology for the environment and sustainable development. In: Naushad M, Lichtfouse E (eds) Green materials for wastewater treatment. Environmental chemistry for a sustainable world, vol 38. Springer, Cham, pp 13–46

Khan MZ, Gul IH, Baig MM, Khan AN (2020) Comprehensive study on structural, electrical, magnetic and photocatalytic degradation properties of Al3+ ions substituted nickel ferrites nanoparticles. J Alloys Compd 848:155795

Khan Y, Sadia H, ALI Shah SZ, Khan MN, Shah AA, Ullah N, Ullah MF, Bibi H, Bafakeeh OT, Khedher NB (2022) Classification, synthetic, and characterization approaches to nanoparticles, and their applications in various fields of nanotechnology: a review. Catalysts 12:1386

Kim J, Chan Hong S, Bae GN, Jung JH (2017) Electrospun magnetic nanoparticle-decorated nanofiber filter and its applications to high-efficiency air filtration. Environ Sci Technol 51:11967–11975

Klaine SJ, Koelmans AA, Horne N, Carley S, Handy RD, Kapustka L, Nowack B, Von DER Kammer F (2012) Paradigms to assess the environmental impact of manufactured nanomaterials. Environ Toxicol Chem 31:3–14

Komvokis VG, Marti M, Delimitis A, Vasalos IA, Triantafyllidis KS (2011) Catalytic decomposition of N_2O over highly active supported Ru nanoparticles ($\leq$ 3 nm) prepared by chemical reduction with ethylene glycol. Appl Catal B Environ 103:62–71

Kumar A, Pandey G (2018) Different methods used for the synthesis of TiO_2 based nanomaterials: a review. Am J Nano Res Appl 6:1–10

Kumar KS, Swygenhoven VAN, H. & Suresh, S. (2003) Mechanical behavior of nanocrystalline metals and alloys. Acta Mater 51:5743–5774

Kumar P, Skouloudis AN, Bell M, Viana M, Carotta MC, Biskos G, Morawska L (2016) Real-time sensors for indoor air monitoring and challenges ahead in deploying them to urban buildings. Sci Total Environ 560:150–159

Kumar A, Sadhu SD, Singh R (2018) Nanotechnology: key for sustainable future. Nano Environ Sci:759–804

Kumar R, Singh DP, Muñoz R, Amami M, Singh RK, Singh S, Kumar V (2023) Graphene-based materials for biotechnological and biomedical applications: drug delivery, bioimaging and biosensing. Mater Today Chem 33:101750

Le TS, Dao TH, Nguyen DC, Nguyen HC, Balikhin IL (2015) Air purification equipment combining a filter coated by silver nanoparticles with a nano-TiO_2 photocatalyst for use in hospitals. Adv Nat Sci Nanosci Nanotechnol 6:015016

Lei M, Zhang L, Lei J, Zong L, Li J, Wu Z, Wang Z (2015) Overview of emerging contaminants and associated human health effects. BioMed Res Int 2015:404796

Li Y, Yin X, Yu J, Ding B (2019) Electrospun nanofibers for high-performance air filtration. Compos Commun 15:6–19

Li X, Zhang L, Yang Z, Wang P, Yan Y, Ran J (2020) Adsorption materials for volatile organic compounds (VOCs) and the key factors for VOCs adsorption process: a review. Sep Purif Technol 235:116213

Li Q, Yan X, Shi M, Wang Q, Liu H, Lin Z, Huang X (2021) Recent advances in metal/ceria catalysts for air pollution control: mechanism insight and application. Environ Sci Nano 8:2760–2779

Lim CT (2017) Nanofiber technology: current status and emerging developments. Prog Polym Sci 70:1–17

Liu L, Li Y, Wang F (2010) Electrochemical corrosion behavior of nanocrystalline materials—a review. J Mater Sci Technol 26:1–14

Lu K (2016) Stabilizing nanostructures in metals using grain and twin boundary architectures. Nat Rev Mater 1:1–13

Macleod HA (2018) Recent developments in deposition techniques for optical thin films and coatings. In: Optical Thin Films and Coatings. Woodhead Publishing, pp 3–23

Mahalik NP (2006) Micromanufacturing and nanotechnology. Springer

Mahesh BV, Raman RKS (2014) Role of nanostructure in electrochemical corrosion and high temperature oxidation: a review. Metall Mater Trans A 45:5799–5822

Malik S, Muhammad K, Waheed Y (2023a) Emerging applications of nanotechnology in healthcare and medicine. Molecules 28:6624

Malik S, Muhammad K, Waheed Y (2023b) Nanotechnology: a revolution in modern industry. Molecules 28:661

Mammadova S, Nasibova A, Khalilov R, Mehraliyeva S, Valiyeva M, Gojayev AS, Zhdanov RI, Eftekhari A (2022) Nanomaterials application in air pollution remediation. Eurasian Chem Commun 4:160–166

Mansoori GA (2017) An introduction to nanoscience and nanotechnology. In: Nanoscience and plant–soil systems. Springer Soil Biology Series, pp 3–20

Marcovich A, Shinn T (2014) Toward a new dimension: exploring the nanoscale. Oxford University Press

Marichy C, Bechelany M, Pinna N (2012) Atomic layer deposition of nanostructured materials for energy and environmental applications. Adv Mater 24:1017–1032

Maroto-Valer MM, Song C, Soong Y (2012) Environmental challenges and greenhouse gas control for fossil fuel utilization in the 21st century. Springer Science & Business Media

Martínez G, Merinero M, Pérez-Aranda M, Pérez-Soriano EM, Ortiz T, Villamor E, Begines B, Alcudia A (2020) Environmental impact of nanoparticles' application as an emerging technology: a review. Materials 14:166

Mehndiratta P, Jain A, Srivastava S, Gupta N (2013) Environmental pollution and nanotechnology. Environ Pollut 2:49

Meyers MA, Mishra A, Benson DJ (2006) Mechanical properties of nanocrystalline materials. Prog Mater Sci 51:427–556

Minnich A, Dresselhaus MS, Ren ZF, Chen G (2009) Bulk nanostructured thermoelectric materials: current research and future prospects. Energy Environ Sci 2:466–479

Minoli D (2005) Nanotechnology applications to telecommunications and networking. Wiley
Mitchell S, Qin R, Zheng N, Pérez-Ramírez J (2021) Nanoscale engineering of catalytic materials for sustainable technologies. Nat Nanotechnol 16:129–139
Mittal M, Kumar A (2014) Carbon nanotube (CNT) gas sensors for emissions from fossil fuel burning. Sensors Actuators B Chem 203:349–362
Mukherjee P, Ahmad A, Mandal D, Senapati S, Sainkar SR, Khan MI, Parishcha R, Ajaykumar P, Alam M, Kumar RJNL (2001) Fungus-mediated synthesis of silver nanoparticles and their immobilization in the mycelial matrix: a novel biological approach to nanoparticle synthesis. Nano Lett 1:515–519
Murray CB, Kagan CR, Bawendi MG (2000) Synthesis and characterization of monodisperse nanocrystals and close-packed nanocrystal assemblies. Annu Rev Mater Sci 30:545–610
Nafisi S, Maibach HI (2017) Nanotechnology in cosmetics. Cosmetic science and technology: theoretical principles and applications, vol 337. Elsevier
Nayak T, Pathan A (2023) Environmental remediation and application of carbon-based nanomaterials in the treatment of heavy metal-contaminated water: a review. Mater Today Proc 92:1659
Nguyen NH, Bai H (2015) Effect of washing pH on the properties of titanate nanotubes and its activity for photocatalytic oxidation of NO and NO_2. Appl Surf Sci 355:672–680
Niederberger M, Garnweitner G (2006) Organic reaction pathways in the nonaqueous synthesis of metal oxide nanoparticles. Chemistry 12:7282–7302
Nile SH, Baskar V, Selvaraj D, Nile A, Xiao J, Kai G (2020) Nanotechnologies in food science: applications, recent trends, and future perspectives. Nano-micro Lett 12:1–34
Nyabadza A, Vázquez M, Coyle S, Fitzpatrick B, Brabazon D (2021) Review of materials and fabrication methods for flexible nano and micro-scale physical and chemical property sensors. Appl Sci 11:8563
Omran BA, Omran BA (2020) Fundamentals of nanotechnology and nanobiotechnology. In: Nanobiotechnology: a multidisciplinary field of science. Nanotechnology in the life sciences. Springer, Cham, pp 1–36
Osman AI, Zhang Y, Farghali M, Rashwan AK, Eltaweil AS, El-Monaem A, Eman M, Mohamed I, Badr MM, Ihara I (2024) Synthesis of green nanoparticles for energy, biomedical, environmental, agricultural, and food applications: a review. Environ Chem Lett 22:1–47
Panchal H, Patel H, Patel J, Shah M (2021) A systematic review on nanotechnology in enhanced oil recovery. Pet Res 6:204–212
Pedersen H, Elliott SDJTCA (2014) Studying chemical vapor deposition processes with theoretical. Chemistry 133:1–10
Pomerantseva E, Bonaccorso F, Feng X, Cui Y, Gogotsi Y (2019) Energy storage: the future enabled by nanomaterials. Science 366:eaan8285
Popok VN (2019) Cluster beam synthesis of polymer composites with nanoparticles. In: Polymer-based multifunctional nanocomposites and their applications. Elsevier
Pramudita D, Suendo V, Azis MM, Indarto A (2022) Nanomaterials for air purification: advances and challenges. In: Nanomaterials for environmental applications. CRC Press, pp 281–315
Prasanna S (2017) Pollution prevention and control using nanotechnology. Int Res J Comput Sci 4:2015–2018
Qu X, Brame J, Li Q, Alvarez PJJ (2013) Nanotechnology for a safe and sustainable water supply: enabling integrated water treatment and reuse. Acc Chem Res 46:834–843
Raizada P, Soni V, Kumar A, Singh P, Khan AAP, Asiri AM, Thakur VK, Nguyen V-H (2021) Surface defect engineering of metal oxides photocatalyst for energy application and water treatment. J Mater 7:388–418
Rajput N (2015) Methods of preparation of nanoparticles-a review. Int. J. Adv. Eng. Technol 7:1806–1811
Ramsden J (2018) Applied nanotechnology: the conversion of research results to products. William Andrew

Rasheed T, Bilal M, Nabeel F, Adeel M, Iqbal HMN (2019) Environmentally-related contaminants of high concern: potential sources and analytical modalities for detection, quantification, and treatment. Environ Int 122:52–66
Rathi BS, Kumar PS (2021) Application of adsorption process for effective removal of emerging contaminants from water and wastewater. Environ Pollut 280:116995
Roco MC, Hersam MC, Mirkin CA, Mirkin CA, Tuominen M (2011) Synthesis, processing, and manufacturing of components, devices, and systems. In: Nanotechnology research directions for societal needs in 2020: retrospective and outlook. Springer, Dordrecht, pp 109–158
Roth H-C, Schwaminger SP, Schindler M, Wagner FE, Berensmeier S (2015) Influencing factors in the co-precipitation process of superparamagnetic iron oxide nano particles: a model based study. J Magn Magn Mater 377:81–89
Rupert TJ, Trenkle JC, Schuh CA (2011) Enhanced solid solution effects on the strength of nanocrystalline alloys. Acta Mater 59:1619–1631
Saeed M, Ilyas N, Bibi F, Shabir S, Mehmood S, Akhtar N, Ali I, Bawazeer S, Al Tawaha AR, Eldin SM (2023) Nanoremediation approaches for the mitigation of heavy metal contamination in vegetables: an overview. Nanotechnol Rev 12:20230156
Salamanca-Buentello F, Persad DL, Court EB, Martin DK, Daar AS, Singer PA (2005) Nanotechnology and the developing world. PLoS Med 2:e97
Santos AC, Morais F, Simões A, Pereira I, Sequeira JAD, Pereira-Silva M, Veiga F, Ribeiro A (2019) Nanotechnology for the development of new cosmetic formulations. Expert Opin Drug Deliv 16:313–330
Selin C (2011) Negotiating plausibility: intervening in the future of nanotechnology. Sci Eng Ethics 17:723–737
Shah MA, Pirzada BM, Price G, Shibiru AL, Qurashi A (2022) Applications of nanotechnology in smart textile industry: a critical review. J Adv Res 38:55–75
Shannon MA, Bohn PW, Elimelech M, Georgiadis JG, Mariñas BJ, Mayes AM (2008) Science and technology for water purification in the coming decades. Nature 452:301–310
Sharma D, Kanchi S, Bisetty KJAJOC (2019) Biogenic synthesis of nanoparticles: a review. Arab J Chem 12:3576–3600
Shen Y (2023) Biomass-derived porous carbons for sorption of volatile organic compounds (VOCs). Fuel 336:126801
Singamaneni S, Bliznyuk VN, Binek C, Tsymbal EY (2011) Magnetic nanoparticles: recent advances in synthesis, self-assembly and applications. J Mater Chem 21:16819–16845
Singh KRB, Solanki PR, Malhotra BD, Pandey AC, Singh RP (2021a) Introduction to nanomaterials: an overview toward broad-spectrum applications. In: Nanomaterials in Bionanotechnology. CRC Press, pp 1–35
Singh RP, Handa R, Manchanda G (2021b) Nanoparticles in sustainable agriculture: an emerging opportunity. J Control Release 329:1234–1248
Sobha K, Surendranath K, Meena V, Jwala KT, Swetha N, Latha KSM (2010) Emerging trends in nanobiotechnology. Biotechnol Mol Biol Rev 5:1–12
Sportelli MC, Izzi M, Volpe A, Clemente M, Picca RA, Ancona A, Lugarà PM, Palazzo G, Cioffi NJA (2018) The pros and cons of the use of laser ablation synthesis for the production of silver nano-antimicrobials. Antibiotics (Basel) 7:67
Srisitthiratkul C, Pongsorrarith V, Intasanta N (2011) The potential use of nanosilver-decorated titanium dioxide nanofibers for toxin decomposition with antimicrobial and self-cleaning properties. Appl Surf Sci 257:8850–8856
Sun Y, Sun M, Xie D (2018) Graphene electronic devices. Elsevier, Graphene
Syduzzaman M, Hassan A, Anik HR, Akter M, Islam MR (2023) Nanotechnology for high-performance textiles: a promising frontier for innovation. Chem Nano Mat 9:e202300205
Tahir MB, Rafique M, Rafique MS, Nawaz T, Rizwan M, Tanveer M (2020) Photocatalytic nanomaterials for degradation of organic pollutants and heavy metals. In: Nanotechnology and photocatalysis for environmental applications. Elsevier

Tan JC, Cheetham AK (2011) Mechanical properties of hybrid inorganic–organic framework materials: establishing fundamental structure–property relationships. Chem Soc Rev 40:1059–1080
Tatarchuk T, Peter A, Al-Najar B, Vijaya J, Bououdina M (2018) Photocatalysis: activity of nanomaterials. In: Hussain CM, Mishra AK (eds) Nanotechnology in environmental science. Wiley-VCH Verlag GmbH & Co. KGaA, pp 209–292
Tavakoli A, Sohrabi M, Kargari AJCP (2007) A review of methods for synthesis of nanostructured metals with emphasis on iron compounds. Chem Pap 61:151–170
Thakur A, Kumar A (2022) Recent advances on rapid detection and remediation of environmental pollutants utilizing nanomaterials-based (bio) sensors. Sci Total Environ 834:155219
Thakur S, Thakur S, Kumar R (2018) Bio-nanotechnology and its role in agriculture and food industry. J Mol Genet Med 12:1747–0862
Thiruvengadam M, Rajakumar G, Chung I-M (2018) Nanotechnology: current uses and future applications in the food industry. 3 Biotech 8:1–13
Tjong S-C (2013) Nanocrystalline materials: their synthesis-structure-property relationships and applications. Newnes
Tjong SC, Chen H (2004) Nanocrystalline materials and coatings. Mater Sci Eng R: Rep 45:1–88
Toberer ES, Baranowski LL, Dames C (2012) Advances in thermal conductivity. Annu Rev Mater Res 42:179–209
Tsuzuki T (2021) Mechanochemical synthesis of metal oxide nanoparticles. Commun Chem 4:1–11
Tsuzuki T, Mccormick PGJJOMS (2004) Mechanochemical synthesis of nanoparticles. J Mater Sci 39:5143–5146
Villa CH, Anselmo AC, Mitragotri S, Muzykantov V (2016) Red blood cells: supercarriers for drugs, biologicals, and nanoparticles and inspiration for advanced delivery systems. Adv Drug Deliv Rev 106:88–103
Vivek P, Rekha M, Steephen A, Kowsalya M, Magesh M (2021) Meticulous review on potential nano–sized catalysts for air and water purifiers. Iran J Catal 11:331–345
Wang HY, Lua AC (2012) Development of metallic nickel nanoparticle catalyst for the decomposition of methane into hydrogen and carbon nanofibers. J Phys Chem C 116:26765–26775
Wang Y, Zhu Y, Wu S (2013) A new nano Cao-based CO2 adsorbent prepared using an adsorption phase technique. Chem Eng J 218:39–45
Wang L, Hasanzadeh Kafshgari M, Meunier M (2020) Optical properties and applications of plasmonic-metal nanoparticles. Adv Funct Mater 30:2005400
Wei J, Li F, Zhou L, Han D, Gong J (2022) Strategies for enhancing peroxymonosulfate activation by heterogenous metal-based catalysis: A review. Chin J Chem Eng 50:12
Wu HB, Lou XW (2017) Metal-organic frameworks and their derived materials for electrochemical energy storage and conversion: promises and challenges. Sci Adv 3:eaap9252
Wu S, Liu J, Wang H, Yan H (2019) A review of performance optimization of Mof-derived metal oxide as electrode materials for supercapacitors. Int J Energy Res 43:697–716
Xu C, Anusuyadevi PR, Aymonier C, Luque R, Marre S (2019) Nanostructured materials for photocatalysis. Chem Soc Rev 48:3868–3902
Yadav TP, Yadav RM, Singh DPJN (2012) Mechanical milling: a top down approach for the synthesis of nanomaterials and nanocomposites. Nanosci Nanotechnol 2:22–48
Yadav S, Sharma AK, Kumar P (2020) Nanoscale self-assembly for therapeutic delivery. Front Bioeng Biotechnol 8:127
Yadav N, Garg VK, Chhillar AK, Rana JS (2021) Detection and remediation of pollutants to maintain ecosustainability employing nanotechnology: a review. Chemosphere 280:130792
Yang X, Shen Z, Zhang B, Yang J, Hong W-X, Zhuang Z, Liu J (2013) Silica nanoparticles capture atmospheric lead: implications in the treatment of environmental heavy metal pollution. Chemosphere 90:653–656
Yetisen AK, Qu H, Manbachi A, Butt H, Dokmeci MR, Hinestroza JP, Skorobogatiy M, Khademhosseini A, Yun SH (2016) Nanotechnology in textiles. ACS Nano 10:3042–3068
Yildiz O, Bradford PD (2013) Aligned carbon nanotube sheet high efficiency particulate air filters. Carbon 64:295–304

Zhang L, Fang M (2010) Nanomaterials in pollution trace detection and environmental improvement. Nano Today 5:128–142

Zhang J, Zhang L (2012) Nanostructures for surface plasmons. Adv Opt Photon 4:157–321

Zhang X, Qu Z, Li X, Zhao Q, Wang Y, Quan X (2011) Low temperature CO oxidation over Ag/Sba-15 nanocomposites prepared via in-situ "ph-adjusting" method. Catal Commun 16:11–14

Zhang Y, Yuan S, Feng X, Li H, Zhou J, Wang B (2016) Preparation of nanofibrous metal–organic framework filters for efficient air pollution control. J Am Chem Soc 138:5785–5788

Zhang J, Qin Z, Zeng D, Xie C (2017) Metal-oxide-semiconductor based gas sensors: screening, preparation, and integration. Phys Chem Chem Phys 19:6313–6329

Zhang H, Zhao H, Khan MA, Zou W, Xu J, Zhang L, Zhang J (2018) Recent progress in advanced electrode materials, separators and electrolytes for lithium batteries. J Mater Chem A 6:20564–20620

Zhao Y-K, Sung W-P, Tsai T-T, Wang H-J (2010) Application of nanoscale silver-doped titanium dioxide as photocatalyst for indoor airborne bacteria control: a feasibility study in medical nursing institutions. J Air Waste Manage Assoc 60:337–345

Zhao W, Adeel M, Zhang P, Zhou P, Huang L, Zhao Y, Ahmad MA, Shakoor N, Lou B, Jiang Y (2022) A critical review on surface-modified nano-catalyst application for the photocatalytic degradation of volatile organic compounds. Environ Sci Nano 9:61–80

Zheng SQ, Lim SS, Foo CY, Haw CY, Chiu WS, Chia CH, Khiew PS (2021) Recent progress on the applications of carbonaceous and metal-organic framework nanomaterials for supercapacitors. Front Mater 8:777149

Zhong L, Haghighat F (2015) Photocatalytic air cleaners and materials technologies–abilities and limitations. Build Environ 91:191–203

Zhou Y, Liu Y, Zhang M, Feng Z, Yu D-G, Wang K (2022) Electrospun nanofiber membranes for air filtration: a review. Nano 12:1077

Zhu M, Han J, Wang F, Shao W, Xiong R, Zhang Q, Pan H, Yang Y, Samal SK, Zhang F (2017) Electrospun nanofibers membranes for effective air filtration. Macromol Mater Eng 302:1600353

Zhu Y, Peng L, Fang Z, Yan C, Zhang X, Yu G (2018) Structural engineering of 2D nanomaterials for energy storage and catalysis. Adv Mater 30:1706347

Chapter 4
Cutting-Edge Nanomaterials for Comprehensive Air Pollution Mitigation and Environmental Restoration

Deepak Punetha, Vivek Pratap Singh, Raghvendra, and Saurabh Kumar Pandey

Abstract In recent years, the detrimental effects of air pollution on human health and the environment have become increasingly evident. Traditional methods of air pollution control often fall short in effectively mitigating the growing challenges posed by pollutants. Nanotechnology has emerged as a promising avenue for tackling air pollution due to the unique properties of nanomaterials. This chapter explores the diverse applications of various nanomaterials in air pollution remediation, including their synthesis, mechanisms of pollutant removal, and potential environmental implications. This research explores the multifaceted application of various nanomaterials in the realm of air pollution remediation, a critical field of environmental science. It delves into the diverse array of nanomaterials, ranging from metal oxides to carbon-based nanomaterials, and their pivotal roles in mitigating air pollutants such as particulate matter, volatile organic compounds, nitrogen oxides, and sulfur dioxide. This study elucidates the underlying mechanisms through which nanomaterials interact with pollutants, including adsorption, catalysis, and photocatalysis, thus elucidating their efficacy in air purification. Furthermore, it scrutinizes the challenges and prospects associated with the implementation of nanomaterial-based remediation strategies, encompassing considerations of scal-

D. Punetha (✉)
Department of Electronics and Communication Engineering, Motilal Nehru National Institute of Technology, Allahabad, Prayagraj, Uttar Pradesh, India
e-mail: deepakpunetha@mnnit.ac.in

V. P. Singh · S. K. Pandey
Sensors and Optoelectronics Research Group (SORG), Department of Electrical Engineering, Indian Institute of Technology Patna, Patna, Bihar, India
e-mail: vivek_2121ee21@iitp.ac.in; saurabh@iitp.ac.in

Raghvendra
Department of Electronics and Communication Engineering, SRM University, Amaravati, Andhra Pradesh, India
e-mail: raghvendra.s@srmap.edu.in

N. Ghosh et al. (eds.), *Nanotechnology in Air Quality Management*, Sustainable Development and Biodiversity 39, https://doi.org/10.1007/978-981-96-8339-0_4

ability, cost-effectiveness, and environmental impact. This chapter culminates in highlighting the promising avenues for future research and development, underscoring the imperative of harnessing nanotechnology to combat the burgeoning menace of air pollution and foster sustainable environmental stewardship.

Keywords Nanomaterials · Air pollution · Remediation · Synthesis · Mechanisms · Environmental implications

4.1 Introduction

Air pollution is a global issue with significant implications for public health, environmental sustainability, and economic development. Various anthropogenic activities, such as industrial emissions, vehicular exhaust, and agricultural practices, contribute to the release of harmful pollutants into the atmosphere (Eranna 2016; Chen et al. 2007; Goldstein 2008; Timmer et al. 2005). Traditional methods of air pollution control, such as filtration systems and chemical scrubbers, have limitations in terms of efficiency, cost-effectiveness, and scalability. Nanotechnology offers innovative solutions to address the challenges associated with air pollution remediation (Eranna 2016; Wang et al. 2010). Nanomaterials, characterized by their nanoscale dimensions and unique physicochemical properties, hold immense potential for capturing, degrading, and neutralizing pollutants in the air. This chapter provides an in-depth exploration of the application of different nanomaterials in air pollution remediation, highlighting their synthesis techniques, mechanisms of pollutant removal, and potential environmental implications (Punetha and Pandey 2018, 2019a, b; Williams 1999; Ghosh and Majumder 2017).

In an era characterized by burgeoning industrialization and rapid urbanization, the menace of air pollution looms large, posing a significant threat to both environmental sustainability and public health. The deleterious consequences of airborne pollutants, ranging from particulate matter to toxic gases, have spurred an urgent quest for innovative remediation strategies (Khan et al. 2019; Jimenez-Cadena et al. 2007). Amid this backdrop, nanotechnology has emerged as a beacon of hope, offering a promising avenue for tackling the multifaceted challenges of air pollution. Nanomaterials, defined by their unique properties at the nanoscale, have garnered substantial attention for their unparalleled potential in environmental remediation. Their high surface area-to-volume ratio, tunable physicochemical properties, and remarkable reactivity render them ideal candidates for mitigating air pollution across diverse contexts. From vehicular emissions choking urban landscapes to industrial effluents tarnishing pristine skies, the application of different nanomaterials presents a multifaceted approach to combatting the pervasive threat of air pollution (Goltsov 2003; Segev-Bar and Haick 2013; Dubinin 1960).

The primary equipment needed to detect the air pollutant in the air is the sensor. A sensor is a device or component that detects or responds to electrical or optical signals, converting a physical parameter into an electrically measurable signal. A

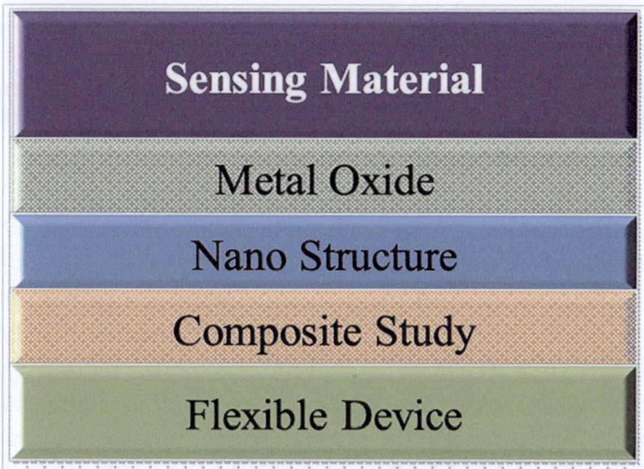

Fig. 4.1 Gas sensing materials (towards improvement in gas sensing performance)

passive sensor is defined as one that exhibits changes in passive electrical quantities—such as capacitance, resistance, or inductance—in response to external stimulation (Punetha et al. 2018a, b, 2020). Gas sensors are a subclass of electrochemical sensors in which either a chemical reaction occurs or charge transport is modulated by the reaction. These sensors operate based on the phenomenon of adsorption, where atoms, ions, or molecules from a substance (which may be a gas, liquid, or dissolved solid) adhere to the surface of an adsorbent. The sensing material or layer plays a critical role in the gas sensing mechanism, where gas adsorption on the surface can trigger a charge-transfer reaction. Gas sensors are generally more effective in high-temperature applications; however, their performance can be enhanced by tailoring physical and surface properties, forming nanocomposites, and developing sensor arrays (Punetha and Singh 2018, 2019; Punetha and Pandey 2020). The potential gas sensing materials responsible for optimized performance parameters are illustrated in Fig. 4.1.

This comprehensive exploration delves into the intricate tapestry of nanomaterials and their versatile applications in remedying air pollution. By harnessing the principles of nanoscience and engineering, researchers have unlocked a myriad of innovative solutions that hold promises for revolutionizing the landscape of environmental sustainability.

4.2 Synthesis of Nanomaterials for Air Pollution Remediation

Air pollution is a pressing environmental concern with significant implications for public health and ecological balance. The advent of nanotechnology offers promising solutions for air pollution remediation due to the unique properties of nanomaterials, such as high surface area-to-volume ratio, tunable surface chemistry, and enhanced reactivity. Nanomaterials also exhibit exceptional catalytic, adsorptive, and photocatalytic properties, making them suitable candidates for addressing different pollutants in the atmosphere. Metal nanoparticles, metal oxides, carbon-based nanomaterials, and hybrid nanocomposites are among the most widely studied materials for air pollution remediation. These materials can effectively target pollutants such as volatile organic compounds (VOCs), nitrogen oxides (NOx), sulfur dioxide (SO_2), particulate matter (PM), and other hazardous airborne contaminants (Chen et al. 2007).

The synthesis of nanomaterials aims to explore the various methods and techniques employed in the fabrication of nanomaterials tailored for air pollution mitigation (Nomura et al. 2023; Fine et al. 2010; Huixia et al. 2015). The nanomaterials utilized for air pollution remediation can be synthesized using various techniques, each imparting specific properties and functionalities to the resulting nanoparticles. Common synthesis methods include chemical vapor deposition, sol–gel synthesis, precipitation, and hydrothermal synthesis. These techniques allow for precise control over the size, shape, composition, and surface properties of nanomaterials, enabling tailored designs for specific pollutant removal applications.

4.2.1 Different Methods for the Synthesis of Nanomaterials

(a) *Chemical Vapor Deposition (CVD)*: CVD is a widely used technique for the synthesis of nanomaterials with precise control over size, morphology, and composition. In the context of air pollution remediation, CVD enables the fabrication of thin films, nanostructures, and catalysts with tailored properties for pollutant removal and degradation. By adjusting the precursor gases and process parameters, researchers can synthesize diverse nanomaterials, including metal oxides, carbon nanotubes, and graphene-based materials (Gautam and Jayatissa 2011).

(b) *Sol–Gel Process*: The sol–gel method offers a versatile approach for the synthesis of metal oxide nanoparticles and nanocomposites. Through controlled hydrolysis and condensation reactions of metal alkoxides or metal salts in a solution, homogeneous colloidal suspensions or gels are formed, which can be subsequently processed into various nanostructures. Sol–gel-derived nanomaterials exhibit high surface areas and surface reactivity, making them effective adsorbents and catalysts for air pollutant removal (Wilson et al. 1994).

(c) *Hydrothermal Synthesis*: Hydrothermal synthesis involves the preparation of nanomaterials under high-temperature and high-pressure aqueous conditions. This method enables the formation of well-defined nanocrystals, nanowires, and nanotubes of metal oxides and other compounds. Hydrothermal synthesis offers advantages such as uniform particle size distribution, crystallinity control, and facile scalability, making it suitable for large-scale production of nanomaterials for air pollution remediation applications (Punetha and Pandey 2019a, b, c, d). Changes in synthesis parameters can optimize the performance charac-

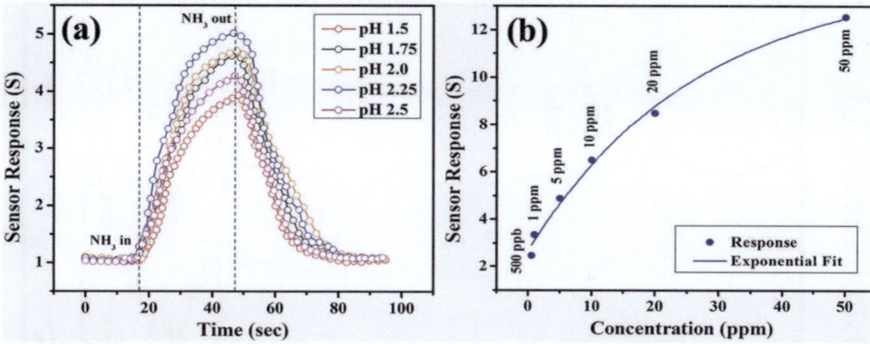

Fig. 4.2 (**a**) Response characteristics of WO_3 nanostructured thin film with pH values (1.5, 1.75, 2, 2.25, and 2.5) toward 5 ppm NH_3 gas concentration, and (**b**) variation in sensor response for different NH_3 gas concentration at pH 2.25 nanostructured WO_3 thin film with 250 °C operating temperatures (Punetha and Pandey 2020)

teristics of gas sensors. As shown in Fig. 4.2a, varying the pH during the synthesis of WO_3 nanomaterials can lead to differences in sensing response. It has also been observed that the sensor response increases exponentially with increasing gas concentration (Fig. 4.2b; Punetha and Pandey 2020).

(d) *Template-Assisted Synthesis*: Template-assisted synthesis involves the use of templates or scaffolds to guide the growth of nanomaterials with specific structures and dimensions. Templates can be organic or inorganic materials with well-defined morphologies, such as micelles, polymers, or porous substrates. By controlling the template properties and synthesis conditions, researchers can fabricate nanostructures tailored for efficient pollutant capture and degradation (Shipu and Lei 2018).

4.2.2 Surface Functionalization and Modification

Surface functionalization plays a crucial role in enhancing the performance and stability of nanomaterials for air pollution remediation. Functional groups such as hydroxyl, carboxyl, amino, and sulfonic acid can be introduced onto the nanomaterial surface to improve adsorption capacity, selectivity, and reactivity toward target pollutants. Additionally, doping with foreign atoms or ions can modify the electronic structure of nanomaterials, thereby enhancing their catalytic activity and photodegradation efficiency (Punetha et al. 2024; Mantri et al. 2023). It has been studied that surface functionalization and modification also affect the sensing response. In Fig. 4.3, the FTIR spectra of tertiary nanocomposites are shown. The bands at 874, 1180, and 1398 cm^{-1} correspond to the C–F stretching vibration, deformation vibrations, and asymmetric stretching of the C–H bond, respectively. The band at 839 cm^{-1} is attributed to a mixed mode of CF_2 asymmetric stretching and CH_2

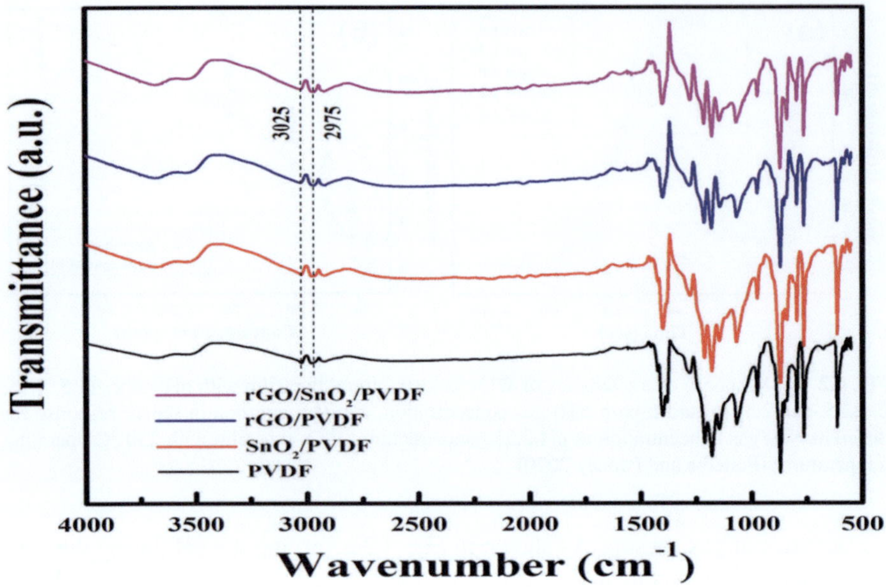

Fig. 4.3 ATR-FTIR spectra of PVDF and its nano-composition with rGO and SnO_2 for the region of 500–4000 cm^{-1} (Punetha et al. 2020)

rocking. The bands near 2975 and 3025 cm^{-1} correspond to the symmetric and asymmetric vibrations of CH_2, respectively. Additional peaks appear at 572 cm^{-1} (Sn–O–Sn stretching), 1067 cm^{-1} (C–O–C stretching), 1215 cm^{-1} (C–O stretching), and 1596 cm^{-1} (O–H bending). A broad peak at 3505 cm^{-1} is associated with O–H stretching of carboxyl groups, while peaks at 760 and 974 cm^{-1} are characteristic of the α-phase of PVDF (Punetha et al. 2020). It has also been observed that varying the nanocomposite composition affects the sensor response, as illustrated in Fig. 4.4.

4.2.3 Characterization Techniques

Characterization of nanomaterials is essential for understanding their structure–property relationships and optimizing their performance for air pollution remediation. Various analytical techniques such as transmission electron microscopy (TEM), scanning electron microscopy (SEM), X-ray diffraction (XRD), Fourier-transform infrared spectroscopy (FTIR), and Brunauer–Emmett–Teller (BET) surface area analysis are commonly employed to investigate morphology, crystal structure, surface chemistry, and porosity of nanomaterials (Punetha and Pandey 2019c).

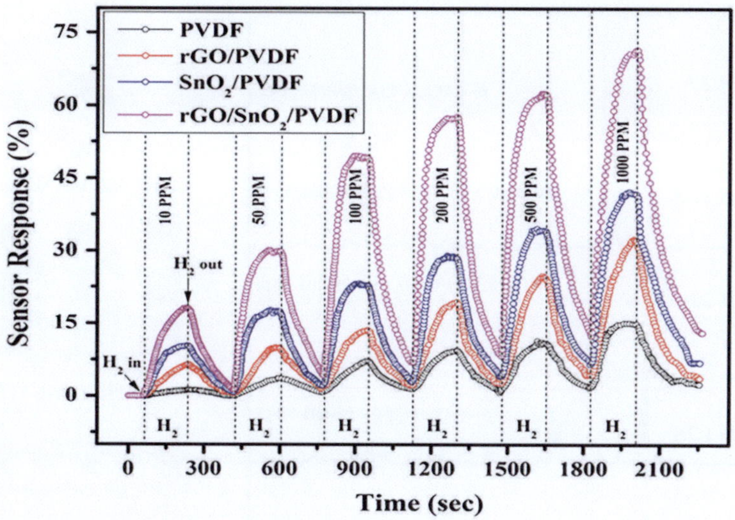

Fig. 4.4 Sensor response of PVDF, rGO/PVDF, SnO_2/PVDF, and rGO/SnO_2/PVDF gas sensor for different H_2 gas concentrations (Punetha et al. 2020)

4.2.4 Applications and Challenges

Nanomaterials hold great potential for diverse applications in air pollution remediation, including catalytic conversion, adsorption, photocatalysis, and filtration. However, several challenges must be addressed to facilitate their practical implementation, including scalability, cost-effectiveness, long-term stability, and environmental safety (Kumar et al. 2023; Punetha and Pandey 2019c, d). Continued research efforts are needed to develop scalable synthesis methods, improve material performance, and ensure the sustainable deployment of nanotechnology-based solutions for air quality improvement. Different crystal structures of WO_3 nanomaterials have been synthesized (Fig. 4.5), and it has been observed that sensor response is also affected by the synthesis mechanism/change in crystal structure (Punetha and Pandey 2020). The variation in corresponding sensor response with respect to different crystal structures is shown in Fig. 4.6.

The synthesis of nanomaterials tailored for air pollution remediation represents a promising avenue for addressing the growing challenges associated with urban and industrial emissions. By leveraging the unique properties of nanomaterials and advancing synthesis techniques, researchers can develop efficient and sustainable solutions for mitigating air pollution and safeguarding human health and the environment. Continued interdisciplinary collaboration and innovation are essential to realize the full potential of nanotechnology in combating air pollution on a global scale (Punetha et al. 2018a, b, 2023).

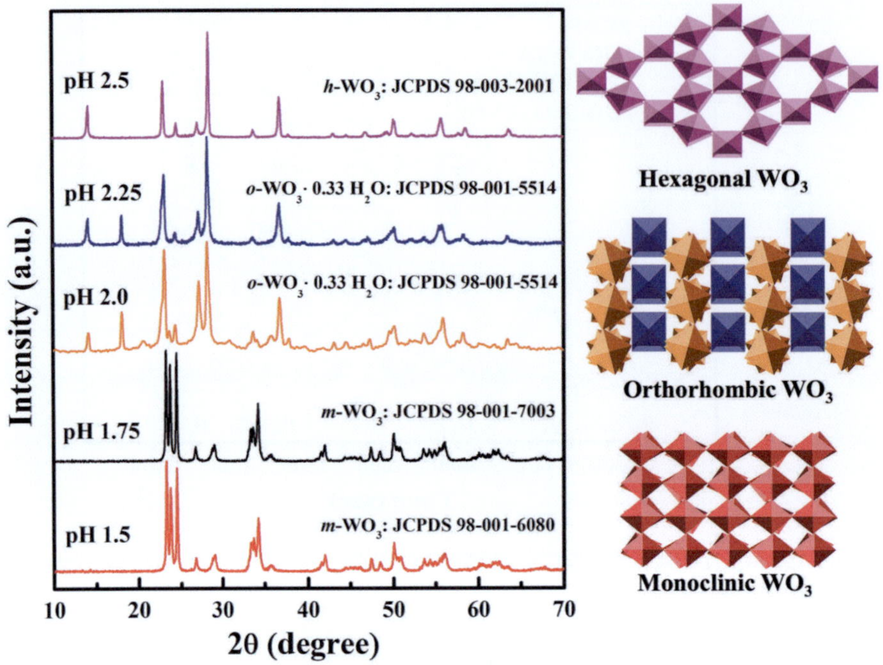

Fig. 4.5 XRD pattern of WO_3 nanoparticles for different pH values (1.5, 1.75, 2, 2.25, and 2.5) and their crystalline structures (Punetha and Pandey 2020)

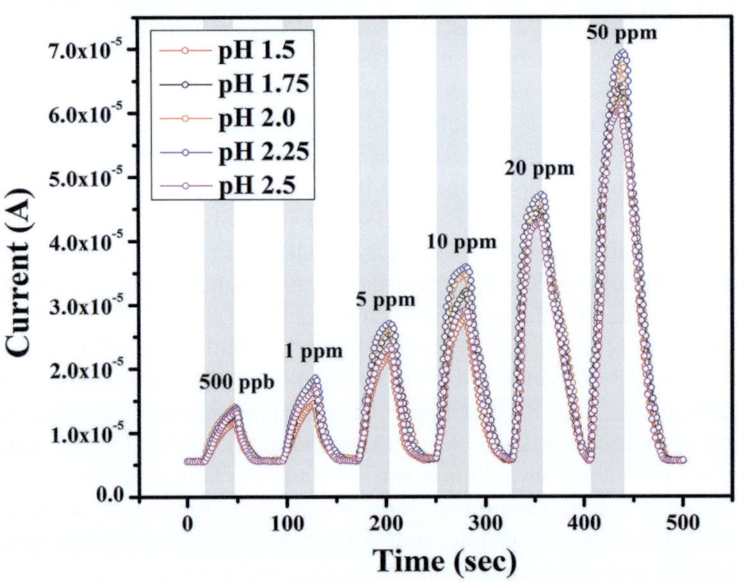

Fig. 4.6 Current configuration of WO_3 nanostructured thin films synthesized at different pH values (1.5, 1.75, 2, 2.25, and 2.5) for NH_3 gas concentrations ranging from 0.5 to 50 ppm, measured at an operating temperature of 250 °C (Punetha and Pandey 2020)

4.3 Mechanisms of Pollutant Removal

Different nanomaterials employ diverse mechanisms for the removal of air pollutants, including physical adsorption, chemical reactions, photocatalysis, and filtration. Carbon-based nanomaterials, such as carbon nanotubes and graphene, exhibit high surface area and adsorption capacity, making them effective adsorbents for gases and volatile organic compounds (VOCs). Metal and metal oxide nanoparticles, such as titanium dioxide (TiO_2) and zinc oxide (ZnO), possess photocatalytic properties that enable the degradation of pollutants under UV irradiation (Punetha and Pandey 2019b; Williams 1999; Ghosh and Majumder 2017). Nanostructured materials, such as zeolites and mesoporous silica nanoparticles, offer selective adsorption and catalytic degradation of specific pollutants.

Gas sensors play a crucial role in various industries, from environmental monitoring to industrial safety. However, the accuracy and reliability of these sensors can be compromised by the presence of pollutants in the environment. Pollutants such as dust, smoke, volatile organic compounds (VOCs), and gases can interfere with sensor readings, leading to inaccurate measurements and potential safety hazards (Eranna 2016). Therefore, understanding the mechanisms of pollutant removal is essential for optimizing the performance of gas sensors.

Before delving into the mechanisms of pollutant removal, it's important to categorize the types of pollutants that gas sensors encounter. These pollutants can be broadly classified into particulate matter, gaseous compounds, and volatile organic compounds (VOCs). Particulate matter includes dust, smoke, and other solid particles suspended in the air. Gaseous compounds encompass various gases such as carbon monoxide (CO), nitrogen dioxide (NO_2), sulfur dioxide (SO_2), and ozone (O_3). VOCs consist of organic chemicals that can easily vaporize into the air, including solvents, fuels, and industrial chemicals (Saxena and Shukla 2023).

4.3.1 Filtration

One of the primary mechanisms for removing pollutants in gas sensor applications is filtration. Filtration involves the physical removal of particulate matter from the air stream before it reaches the sensor. This can be achieved through various filtration techniques such as mechanical filters, electrostatic precipitators, and high-efficiency particulate air (HEPA) filters. Mechanical filters utilize porous materials to trap particles based on size, while electrostatic precipitators charge particles and collect them on oppositely charged plates. HEPA filters, known for their high efficiency, can capture particles as small as 0.3 micrometers (Afrin and Shah 2015).

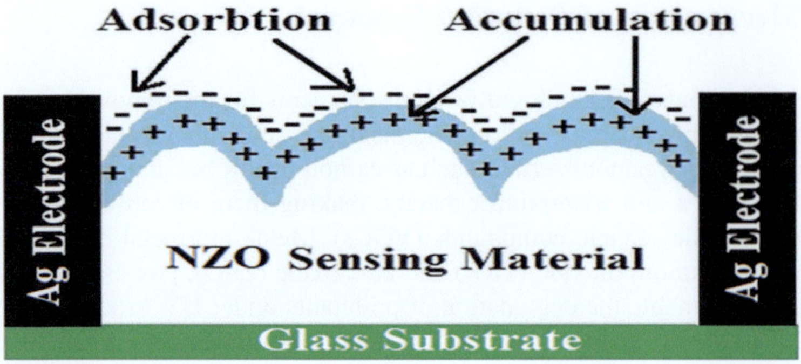

Fig. 4.7 The adsorption and accumulation phenomena during the gas sensing process (Punetha et al. 2018a, b)

4.3.2 Adsorption

Adsorption is another effective mechanism for pollutant removal in gas sensor applications, particularly for gaseous compounds and VOCs. Adsorption involves the adhesion of gas molecules onto the surface of a solid adsorbent material. Common adsorbents used in gas sensor applications include activated carbon, zeolites, and molecular sieves. These materials have high surface areas and affinity for specific gases, allowing them to selectively remove pollutants from the air stream (Punetha et al. 2018a, b; Punetha and Singh 2018, 2019). The adsorbed pollutants can later be desorbed from the adsorbent surface for disposal or regeneration (Fig. 4.7).

4.3.3 Chemical Reactions

Chemical reactions play a vital role in pollutant removal, especially for reactive gases and VOCs. Gas sensors can incorporate catalytic materials that facilitate chemical reactions to convert harmful gases into less toxic or inert substances. For instance, catalytic converters in automotive exhaust systems utilize noble metal catalysts such as platinum and palladium to catalyze the conversion of carbon monoxide and nitrogen oxides into carbon dioxide and nitrogen gas, respectively. Similarly, gas sensors can employ catalysts to oxidize or reduce VOCs and other reactive gases, improving sensor accuracy and reducing environmental impact (Fig. 4.8) (Punetha and Pandey 2019d; Punetha et al. 2018a, b).

4.3.4 Thermal Treatment

Thermal treatment is another effective method for pollutant removal in gas sensor applications, particularly for volatile organic compounds (VOCs). Thermal treatment involves heating the contaminated air stream to high temperatures to thermally

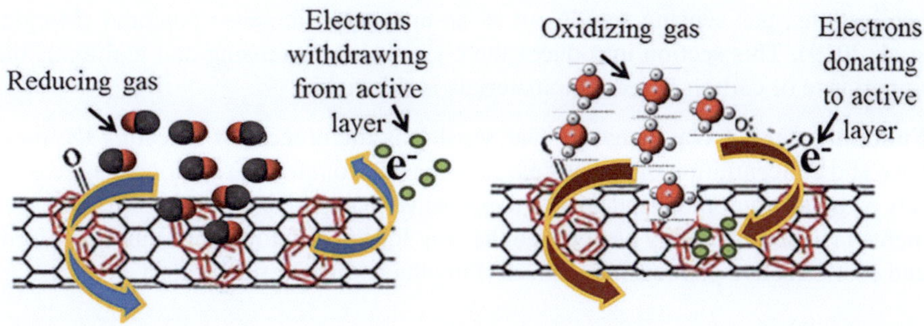

Fig. 4.8 The illustrated mechanism of charge transfer reaction associated with gas sensing (Afrin and Shah 2015)

decompose or vaporize the pollutants. This process can be achieved through thermal oxidizers, catalytic incinerators, or thermal desorption units. By subjecting the pollutants to high temperatures, thermal treatment can effectively eliminate VOCs and other volatile contaminants, ensuring clean air reaches the gas sensor for accurate measurements (Segev-Bar and Haick 2013; Dubinin 1960).

Understanding the mechanisms of pollutant removal is essential for optimizing the performance and reliability of gas sensors in various applications. Filtration, adsorption, chemical reactions, and thermal treatment are among the key mechanisms employed to remove pollutants and ensure accurate sensor readings. By integrating these pollutant removal techniques into gas sensor design and operation, researchers and engineers can enhance sensor performance, mitigate environmental impact, and improve overall air quality monitoring capabilities. Continued research and innovation in pollutant removal technologies will further advance the effectiveness and efficiency of gas sensor systems in the future.

4.4 Applications for Different Nanomaterials

4.4.1 Carbon-Based Nanomaterials

Carbon-based nanomaterials have garnered significant attention in recent years due to their unique properties and wide-ranging applications in various fields, including gas sensing. This comprehensive review explores the role of carbon-based nanomaterials as gas sensors, focusing on their synthesis methods, structural properties, sensing mechanisms, and applications. The review covers carbon nanotubes (CNTs), graphene, carbon nanofibers (CNFs), carbon quantum dots (CQDs), and other emerging carbon-based materials. Additionally, it discusses the challenges and prospects in the development of carbon-based nanomaterials for gas sensing applications. Carbon-based nanomaterials have garnered significant attention in recent years due to their unique properties and wide range of applications. Among these applications, gas sensing stands out as an area with immense potential (Mingfei et al. 2019). This section introduces the concept of gas sensing and highlights the importance of carbon-based nanomaterials in this field.

Fundamentals of Gas Sensing Gas sensing is the process of detecting the presence and concentration of various gases in the environment. Traditional gas sensors rely on bulky equipment and are often expensive to manufacture. However, advancements in nanotechnology have paved the way for the development of more efficient and cost-effective gas sensing devices (Punetha and Pandey 2019c, d).

Properties of Carbon-Based Nanomaterials Carbon-based nanomaterials, including carbon nanotubes (CNTs), graphene, and carbon nanofibers (CNFs), possess unique properties that make them ideal candidates for gas sensing applications. These properties include high surface area, excellent conductivity, and sensitivity to gas molecules (Punetha and Pandey 2019b).

Mechanisms of Gas Sensing This section delves into the mechanisms through which carbon-based nanomaterials detect gas molecules. The interaction between gas molecules and the surface of the nanomaterials leads to changes in conductivity or other physical properties, which can be measured and analyzed to determine the presence and concentration of the gas. The transport mechanism of charge species is shown in Fig. 4.9 by using an energy band diagram, in which Fig. 4.9 corresponds

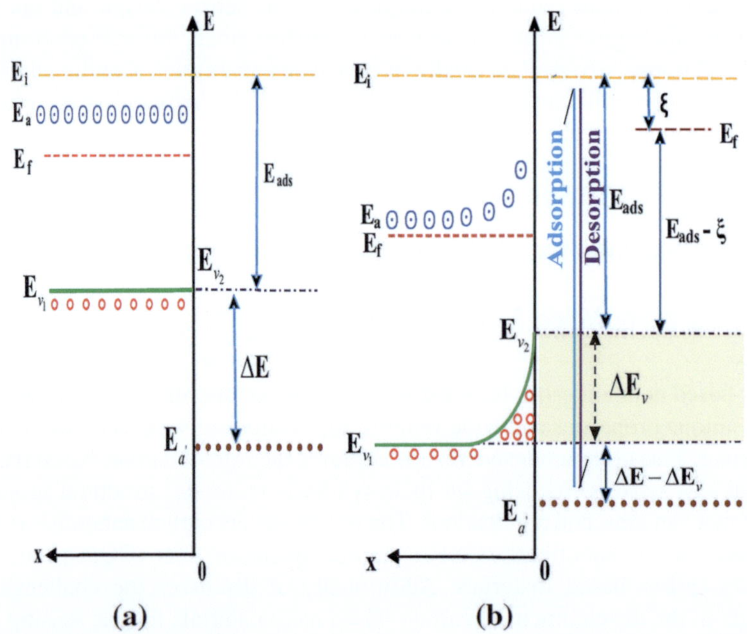

Fig. 4.9 Energy band diagram for semiconductor thin film (**a**) without adsorption, and (**b**) during adsorption and desorption (Punetha et al. 2018a, b)

to a semiconductor thin film (a) without adsorption, and (b) during adsorption and desorption (Punetha et al. 2018a, b).

Types of Gas Sensors Utilizing Carbon-Based Nanomaterials There are various types of gas sensors that utilize carbon-based nanomaterials, including resistive, capacitive, and field-effect transistor (FET) sensors. Each type has its own advantages and limitations, depending on factors such as sensitivity, selectivity, and response time.

Applications of Carbon-Based Nanomaterial Gas Sensors The versatility of carbon-based nanomaterial gas sensors makes them suitable for a wide range of applications. These include environmental monitoring, industrial safety, healthcare diagnostics, and even military and defense purposes. The ability to detect gases with high sensitivity and selectivity makes them invaluable in ensuring public health and safety (Punetha and Pandey 2020; Nomura et al. 2023; Fine et al. 2010). The illustrated interaction mechanism of carbon-based nanomaterial with metal oxide and piezoelectric polymer is shown in Fig. 4.10. It signifies that carbon-based nanomaterial can also be used as a flexible sensor (E-skin) and self-powered gas sensing device (Punetha et al. 2020, 2024). It has also been observed (Fig. 4.11) that sensor response changes by changing the composition of carbon-based nanomaterial, which signifies the importance of carbon nanomaterial in gas sensing.

Challenges and Future Perspectives Despite their promising potential, carbon-based nanomaterial gas sensors still face certain challenges, such as scalability, reproducibility, and selectivity toward specific gas molecules. Addressing these challenges requires further research and development efforts. However, with ongo-

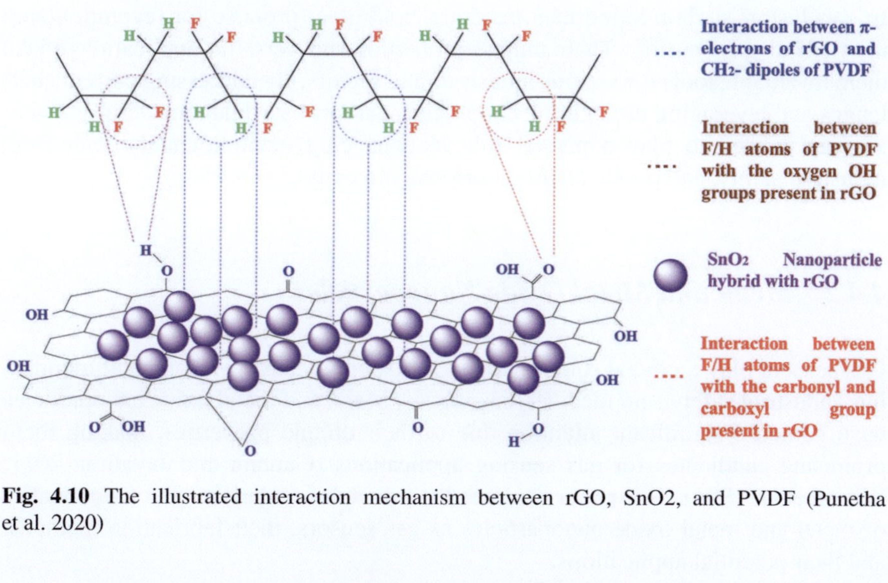

Fig. 4.10 The illustrated interaction mechanism between rGO, SnO2, and PVDF (Punetha et al. 2020)

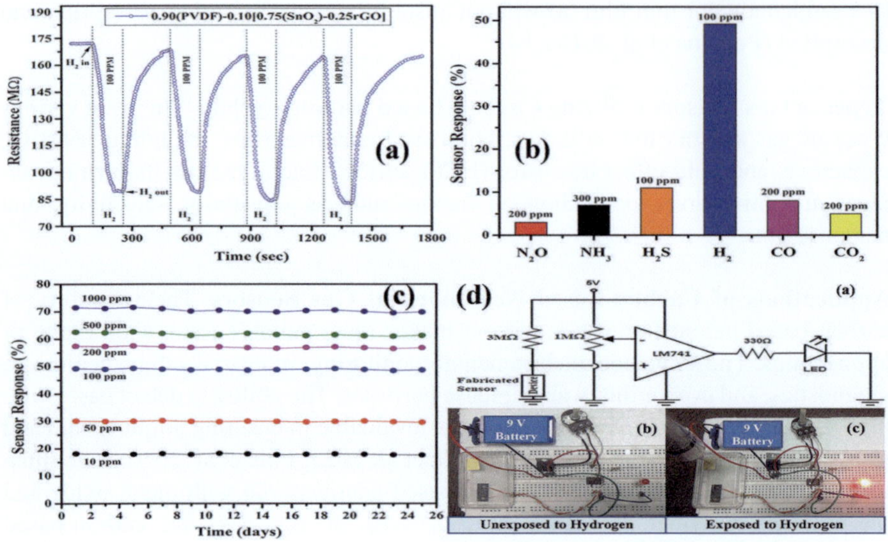

Fig. 4.11 (**a**) Baseline resistance of 0.90(PVDF)–0.10[0.75(SnO2)-0.25rGO] nanocomposite gas sensor at 100 ppm hydrogen gas concentration, (**b**) selectivity of the 0.90(PVDF)–0.10[0.75(SnO 2)-0.25rGO] nanocomposite gas sensor toward various target gases, (**c**) time-dependent stability of the 0.90(PVDF)–0.10[0.75(SnO2)-0.25rGO] nanocomposite sensor for 10, 50, 100, 200, 500, and 1000 ppm H2 gas concentration, and (**d**) real-time demonstration of the proposed sensor, circuit diagram, and the sensor prototype in the absence and presence of the hydrogen gas, respectively (Punetha et al. 2020)

ing advancements in nanotechnology and material science, the future of carbon-based nanomaterial gas sensors looks promising.

In conclusion, carbon-based nanomaterials hold great promise for revolutionizing the field of gas sensing. Their unique properties and versatile applications make them invaluable tools for various industries and sectors. By addressing current challenges and leveraging emerging technologies, carbon-based nanomaterial gas sensors are poised to play a pivotal role in enhancing environmental monitoring, industrial safety, and public health in the years to come.

4.4.2 Metal and Metal Oxide Nanoparticles

Gas sensors play a crucial role in various fields, including environmental monitoring, industrial safety, and medical diagnostics. Metal and metal oxide nanoparticles have garnered significant attention due to their unique properties, making them promising candidates for gas sensing applications (Gautam and Jayatissa 2011; Wilson et al. 1994). This paper explores the principles underlying the functionality of metal and metal oxide nanoparticles as gas sensors, their fabrication methods, and their potential applications.

Fundamentals of Gas Sensing Gas sensing involves the detection and quantification of target gases in the surrounding environment. This process typically relies on the interaction between the gas molecules and the sensing material, leading to measurable changes in electrical, optical, or mechanical properties. In metal oxides, charged species such as O_2^-, O^- and O^{2-} are adsorbed on the metal oxide, charging it negatively. Other gases adsorbing on the surface can give rise to a charge-transfer reaction (Punetha and Pandey 2019a).

If a reducing species, such as CO, is present, it oxidizes and releases a negative charge, which increases the conductivity of the material (Punetha and Pandey 2018). For example:

$$CO_{(g)} + O^-_{(ad)} \rightarrow CO_{2(g)} + e^-$$

$$CO + 2O^- \rightarrow CO_3^{2-} \rightarrow CO_2 + ½O_2 + 2e^-$$

In the presence of an electronegative species such as NO_2, the charges are attracted to the adsorbed molecule, which decreases the conductivity of the material:

$$NO_2 + e^- \rightarrow NO_2^-$$

Properties of Metal and Metal Oxide Nanoparticles Metal nanoparticles exhibit distinct electronic, optical, and catalytic properties compared to their bulk counterparts. These properties are highly sensitive to the presence of gas molecules, making them ideal candidates for gas sensing applications. Similarly, metal oxide nanoparticles possess a high surface-to-volume ratio, which enhances their sensitivity to gas molecules due to increased surface interactions (Williams 1999; Ghosh and Majumder 2017).

Mechanisms of Gas Sensing with Nanoparticles Gas sensing mechanisms involving metal and metal oxide nanoparticles can be categorized into chemisorption-based and physisorption-based mechanisms. Chemisorption involves the formation of chemical bonds between gas molecules and nanoparticle surfaces, leading to changes in electrical conductivity or optical properties. Physisorption, on the other hand, involves the adsorption of gas molecules onto the nanoparticle surface through weak physical interactions (Fig. 4.12), resulting in changes in surface morphology or mass (Punetha and Pandey 2019a).

Fabrication Methods Several techniques are employed for the fabrication of metal and metal oxide nanoparticles suitable for gas sensing applications. These include chemical synthesis methods such as sol–gel, hydrothermal, and chemical vapor deposition (CVD), as well as physical techniques such as laser ablation and sputtering. Each method offers unique advantages in terms of particle size control, uniformity, and scalability (Fig. 4.13) (Huixia et al. 2015; Gautam and Jayatissa 2011; Wilson et al. 1994).

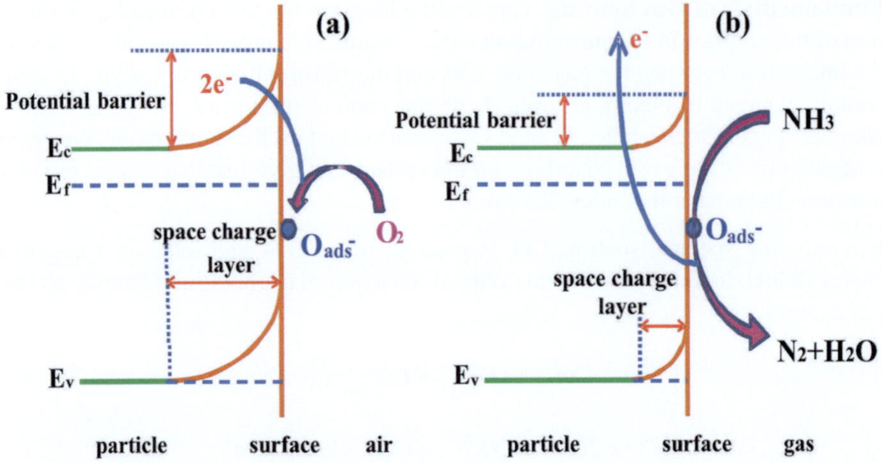

Fig. 4.12 Gas sensing mechanism for WO_3-based sensor in the (a) absence and (b) presence of NH_3 gas

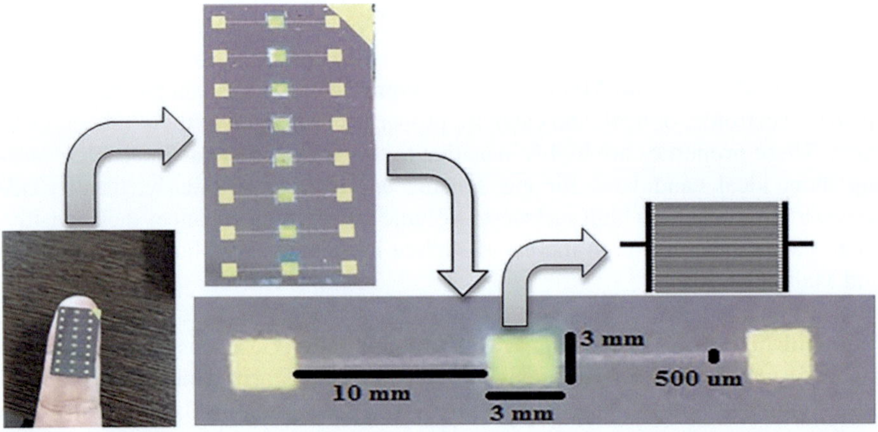

Fig. 4.13 The pictorial image of a fabricated gas sensor by using photolithography (Punetha and Pandey 2019a)

Surface Functionalization Surface functionalization plays a crucial role in enhancing the selectivity and sensitivity of metal and metal oxide nanoparticle-based gas sensors. Functionalization can be achieved through the deposition of organic or inorganic coatings, doping with foreign atoms, or modifying surface morphology (Saxena and Shukla 2023; Afrin and Shah 2015). These strategies enable the tailored design of sensors for specific gas detection applications (Fig. 4.14).

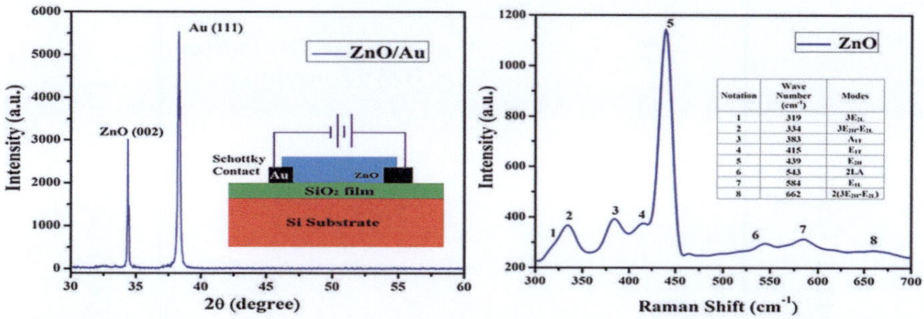

Fig. 4.14 The structural parameters of ZnO thin film, such as crystallite size and FWHM (full width at half maximum), have been calculated by using Scherrer's formula, and values are 71.37 nm and 0.13495°, respectively (Punetha and Pandey 2019a)

Gas Sensing Performance The performance of metal and metal oxide nanoparticle-based gas sensors is evaluated based on parameters such as sensitivity, selectivity, response time, and stability. Sensitivity refers to the ability of the sensor to detect low concentrations of target gases, while selectivity determines its ability to discriminate between different gas species. Response time and stability are critical for real-time monitoring applications and long-term reliability, respectively (Saxena and Shukla 2023; Afrin and Shah 2015).

Applications Metal and metal oxide nanoparticle-based gas sensors find applications in various fields, including environmental monitoring, industrial safety, healthcare, and homeland security. They are used for detecting a wide range of gases such as carbon monoxide, nitrogen dioxide, hydrogen sulfide, and volatile organic compounds (Huixia et al. 2015; Gautam and Jayatissa 2011; Wilson et al. 1994). Their miniaturized size, low power consumption, and compatibility with integrated circuitry make them suitable for portable and wearable sensor platforms.

Metal and metal oxide nanoparticles hold great promise for the development of highly sensitive and selective gas sensors. Their unique properties, coupled with advances in fabrication techniques and surface functionalization strategies, enable the design of tailored sensors for diverse applications (Eranna 2016). Continued research efforts in this field are expected to lead to further improvements in gas sensing performance and the widespread adoption of nanoparticle-based sensor technologies.

4.4.3 Nanostructured Materials

Gas sensors play a vital role in various industries and applications, including environmental monitoring, industrial safety, and healthcare. Nanostructured materials have garnered significant attention in recent years for their potential in enhancing

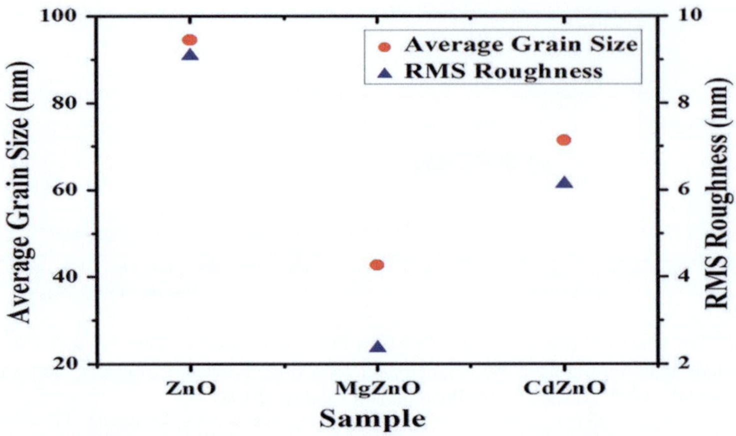

Fig. 4.15 Average grain size and RMS roughness for different samples having a surface area of 1 μm (Punetha and Pandey 2018)

gas sensing performance (Williams 1999; Ghosh and Majumder 2017). Figure 4.15 shows the average grain size and RMS roughness of different samples, all measured over a fixed surface area of 1 μm. This article explores the principles behind gas sensing, the advantages of nanostructured materials, their synthesis methods, and their applications in gas sensor technology.

Gas sensing involves the detection and measurement of specific gases or vapors in the surrounding environment. This process relies on the interaction between the gas molecules and the sensing material, resulting in measurable changes in electrical, optical, or mass-based properties (Punetha and Pandey 2019d). Typically, gas sensors operate based on principles such as chemiresistive, conductometric, optical, or surface acoustic wave (SAW) sensing.

Advantages of Nanostructured Materials Nanostructured materials offer several advantages over conventional sensing materials. Their high surface-to-volume ratio provides a large active surface area for gas interaction, leading to enhanced sensitivity and selectivity. Additionally, their tunable properties at the nanoscale allow for precise control over sensing performance, including response time, recovery time, and detection limits. Moreover, nanostructured materials exhibit improved stability and robustness, making them suitable for long-term sensing applications. In Fig. 4.16, the FESEM micrographs and corresponding EDS spectra of rGO/metal oxide nanocomposite are shown, and it is shown (Fig. 4.17) that sensor response toward high surface to volume ratio results in the maximum sensitivity. The sensing device has been tested for different time intervals, resulting in good stability (deviation <4%), as shown in Fig. 4.18.

Nanostructured materials utilized in gas sensors encompass a wide range of materials, including metal oxides (e.g., ZnO, SnO_2, TiO_2), carbon-based materials (e.g., carbon nanotubes, graphene), metal nanoparticles (e.g., Pt, Pd, Au), and 2D

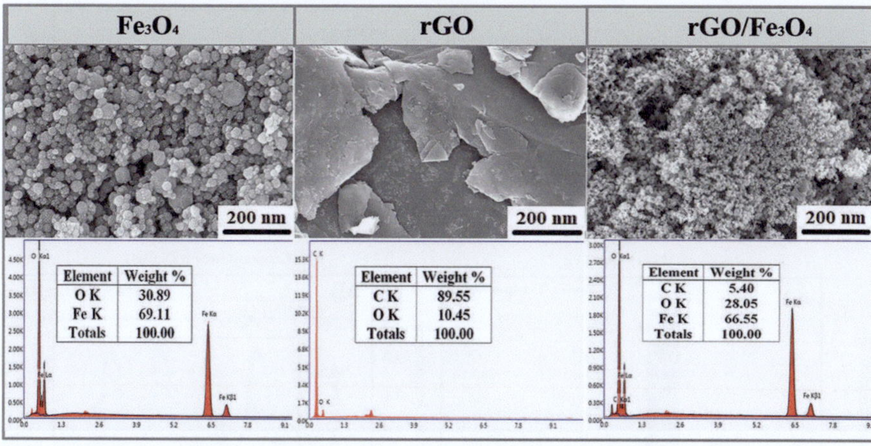

Fig. 4.16 FESEM micrographs and corresponding EDS spectra of Fe_3O_4, rGO, and rGO/Fe3O4 nanocomposite at a scale bar of 200 nm (Punetha and Pandey 2019c)

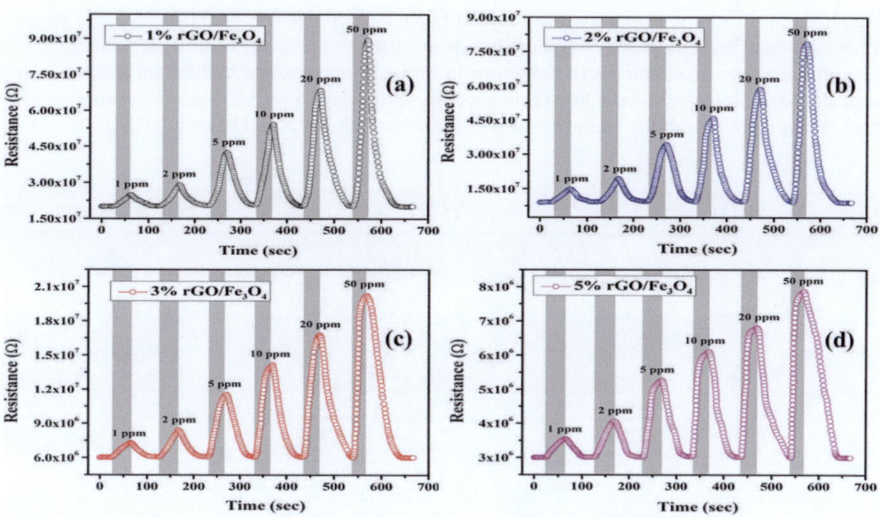

Fig. 4.17 Baseline resistance of (a) 1% rGO, (b) 2% rGO, (c) 3% rGO, and (d) 5% rGO-based rGO/Fe_3O_4 nanocomposite gas sensor toward 1–50 ppm CO gas concentration at room temperature

materials (e.g., MoS_2, WS_2). Each material type exhibits distinct sensing mechanisms and properties, offering versatility in sensor design and application (Eranna 2016). The structural analysis of a nanocomposite of metal oxide and 2D material is shown in Fig. 4.19 through FESEM micrographs (Punetha and Pandey 2019b).

Gas Sensing Mechanisms The gas sensing mechanisms of nanostructured materials vary depending on their composition and structure. For metal oxide-based sen-

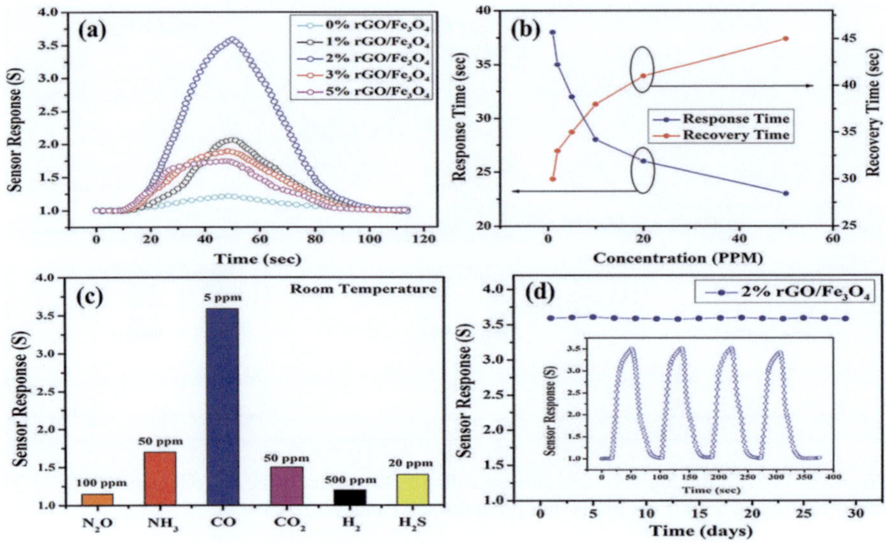

Fig. 4.18 (**a**) Response characteristics for rGO/Fe_3O_4 nanocomposites with 0%, 1%, 2%, 3%, and 5% rGO weight percentage, (**b**) variation in response time and recovery time toward change in CO gas concentration, (**c**) selectivity of the sensor toward different gases with different gas concentration, and (**d**) stability of the rGO/Fe_3O_4 nanocomposite toward 5 ppm CO gas concentration. The inset shows the repeatability characteristics of the sensor (Punetha and Pandey 2019c)

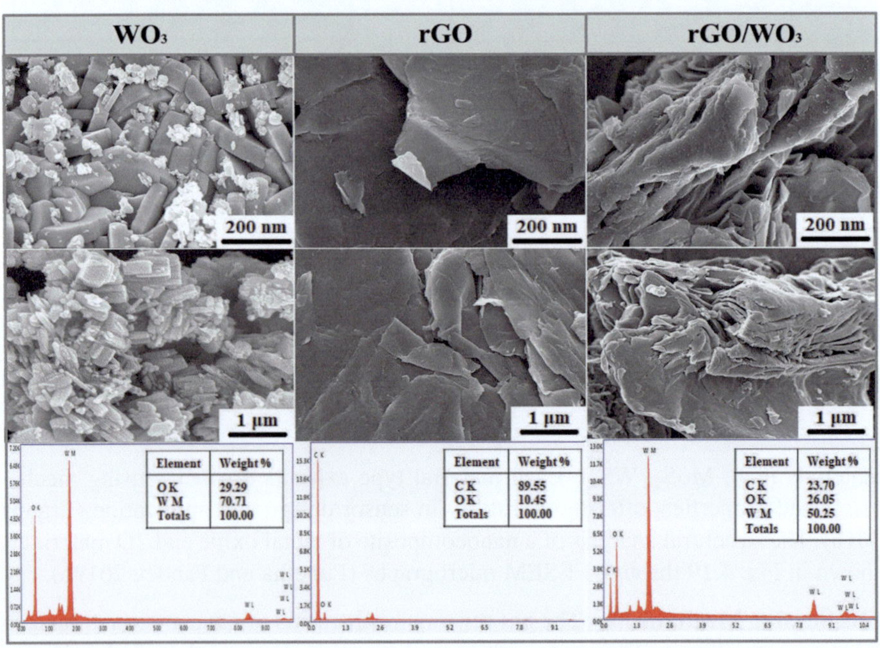

Fig. 4.19 FESEM micrographs and corresponding EDS spectra of WO_3, rGO, and rGO/WO_3 nanocomposite at a scale bar of 200 nm and 1 µm (Punetha and Pandey 2019b)

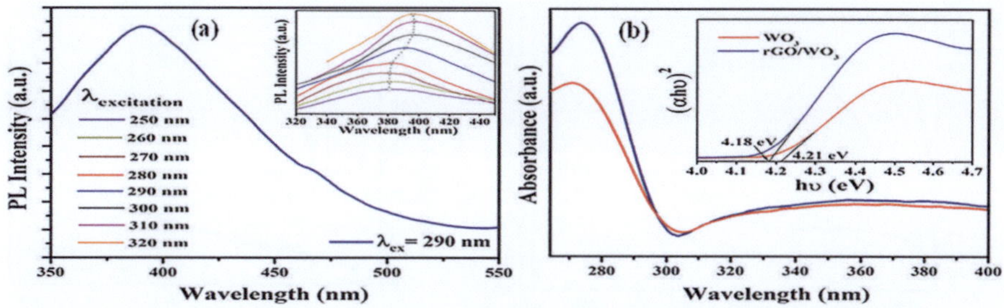

Fig. 4.20 (**a**) Room-temperature PL spectra of WO_3 nanomaterial corresponding to 290 nm excitation wavelength, and the inset shows the shifting of NBE corresponding to variation in excitation wavelength; (**b**) UV-vis absorbance spectra, and the inset shows the Tauc plots to determine the optical band gap of WO_3 nanomaterial and rGO/WO_3 nanocomposite (Punetha and Pandey 2019b)

sors, gas adsorption leads to changes in the material's conductivity or surface potential due to charge transfer processes. In carbon-based sensors, gas molecules interact with the graphene or carbon nanotube surface, altering the electrical conductivity through charge transfer or chemical reactions. Understanding these mechanisms is crucial for optimizing sensor performance and selectivity. The optical characteristics of carbon-based sensors have been shown through PL spectra and UV-vis analysis in Fig. 4.20. Furthermore, the thermal stability of the nanocomposite in the air by using TGA, DTG, and DTA analysis for decomposition in the temperature range of 30–800 °C, and nitrogen adsorption–desorption isotherm by the BJH (Barrett, Joyner, and Halenda) method from the desorption branch of carbon-based nanocomposite are shown in Fig. 4.21 (Punetha and Pandey 2019b). It is observed from Fig. 4.22 that the sensor shows good selectivity toward different gases with different gas concentrations. The variation in response time and recovery time toward change in gas concentration and humidity response toward different operating temperature has also been studied.

Applications of Nanostructured Gas Sensors Nanostructured gas sensors find applications in various fields, including environmental monitoring, industrial safety, healthcare, and automotive systems. They are used for detecting a wide range of gases, such as nitrogen dioxide (NO_2), carbon monoxide (CO), hydrogen (H_2), methane (CH_4), and volatile organic compounds (VOCs). Additionally, nanostructured gas sensors hold promise for emerging applications, such as wearable sensors and internet-of-things (IoT) devices, due to their miniaturization and low power consumption (Figs. 4.23 and 4.24). Furthermore, the sensing of nanostructure-based gas sensors can increase by forming an array (Figs. 4.25 and 4.26). Array configurations (series and parallel) have been made to study the impact of nanostructured-based array devices on gas sensing characteristics.

Nanostructured materials have revolutionized gas sensor technology by offering enhanced sensitivity, selectivity, and stability. Through precise synthesis methods and understanding of gas sensing mechanisms, researchers continue to develop

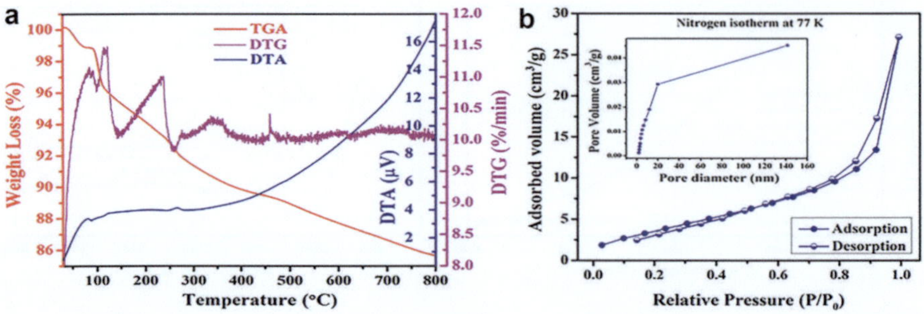

Fig. 4.21 (**a**) Thermal stability of the nanocomposite in the air by using TGA, DTG, and DTA analysis for decomposition in the temperature range of 30–800 °C; (**b**) nitrogen adsorption–desorption isotherm. The inset shows the pore-size distribution calculated by the BJH (Barrett, Joyner, and Halenda) method from the desorption branch of rGO/WO_3 nanocomposite (Punetha and Pandey 2019b)

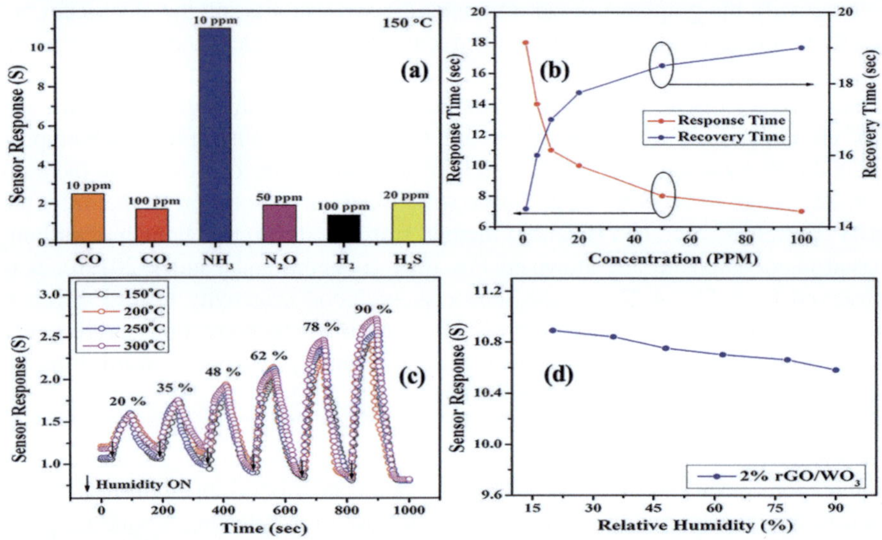

Fig. 4.22 (**a**) Selectivity of the sensor toward different gases with different gas concentration, (**b**) variation in response time and recovery time toward change in NH_3 gas concentration at 150 °C operating temperature, (**c**) humidity response toward different operating temperature, and (**d**) sensing performance of the sensor toward 10 ppm NH_3 exposure under different humidity conditions (Punetha and Pandey 2019b)

novel nanostructured materials tailored for specific sensing applications (Mingle et al. 2019; Zhang et al. 2016). As advancements in nanotechnology and material science progress, nanostructured gas sensors are poised to play an increasingly important role in addressing societal and environmental challenges.

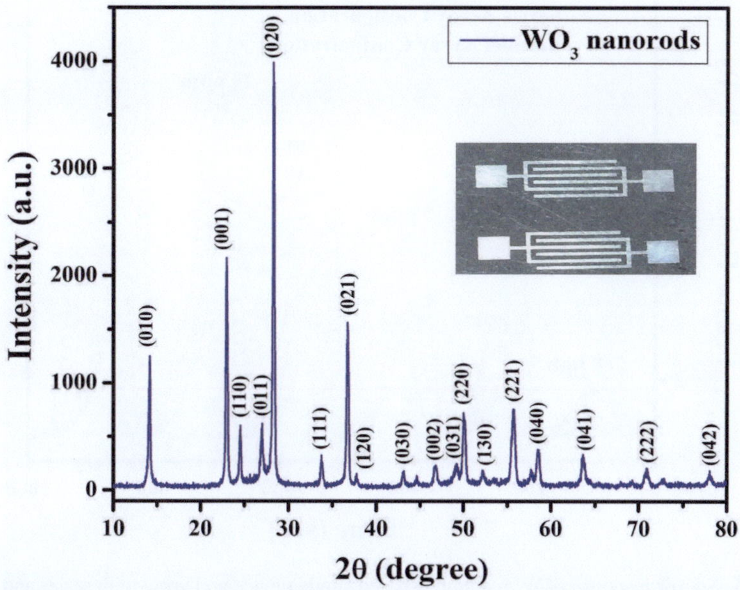

Fig. 4.23 XRD pattern of WO_3 nanorods. The inset shows the fabricated sensor (Punetha and Pandey 2019d)

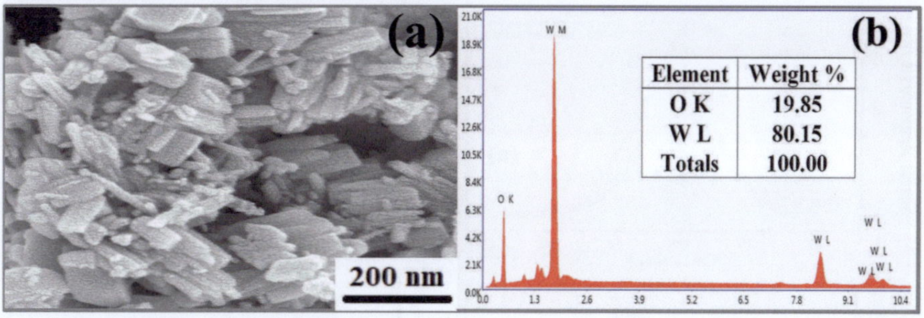

Fig. 4.24 (a) FESEM micrographs and (b) EDS spectra of WO_3 nanorod at a scale bar of 200 nm (Punetha and Pandey 2019d)

4.5 Environmental Implications

While nanomaterials offer promising solutions for air pollution remediation, their widespread application raises concerns regarding potential environmental impacts. The release of nanoparticles into the environment may have unintended consequences on ecosystems and human health. Nanotoxicity, bioaccumulation, and long-term environmental persistence are among the key challenges associated with the use of nanomaterials in air pollution control (Nanshu and Zhou 2017). Therefore, comprehensive risk assessments and mitigation strategies are essential to ensure the safe and sustainable deployment of nanotechnology for environmental remediation.

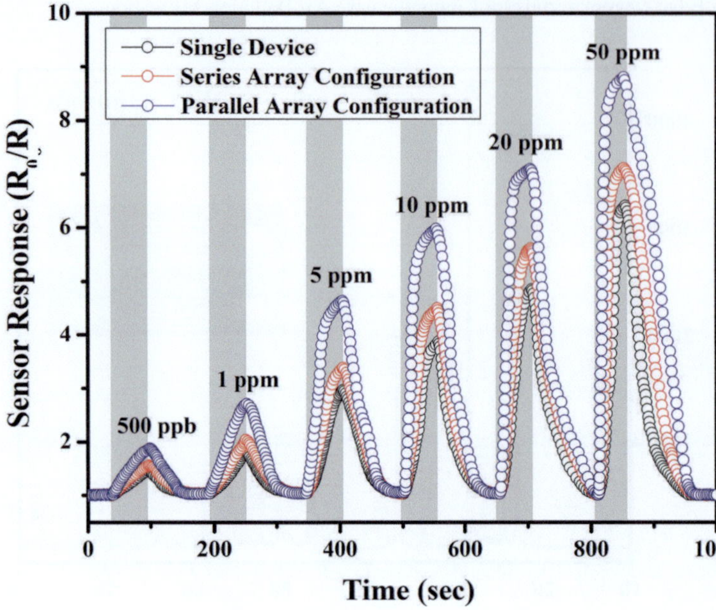

Fig. 4.25 Sensor response of WO_3 nanorod-based single device and array with series and parallel configurations gas sensor toward 500 ppb–50 ppm NH_3 gas concentration at 250 °C operating temperature (Punetha and Pandey 2019d)

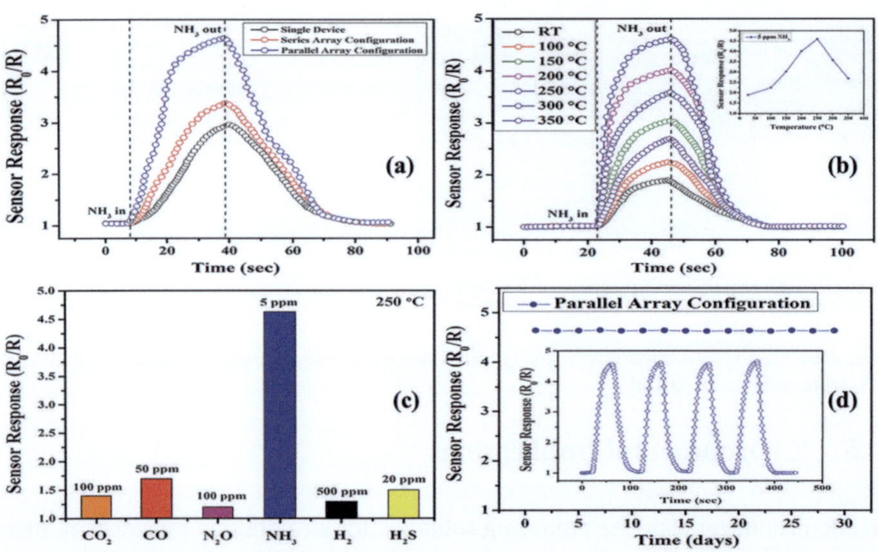

Fig. 4.26 (**a**) Response characteristics for WO_3 nanorod-based sensor array, (**b**) variation in sensor response for different operating temperature toward 5 ppm NH_3 gas concentration for parallel array configuration, (**c**) selectivity of the sensor toward different gases with different gas concentration, and (**d**) stability of the WO_3 nanorod-based parallel array configuration toward 5 ppm NH_3 gas concentration. The inset shows the repeatability characteristics of the sensor (Punetha and Pandey 2019d)

Gas sensors play a crucial role in monitoring air quality, detecting pollutants, and ensuring safety in various environments. However, the environmental implications associated with gas sensor technology are multifaceted. This comprehensive discussion delves into the environmental considerations related to the production, usage, and disposal of gas sensors, highlighting both their benefits and potential drawbacks.

4.5.1 Manufacturing Processes

The environmental impact of gas sensors begins with their manufacturing processes. These processes typically involve the use of various materials, including metals, semiconductors, and polymers. Mining and processing these raw materials can lead to habitat destruction, water pollution, and greenhouse gas emissions. Additionally, the energy-intensive nature of manufacturing contributes to carbon emissions and exacerbates climate change. To mitigate these environmental impacts, manufacturers must adopt sustainable practices, such as using recycled materials, reducing energy consumption, and implementing efficient waste management systems (Korotcenkov 2013; Arshak and Gaidan 2005; Dai et al. 2020; Chan et al. 2002).

4.5.2 Energy Consumption

Gas sensors require energy to operate, whether they are deployed in industrial settings, vehicles, or smart home devices (Punetha et al. 2024). The continuous operation of gas sensors consumes electricity, which predominantly comes from fossil fuel-based power plants. This energy consumption contributes to air pollution, carbon emissions, and resource depletion. However, advancements in sensor technology have led to the development of low-power sensors and energy-efficient algorithms, reducing their environmental footprint. Furthermore, integrating renewable energy sources, such as solar or wind power, can further mitigate the environmental impact of gas sensor operation (Baranov et al. 2015; Nazemi et al. 2019).

4.5.3 Chemicals and Hazardous Materials

Gas sensors often contain chemicals and hazardous materials that pose risks to human health and the environment. For instance, some sensors use toxic gases, such as lead, mercury, or arsenic, in their construction. Improper handling or disposal of these sensors can result in soil and water contamination, as well as adverse health effects for workers and local communities (Ly et al. 2021; Huang et al. 2022). To address this issue, manufacturers must adhere to strict regulations regarding the use and disposal of hazardous materials. Additionally, research efforts are underway to develop alternative sensor materials that are both effective and environmentally benign.

4.5.4 Lifespan and Maintenance

The lifespan of gas sensors varies depending on factors such as usage patterns, environmental conditions, and quality of materials. Over time, sensors may degrade or become less accurate, requiring maintenance or replacement. The disposal of old or malfunctioning sensors presents environmental challenges, as they may contain non-biodegradable components or hazardous substances (Punetha and Singh 2018, 2019). Extending the lifespan of gas sensors through regular maintenance, calibration, and refurbishment can reduce the frequency of disposal and minimize environmental impact. Furthermore, incorporating modular designs and recyclable materials into sensor construction facilitates easier repair and recycling at the end of their lifecycle (Yadav and Indurkar 2021).

4.5.5 End-of-Life Disposal

Proper disposal of gas sensors is crucial to prevent environmental contamination and reduce waste. However, many sensors end up in landfills, where they can leach toxic substances into the soil and groundwater. To address this issue, manufacturers and regulatory agencies must promote responsible disposal practices, such as recycling or incineration with energy recovery (Leanne et al. 2014). Recycling programs can recover valuable metals and materials from sensors, reducing the demand for virgin resources and minimizing environmental degradation. Additionally, designing sensors for easy disassembly and recycling can incentivize consumers to participate in recycling initiatives.

Gas sensors play a vital role in environmental monitoring and safety applications, but their production, usage, and disposal have significant environmental implications. By implementing sustainable manufacturing practices, reducing energy consumption, minimizing the use of hazardous materials, extending sensor lifespan, and promoting responsible disposal, stakeholders can mitigate the environmental impact of gas sensors. Continued research and innovation are essential to developing environmentally friendly sensor technologies that effectively safeguard both human health and the environment.

4.6 Conclusion

The application of nanomaterials in air pollution remediation represents a rapidly evolving field with significant potential to address the complex challenges posed by urban and industrial emissions. By harnessing the unique properties of nanomaterials, such as high surface area, catalytic activity, and selective adsorption, innovative air purification technologies can be developed to improve air quality and safeguard public health. However, it is imperative to consider the environmental implications and risks associated with the widespread use of nanotechnology, emphasizing the importance of responsible innovation and regulatory oversight in ensuring the safe and sustainable deployment of nanomaterials for environmental remediation. The application of various nanomaterials in air pollution remediation holds immense promise and potential for addressing the pressing global challenge of deteriorating air quality. Through meticulous research and experimentation, it has been demonstrated that nanomaterials possess unique properties such as high surface area to volume ratio, exceptional catalytic activity, and tunable physicochemical characteristics, which enable them to effectively capture, degrade, or neutralize a wide range of air pollutants including particulate matter, volatile organic compounds, nitrogen oxides, and heavy metals. Moreover, advancements in nanotechnology have facilitated the development of innovative nano-enabled air purification systems and devices that exhibit superior efficiency, durability, and scalability compared to conventional counterparts. However, the successful implementation of nanomaterial-based solutions necessitates thorough consideration of factors such as material toxicity, environmental impact, cost-effectiveness, and regulatory compliance to ensure their safe and sustainable deployment. Therefore, while significant strides have been made in harnessing the potential of nanomaterials for air pollution remediation, further interdisciplinary research, collaborative efforts, and technological innovations are essential to optimize their efficacy, mitigate potential risks, and facilitate their widespread adoption for achieving cleaner and healthier air quality standards globally.

References

Afrin R, Shah NA (2015) Room temperature gas sensors based on carboxyl and thiol functionalized carbon nanotubes buckypapers. Diam Relat Mater 60:42–49

Arshak K, Gaidan I (2005) Development of a novel gas sensor based on oxide thick films. Mater Sci Eng B 118(1–3):44–49

Baranov A, Spirjakin D, Akbari S, Somov A (2015) Optimization of power consumption for gas sensor nodes: a survey. Sensors Actuators A Phys 233:279–289

Chan PCH, Yan G-Z, Sheng L-Y, Sharma RK, Tang Z, Sin JKO, Hsing I-M, Wang Y (2002) An integrated gas sensor technology using surface micro-machining. Sensors Actuators B Chem 82(2–3):277–283

Chen TM, Kuschner WG, Gokhale J, Shofer S (2007) Outdoor air pollution: nitrogen dioxide, sulfur dioxide, and carbon monoxide health effects. Am J Med Sci 333(4):249–256

Dai J, Ogbeide O, Macadam N, Sun Q, Yu W, Li Y, Su B-L, Hasan T, Huang X, Huang W (2020) Printed gas sensors. Chem Soc Rev 49(6):1756–1789

Dubinin MM (1960) The potential theory of adsorption of gases and vapors for adsorbents with energetically nonuniform surfaces. Chem Rev 60:235–241

Eranna G (2016) Metal oxide nanostructures as gas sensing devices. CRC Press, Boca Raton

Fine GF, Cavanagh LM, Afonja A, Binions R (2010) Metal oxide semi-conductor gas sensors in environmental monitoring. Sensors 10(6):5469–5502

Gautam M, Jayatissa AH (2011) Gas sensing properties of graphene synthesized by chemical vapor deposition. Mater Sci Eng C 31(7):1405–1411

Ghosh A, Majumder SB (2017) Modeling the sensing characteristics of chemi-resistive thin film semi-conducting gas sensors. Phys Chem 19(34):23431–23443

Goldstein M (2008) Carbon monoxide poisoning. J Emergency. Nursing 34(6):538–542

Goltsov VA (2003) Step on the road to hydrogen civilization. Int J Hydrog Energy 27:719–723

Huang X, Gong Z, Lv Y (2022) Advances in metal-organic frameworks-based gas sensors for hazardous substances. TrAC Trends Anal Chem 153:116644

Huixia L, Yong L, Yanni T, Lanlan L, Qing Z, Kun L (2015) Room temperature gas sensing properties of tubular hydroxyapatite. New J Chem 39(5):3865–3874

Jimenez-Cadena G, Riu J, Rius FX (2007) Gas sensors based on nanostructured materials. Analyst 132(11):1083–1099

Khan MAH, Rao MV, Li Q (2019) Recent advances in electrochemical sensors for detecting toxic gases: NO2, SO2 and H2S. Sensors 19(4):905

Korotcenkov G (2013) Handbook of gas sensor materials. Conventional approaches, vol 1. Springer, New York

Kumar A, Punetha D, Pandey SK (2023) Computational modeling and analysis of a chemical gas sensor utilizing WO3 thin films for NO2 detection. J Comput Electron 22:760–767

Leanne G, Busnaina AA, Isaacs JA, Zimmerman JB, Eckelman MJ (2014) Life cycle impacts and benefits of a carbon nanotube-enabled chemical gas sensor. Environ Sci Technol 48(19):11360–11368

Ly NH, Kim HH, Joo S-W (2021) On-site detection for hazardous materials in chemical accidents. Bull Korean Chem Soc 42(1):4–16

Mantri MR, Panda DP, Punetha D, Pandey SK, Singh VP, Pandey SK, Chakrabarti S (2023) Improvement in performance of InAs surface quantum dot heterostructure based H2S gas sensor by introducing buried quantum dot layer. IEEE Sensors J 23(14):15369–15375

Mingfei P, Yin Z, Liu K, Du X, Liu H, Wang S (2019) Carbon-based nanomaterials in sensors for food safety. Nanomaterials 9(9):1330

Nanshu L, Zhou S (2017) Gas adsorption on monolayer blue phosphorus: implications for environmental stability and gas sensors. Nanotechnology 28(17):175708

Nazemi H, Joseph A, Park J, Emadi A (2019) Advanced micro- and nano-gas sensor technology: a review. Sensors 19(6):1285

Nomura K, Ohta H, Ueda K, Kamiya T, Hirano M, Hosono H (2023) Thin-film transistor fabricated in single-crystalline transparent oxide semiconductor. Science 300(5623):1269–1272

Punetha D, Pandey SK (2018) CO gas sensor based on E-beam evaporated ZnO, MgZnO, and CdZnO thin films: a comparative study. IEEE Sensors J 19(7):2450–2457

Punetha D, Pandey SK (2019a) Optimization in NH3 gas response of WO3 nanorods based sensor array. In: IEEE sensors. McGill University, Montreal, QC, Canada, Oct 27–30

Punetha D, Pandey SK (2019b) Sensitivity enhancement of ammonia gas sensor based on hydrothermally synthesized rGO/WO3 nanocomposites. IEEE Sensors J 20(4):1738–1745

Punetha D, Pandey SK (2019c) Ultrafast and highly selective CO gas sensor based on rGO/Fe3O4 nanocomposite at room temperature. In: IEEE sensors. McGill University, Montreal, QC, Canada, Oct 27–30

Punetha D, Pandey SK (2019d) Ultrasensitive NH3 gas sensor based on Au/ZnO/n-Si heterojunction Schottky diode. IEEE Trans Electron Devices 66(8):3560–3567

Punetha D, Pandey SK (2020) Enhancement and optimization in sensing characteristics of ammonia gas sensor based on light assisted nanostructured WO3 thin film. IEEE Sensors J 20(24):14617–14623

Punetha D, Singh SV (2018) A gas detection device. Indian Patent, Application No 201811022863, June 2018

Punetha D, Singh SV (2019) A gas sensor analyzing apparatus. Indian Patent, Application No 201911019803, May 2019

Punetha D, Dixit H, Pandey SK (2018a) Modeling and analysis of an Ni: ZnO-based Schottky pattern for NO2 detection. J Comput Electron 18(1):300–307

Punetha D, Ranjan R, Pandey SK (2018b) Numerical modeling and performance analysis of zinc oxide (ZnO) thin-film based gas sensor. AIP Conf Proc 1966(1):020008

Punetha D, Kar M, Pandey SK (2020) A new type low-cost, flexible and wearable tertiary nanocomposite sensor for room temperature hydrogen gas sensing. Sci Rep 10:2151

Punetha D, Kumar A, Pandey SK, Chakrabarti S (2023) Wearable piezoelectric nanogenerator-based hazardous gas monitoring gadget for self-powered ammonia early warning. In: Organic and hybrid sensors and bioelectronics XVI. SPIE, San Diego, CA, USA, Aug 20–24

Punetha D, Kumar A, Pandey SK, Chakrabarti S (2024) Tertiary nanocomposite-based self-powered E-skin as energy harvester and electronic nose. J Mater Sci: Mater Electron 35:160

Saxena P, Shukla P (2023) A review on recent developments and advances in environmental gas sensors to monitor toxic gas pollutants. Environ Prog Sustain Energy 42(5):e14126

Segev-Bar M, Haick H (2013) Flexible sensors based on nanoparticles. ACS Nano 7:8366–8378

Shipu X, Lei Y (2018) Template-assisted fabrication of nanostructured arrays for sensing applications. ChemPlusChem 83(8):741–755

Timmer B, Olthuis W, Berg AVD (2005) Ammonia sensors and their applications—a review. Sensors Actuators B Chem 107(2):666–677

Wang C, Yin L, Zhang L, Xiang D, Gao R (2010) Metal oxide gas sensors: sensitivity and influencing factors. Sensors 10(3):2088–2106

Williams DE (1999) Semiconducting oxides as gas-sensitive resistors. Sensors Actuators B Chem 57(1–3):1–16

Wilson A, Wright JD, Murphy JJ, Stroud MAM, Thorpe SC (1994) Sol-gel materials for gas-sensing applications. Sensors Actuators B Chem 19(1–3):506–510

Yadav A, Indurkar PD (2021) Gas sensor applications in water quality monitoring and maintenance. Water Conserv Sci Eng 6(3):175–190

Zhang J, Liu X, Neri G, Pinna N (2016) Nanostructured materials for room-temperature gas sensors. Adv Mater 28(5):795–831

Chapter 5
Application of Different Nanomaterials in Air Pollution Remediation

Hassan Shokry Hassan, Mahmoud Samy, and Marwa Farouk Elkady

Abstract Generally, the term pollution is derived from the production and consumption of resources, where the produced streams are very wasteful in their current state. Most of these waste streams cannot be reintegrated into the environment efficiently or economically. Environmental pollution is classified mainly into air, water, and soil pollution. These pollution types not only destroy biodiversity but also hurt the living's health. On the other hand, pollution levels that are increasing day by day need better developments or technological discoveries immediately. The concept of molecular level pollution control is either the detection or separation of definite molecules from atoms and molecules mixture.

There are many technologies developed for effectively treating environmental pollutants. However, the science and engineering area that is responsible for tomorrow's breakthroughs is "Nanotechnology." Nanotechnology can develop the current environmental technologies to generate novel technology, which is more advanced compared with the existing technologies. Consequently, environmental nanotechnology has a vital role in the development of current environmental science and engineering. In this sense, nanotechnology has two main capabilities that can be applied in the field of environment protection, including the clean up or remediation or purification and the detection or sensing of contaminants.

H. S. Hassan
Environmental Engineering Department, Egypt-Japan University for Science and Technology (E-JUST), New Borg El-Arab City, Alexandria, Egypt
e-mail: hassan.shokry@ejust.edu.eg

M. Samy
Public Works Engineering Department, Faculty of Engineering, Mansoura University, Mansoura, Egypt
e-mail: msa203050@mans.edu.eg

M. F. Elkady (✉)
Chemical and Petrochemical Engineering Department, Egypt-Japan University for Science and Technology (E-JUST), New Borg El-Arab City, Alexandria, Egypt
e-mail: marwa.elkady@ejust.edu.eg

N. Ghosh et al. (eds.), *Nanotechnology in Air Quality Management*, Sustainable Development and Biodiversity 39, https://doi.org/10.1007/978-981-96-8339-0_5

This book chapter highlights the recent nanotechnology progresses and its impact on the efficient detection and cleaning the hazardous air contaminates with low-cost efficient and fast process. Essentially, the main features of this justification are to briefly outline the nanotechnology advantages compared with conventional air purification technologies and then highlight the applications of some nanomaterials for detecting air contaminants, as well as cleaning, maintaining, and enhancing air quality. Finally, the future horizons of such nanomaterials for gas sensing and remediation applications are also emphasized.

Keywords Nanotechnology · Environmental pollution · Air pollution · Detectors · Air remediation · Filtration · Nanomaterials

Abbreviations

AC	Activated carbon
CFCs	Chlorofluorocarbon compounds
CFCs	Chlorofluorocarbons
CNTs	Carbon nanotubes
GIS	Geographical Information Systems
MF	Microfiltration
MOFs	Metal organic framework
NF	Nanofiltration
NMVOCs	Non-methane volatile organic compounds
PDA	Personal Digital Assistant
PM	Particulate matter
TNTs	Titanate nanotubes
UF	Ultrafiltration
VOC	Volatile organic compounds

5.1 Introduction

Environmental pollution represents the contaminant's introduction into the environment that yields disorder, instability, and destruction of the ecology. There are three phases of pollution: water, air, and soil pollution (Abozeid et al. 2019). Pollution control term represents the decreasing or removal of pollutants from their source through process modification, utilization of nontoxic matters, applying preservation techniques, and applying the concept of materials reusability rather than considering them as waste streams (Zhang and Fang 2010). Air pollution control means the control of air effluents and emissions into the environment. The priority for keeping the environment safe is pollution prevention through waste minimization rather than pollution control.

The continuous spread of the development and growth processes, including transportation, construction, petroleum refining, mining, and so on, reduces the natural resources and has large production from harmful wastes that cause air, water, and soil pollution and subsequently threaten human public health and the safe environment (Singh and Sachan 2016). Particularly for air pollution, the created wastes are discharged to the environment in various forms, for example, the air pollutants comprise toxic gases such as nitrogen and sulfur oxides, carbon oxides, and so on, as well as volatile organic compounds (VOCs). These environmental contaminates have a great impact to adversely impact on human health, where they have their way into the human body either through breath or absorption attitudes high human risk (Fereidoun et al. 2007). Recently, many studies revealed an increase in the rate of morbidity due to the high risk of air pollution. Consequently, there is an urgent need for the upgrading of effective, sustainable, and low-cost technologies to screen and correctly treat toxic ecological contaminates (Bhawana and Fulekar 2012).

Nanotechnology represents the applied science, which is related to the creation, characterization, design, and application of materials at the nanoscale. During the last century, most scientists have searched for answers to most research problems through the implementation of nanotechnology in various fields such as that concerned by ecology, computer science, medicine, and sports. Novel technologies were applied and developed for air pollution control to implement new and low-cost-efficient processes for air pollution prevention and detection (Karn et al. 2009). The application of nanotechnology in the air pollution control field is mainly based on the physical and chemical relations between nanomaterials of specific size and shape and air contaminants (Abdelraheem 2019). Both air remediation and sensing processes rely mainly on using various nanotechnologies-based approaches such as adsorption, chemical reactions, catalysis, and filtration (Fig. 5.1) for the removal

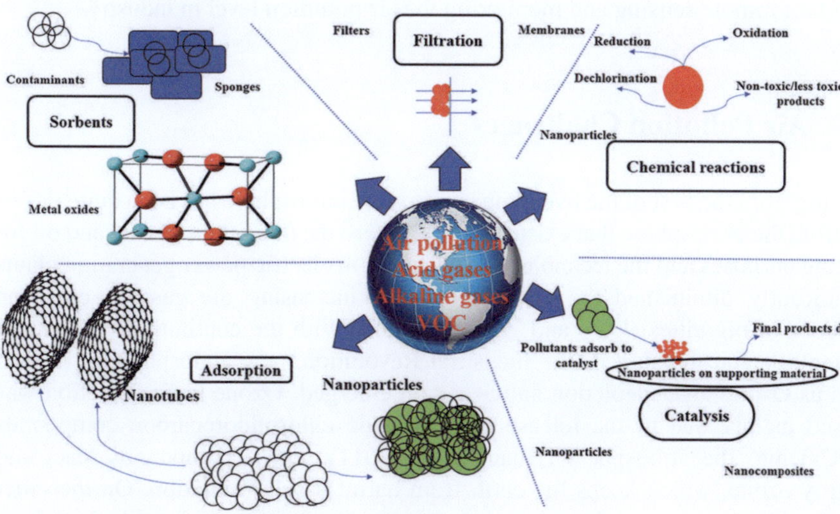

Fig. 5.1 Air pollution control approaches based on nanotechnology

and detection of air pollutants. Among these techniques, the adsorption approach using advanced nanoadsorbents represents the most promising and facile technique for the remediation of harmful emitted gases (Abdelraheem 2019).

Recently, great consideration has been given to the application of nanocatalysts characterized by high surface area for gas reactions. Catalytic nanomaterials are based on their work on enhancing chemical reactions through transformation of the harmful emissions from industrial plants into harmless gases. Another working approach is the utilization of nanomembranes as nanofilters, which are characterized by their small pores that are suitable for polluted gas separation from polluted air. This original technology can separate large volumes of gas effectively (Hilal et al. 2004). On the other hand, the side of air pollution control is concerned with pollution detection; the solid-state gas sensors (which control the ventilation) that are based on absorption/adsorption processes are widely used to monitor and detect fire and toxic gases (Fig. 5.1).

To recognize the application of nanomaterials as gas sensors, one must realize the features of the installation and communication techniques during the gas sensing action. Recently, the attention of scientists and engineers to gas-sensitive materials has grown extraordinarily because of the advances in nanotechnology (Ali et al. 2020). This attention is fundamentally based upon the talented electronic characteristics of nanomaterials, their large surface area, and the availability for controlling the morphological structure using novel synthesis techniques (Font and Artola 2011). Among the major trends in material–gas sensor nanotechnology, the invention of electronic noses was developed as an advanced gas-sensing technology. Such multisensor systems known as hybrid sensors can be invented on a single substrate, which can include gas sensors of various types. Recently, this hybrid sensor (multisensor systems) may relate to Personal Digital Assistant (PDA) as well as communication tools such as Bluetooth and Geographical Information Systems (GIS) for remote sensing and monitoring the air pollution level in industry.

5.2 Air Pollution Challenges

Starting from the first of the twentieth century, the atmosphere has been quite different from the atmosphere that existed before, where the utilization of coal and oil for heating purposes and the technological advances in electric power generation plants significantly diminished the air quality, with increasing the gaseous emission recorded at big cities (Daly and Zannetti 2007). With the continuous progress of technology and the start of the Industrial Revolution, new atmospheric problems such as Ozone layer depletion and acid rain emerged. Ozone layer depletion was caused mainly due to the release of man-made chlorofluorocarbon compounds (CFCs) into the atmosphere (Huang et al. 2011). Those compounds react and destroy ozone, which keeps the earth from harmful solar radiation. On the other hand, acid rain was mainly ascribed to sulfur and nitrogen oxide emissions from fossil fuel combustion and represented a critical pollution that demanded an immediate and integral response from the international community (Huang et al. 2011).

Nowadays, in the second decade of the twenty-first century, air pollution control represents challenges; so, a single approach to control the atmospheric pollution associated with our fast-developing society will probably not be enough to challenge worldwide environmental issues (Chen et al. 2008). Hence, the combination and integration of different technologies will be necessary to deal with the recent air quality problems.

5.3 Main Sources of the Contaminants in Air

Industrial applications have a large contribution to air pollution due to their operational emissions. The nature and amount of air pollutants depend very much on the practice of the industry (Lanjouw and Mody 1996). Most chemical processes such as the fertilizer production industry, petroleum refineries, cement and iron ore industries, and so on have a major contribution to air pollution emissions. These processes emitted either solid particulate matter, liquid droplets, or gases as air contaminants. Air contaminates can be divided into primary or secondary pollutants. The primary air contaminants are matters that are directly emitted from the process, such as emitted carbon monoxide gas from fuel burning at a motor car. However, the secondary air contaminates are not released directly into the atmosphere, but they are induced due to the interaction or reaction of the primary air contaminates. As an example of the most important secondary air, pollutant is the ground level ozone, which is responsible for photochemical smog (Kesselmeier and Staudt 1999). So, the main primary air contaminates produced by human activity include:

1. Acidic gases such as nitrogen oxides (NO_x) and sulfur oxides (SO_x);
2. Alkaline gases such as carbon monoxide (CO) carbon dioxide (CO_2) and ammonia (NH_3);
3. Volatile organic compounds (VOC_S) that are classified into two classes of methane (CH_4) and non-methane (NMVOCs). All the aromatic compounds including benzene derivatives and organic acids, aldehydes, and alcohols are classified as NMVOCs;
4. Chlorofluorocarbons (CFCs);
5. Particulate matter (PM) that are small particles of solid or liquid presence in a gas;
6. Odors such as smells emitted from trash, sewage, and industrial processes.

5.4 Recent Advances in Nanotechnology for Air Pollution Remediation and Control

Nanotechnology, condensed to "nanotech," represents the science of matter control at the atomic and molecular level (Ranveer et al. 2016). Commonly, nanotechnology interacts with size structures of 100 nm or smaller that includes developing materials within this scale. The unique and potential nanomaterial properties comprise dramatically increased surface areas and reactivities, enlarged electrical properties, and enhanced strength–weight ratios. All these promising properties qualified the designed materials to have great potential in many applications especially that concerned with environmental protection, particularly the air pollution control field (Karn et al. 2009). In terms of nanomaterials applications in air pollution control, the promising properties of nanomaterials improve the accuracy and sensitivity of nanosensor devices. On the other side, the designed nanomaterials are effectively reacting with air contaminants and degrading them into nontoxic products. Therefore, nanotechnology delivers a powerful way for both detection and treatment of pollutants for air pollution control (Ranveer et al. 2016). Recent research at the level of nanoscale develops innovative, effective, and cheap methodologies for air pollution remediation and detection. Currently, the nanotechnology progress offers a talented resolution to detect and treat air pollution through the development of costly, fast, and effective processes.

Air pollution control means two different concepts: air remediation (clean-up) and detection (sensor) of air pollutants. Remediation or clean-up is the science of removal or reduction of contaminates from polluted air using either physical or chemical techniques. However, air detection or sensing is the science of monitoring the presence of contaminants in polluted air. The conventional air pollution techniques are designed to be operated based on the remediation concept through the isolation of air contaminants from polluted air. Most conventional air pollution control methods are concentrated mainly on the separation of pollutants whether these pollutants are particulate matters or gaseous contaminates. The most common control devices utilized for the separation of particulate contaminants from polluted air are the gravitational settling chamber, cyclone separators, fabric filters, and electrostatic precipitators (Zhang and Fang 2010). However, the gaseous contaminates in polluted air are separated by absorption approaches using either wet or dry absorption according to the characteristics of polluted gas. Most of these techniques produce undesirable polluted by-products wastes that require further treatment processes such as the disposal of the separated air contaminants to landfill (Yunus et al. 2012). So, further treatment processes are required to achieve the final targeted goal of environmental protection, resulting in high operation cost and time. All these limitations can be effectively targeted through nanotechnology.

5.4.1 Air Remediation Using Nanotechnology

There are three principal techniques in which nanotechnology can be employed for the treatment and reduction of various air pollutants as follows (Ranveer et al. 2016):

1. Adsorption using nanoadsorbent materials
2. Degradation using nanocatalysts
3. Filtration and separation using nanofilters

5.4.1.1 Adsorption Technique for Air Remediation

Nanotechnology is hypothetical in that most of our life's difficulties including air pollution can be resolved using the nanoscale adsorbent materials that are known as nanoadsorbents. The behavior of adsorption processes is highly dependent on the adsorbent nature, its surface areas, and its porosity. Before investigating the available types of nanoadsorbents applied for air remediation, it is important to highlight briefly the process definition and types of adsorption processes that take place in the air remediation process. Adsorption represents a mass transfer process in which a porous solid material is contacted with a liquid or gaseous stream for pollutants decontamination by depositing (adsorbing) them onto the solid. Two main types of adsorption occur in air remediation using nanoadsorbent, which are the physisorption and chemisorption adsorption processes. At the physisorption process, the air pollutants are collected onto the surface of the nanoadsorbent material by the action of van der Waals forces and other relatively weak binding forces and remain hosted onto the material's surfaces. Therefore, both the air pollutants and the surface of nanoadsorbent material remain the same with no irreversible chemical changes occurring (Mensah et al. 2022a). However, during the chemisorption process, the air contaminated at the polluted air streams reacted with the nanoadsorbent material surfaces irreversibly. The concept of chemisorption is almost applied to the catalytic processes. Recently, the most commonly used nanoadsorbent materials for air remediation applications are nanocarbon-based nanomaterials, and metal and metal oxide nanomaterials as an example of the most popular inorganic nanomaterials and polymeric composite nanomaterials.

5.4.1.1.1 Carbon-Based Nanoadsorbent Material for Air Remediation

Carbon nanomaterials such as activated carbon, carbon nanotubes (CNTs), graphene, fluorine, carbon fibers, and so on (Fig. 5.2) have enormously physical properties such as high pore diameter, pore volume, and surface area assembly the material distinguishes them for many applications. They can act as nanoadsorbents with high selectivity and capacity. Additionally, the highly sensitive surface sites have a significant impact on the adsorption. Activated carbon (AC) nanomaterial represents one of the most promising and earliest materials used for air cleaning to deliver pure air to the environment. Activated carbon is an extremely porous material that is utilized for decontamination of the pollutants from the air that depend on its operation on surface adsorption (Moreno-Castilla 2004). It seems to be efficient for high amounts of organic and inorganic air contaminant decontamination. So recently, there has been important progress toward enhancing air treatment processes using activated carbon nanomaterial through either chemical or biological modifications of the already existing AC nanomaterials (Bhatnagar et al. 2013).

Lately, advanced nanomaterials that are known as graphite-like carbon framework materials have been developed for high-performance air purification (Moreno-Castilla 2004). The high reactivity of these advanced carbon-based nanomaterials is

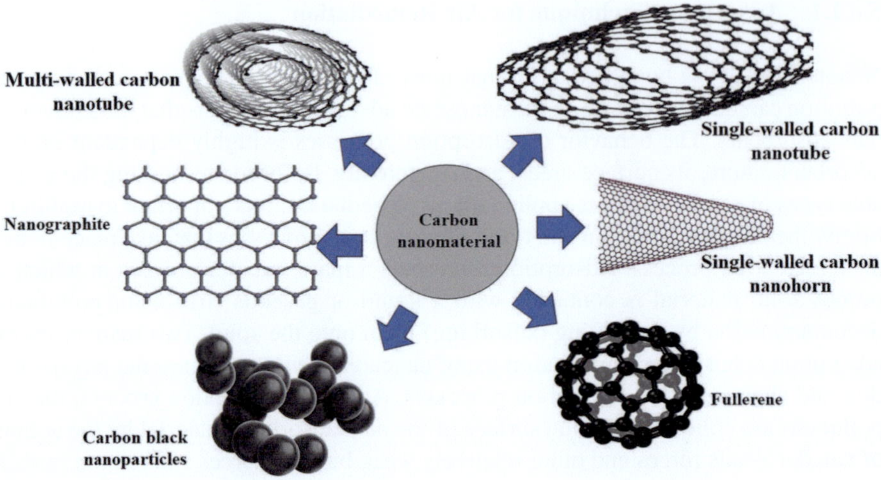

Fig. 5.2 Carbon-based nanomaterials

returned to the induction of extremely reactive surface locations or modified structure bonds that show a vital contribution to the adsorption process. Also, the addition of other oxygenated functional groups at the carbon framework materials can offer new reactive sites available for the adsorption process. In this regard, Fig. 5.2 delivers examples of the most advanced carbon-based nanostructured materials such as fullerene, carbon nanotubes, graphene, and graphite (Bergmann and Machado 2015).

Most graphite-like carbonaceous structure nanomaterials are preferably suitable for the adsorption of organic contaminants from polluted air (Gupta and Saleh 2013). Generally, the adsorption process takes place through π–π interactions among the air pollutants and the surface of carbon-based material. Also, this interaction may occur through hydrogen and Van der Waals bonding. These materials also have the capability for cation/anion–π interactions with various air contaminates. Accordingly, these materials have a vital role in solving the recently suffering problem of global warming due to their promising properties and their ability to adsorb various types of air pollutants.

Global warming represents the most serious problem of outdoor air pollution, which yield large variation in land, water sources, and climate all over the world as a result of greenhouse gases. These greenhouse gases including CH_4, CO_2, N_2O, and fluorinated gases have a negative impact on the atmosphere due to their sustainability in the climate for many years. Due to the superior adsorption properties of the carbon-based nanomaterials and their ability to adsorb various air pollutants types as discussed previously, they are considered the most promising adsorbent materials for air pollution control and they have a great contribution to solving the global warming problem. In this regard, scientists try to develop and synthesize different carbon-based nanomaterials with high efficiency for greenhouse gas adsorption (Ren et al. 2011).

Recently, there have been many investigations that deal with the methodologies utilized to modify and enhance the adsorption properties of carbon-based nanomaterials to permit a great interaction between the air pollutants and the material. So, it was investigated that the main interaction between the organic pollutants and the surface of most carbon-based materials takes place through noncovalent interactions such as hydrogen bonding, electrostatic forces, π–π interactions, and van der Waals forces. Moreover, CNTs nanomaterial allows the grouping of one or more surface groups such as OH, COOH, and C=O that enhance the adsorption capacity of all material. CNTs nanomaterials have one-dimensional morphological structures that result from the rolling up of graphene sheets in a concentric formula and may exist in either single-walled or multiwalled carbon nanotubes as indicated in Fig. 5.3 (Wang et al. 2014). As indicated from this figure, it was elucidated that these materials are characterized by their cylindrical pores; so, the air contaminates molecules will react with their wall's carbon molecules. This interaction depends on the pore's size and shape. So, it was evident that CNTs nanomaterials are highly graphitic materials compared with activated carbon nanomaterials that confirm their higher adsorption capacity compared with activated carbon nanomaterials (Wang et al. 2014).

These promising adsorption capacity of CNTs nanomaterials and the availability for preparing various molecular structures from these compounds besides their ability to be chemically modified to develop many carbon-based nanomaterials encouraged the researcher for more investigations in the field of utilization of carbon-based nanomaterials for air remediation and clean-up. So, recently, Su et al. (2009) studied the performance of both single and multiwall carbon nanotubes for decontamination of CO_2 gases from polluted air (Su et al. 2009). The adsorption performance of CNTs deposited on quartz filters for VOCs was investigated by Asyadi et al. (2017). Moreover, the developed Si-doped and boron-doped single-walled CNTs were tested for adsorption of both CO and methanol gases (Wu et al. 2013). Rather than

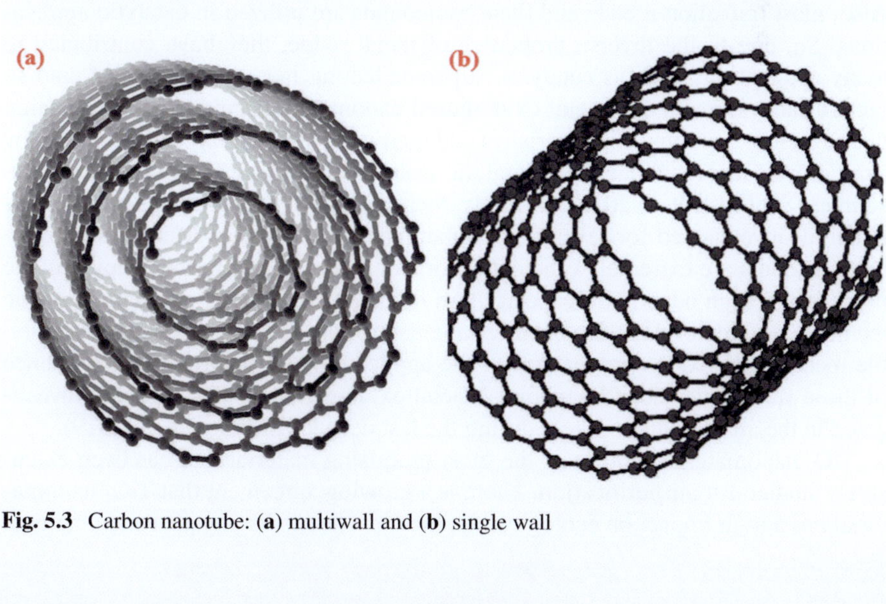

Fig. 5.3 Carbon nanotube: (**a**) multiwall and (**b**) single wall

carbon nanotube materials, both graphene and fullerene have high applications as carbon-based adsorbed materials for air pollution treatment. Graphene is a single flat sheet (layer) of carbon atoms arranged in a repeating hexagonal lattice, but fullerene is composed of at least 60 carbon atoms that form a closed cage known as buckyball. Graphene-based composite nanomaterial was evaluated for the adsorption of various air pollutants such as CO_2, SO_2, H_2S, NH_3, and N_2 (Zhang et al. 2010). It was elucidated that the functional groups present onto the graphene oxide are responsible mainly for the gas adsorption process. Moreover, Zhang et al. evaluated the efficiency of TiO_2-graphene nanocomposites for benzene decontamination (Zhang et al. 2010). On the other hand, the efficiency of fullerene for CO_2 adsorption was discussed by Dong et al. (2015). However, the adsorption performance of chemically modified fullerene, which is fullerene-like boron nitride nanocage, was assisted against N_2O by Esrafili (2017).

5.4.1.1.2 Metals and Metal Oxides-Based Nanoadsorbent for Air Remediation

Metal and metal oxide nanomaterials have a vital role in many areas of chemistry, physics, and materials science. The formation of metal oxides is formed due to the tendency of metal ions for consequence coordination tendency. Accordingly, the oxides procedure sphere around metal ions and close-packed metal oxide structure was produced (Arora et al. 2016). The deviation in the physicochemical properties of metal oxides has great chemists' attraction due to their sensitivity to any change in material composition and structure. Special attention was given to metal oxides by scientists because of their facile formation mode and their multifunctional behavior compared with metal. Especially, the transition metals are characterized by the formation of colored compounds and their demonstration with magnetic properties. Also, most transition metals and their compounds are utilized in catalytic applications. So, due to the diverse properties of metal oxide, they have contributed to many applications such as catalysts, superconducting materials, ceramics, and so on. Nowadays, metal and metal oxide-based nanomaterials gained more attraction due to their exceptional characteristics and their wide application range, where they can work as adsorbent agents or catalytic materials in the presence of heat or light sources (Santhosh et al. 2016). Recently, there are many various metal-based nanomaterials investigated for several air contaminants remediation. Metal/metal oxide nanomaterials are extremely efficient adsorbent materials due to their rapid kinetic process and high adsorption capacity (An et al. 2015). Various efficient synthetic techniques were developed to obtain shape-controlled, monodisperse, and very stable metal/metal oxide nanomaterials. The application and adsorption performance of these various developed nanometal/metal oxide materials were broadly investigated in the air remediation field during the last decade (El Fawal et al. 2019).

TiO_2 nanomaterial represents the most promising material that has been extensively studied for air purification. There is a growing agreement that TiO_2 nanomaterial has a high impact on ecological applications due to its physical and chemical strength, nontoxicity, low cost, and corrosion resistance. So, many researchers investigated many techniques to develop titanium dioxide at various morphological structures as indicated in Fig. 5.4. These developed nanostructures were evaluated for different applications based on the adsorption process, especially their adsorption capabilities for both water and air contaminants from the polluted streams. Here, we will highlight some research that investigates the performance of TiO_2 nanomaterials for air pollutants decontamination. Some research works elucidated the adsorption properties of titanium-based materials for carbon dioxide adsorption. An et al. (2015) investigated the adsorption performance of the synthesized Au/Fe_2O_3/TiO_2 hybrid nanotube material for CO_2 adsorption (An et al. 2015). Also, the adsorption of synthesized titanium dioxide nanotubes for carbon oxide was evaluated by Paltanea et al. (2018). Liu recommended the polyethylene imine (PEI)-modified TiO_2 nanotubes for capturing CO_2 from power plant flue gas, where this developed material showed high CO_2 adsorption capacity (Cheng et al. 2015).

ZnO nanomaterials have gained extensive research attention due to their outstanding physical and chemical properties, leading to numerous opportunities in environmental, energy, and electrical applications. Out of these applications, the usage of ZnO nanostructured materials for H_2S removal from the polluted air, mostly at high temperatures through an adsorption mechanism, because of their thermal stability for the non-catalytic suffixation reaction: (Portela et al. 2014). Huy et al. (2019) described a simple single-step technique via ultrasonic-assisted precipitation with no surfactants and post-heat treatments for preparing ZnO nanoparticles and their application for removing H_2S in the air as a nanoadsorbent (Huy et al. 2019). The assessment results of the adsorption experiments confirmed that the prepared ZnO nanoadsorbent had a higher adsorption ability for H_2S due to the high air volume treated and longer adsorption time.

Several studies relying on metal organic framework (MOFs) as nanoadsorbents to capture sulfur and nitrogen compounds from gas streams were previously reported

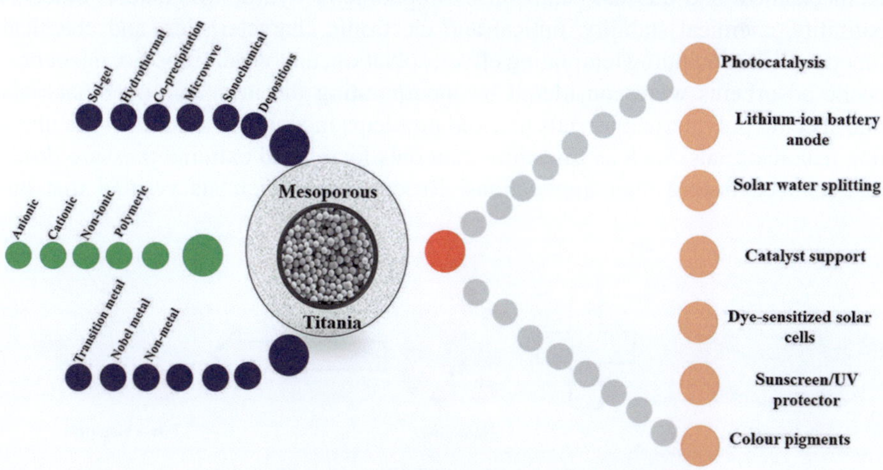

Fig. 5.4 Various synthesis techniques and applications of TiO_2 nanomaterials

(Li et al. 2016). Among these MOF-based nanoadsorbents, NH_2-substituted UiO-66 showed a higher ability to adsorb acid gases (e.g., SO_2, CO_2, NO_2, and H_2S). It is a potential detoxification material since harmful gases can be adsorbed with UiO-66-NH_2 within a few minutes. The high porosity, variable pore size, and increased concentrations of UiO-66-NH_2 make it an ideal material for acid gas removal (DeCoste et al. 2015).

5.4.1.1.3 Polymeric Composite Nanomaterial Adsorbent Agent for Air Remediation

The application of most inorganic nanomaterials and the ecological remediation applications gave extreme pressure drops during their operations of fixed-bed nanomaterials or at any other flow systems. Moreover, these inorganic nanomaterials have many other limitations during their operation such as difficulty for both separation and reuse, besides their possible ecosystems dangerous due to their release into the environment (William et al. 2008). Moreover, the utilization of these inorganic nanomaterials as aqueous suspensions restricts their applications due to the separation problems of these fine inorganic particles. Most of these limitations may be solved through the immobilization of these inorganic nanomaterials onto polymeric matrices, such as porous resin as indicated in Fig. 5.5.

This polymeric matrix can solve all the discussed limitations to a considerable extent, helping the particle loss reduction, inhibition of agglomeration of particles, and possible application of convective flow happening by free-standing particles. The most widely utilized host polymeric materials for nanocomposite fabrication comprise cellulose acetate, alginate, and chitosan. These polymeric host materials must be characterized by their brilliant mechanical strength for large periods of composite utilization. The selection of the polymeric support material is based on its mechanical and thermal behavior, hydrophobic or hydrophilic nature, biocompatibility, chemical stability, optical and electronic characteristics, and chemical functionalities including templating effect, solvation, and wettability. So, nanocomposite adsorbents were considered by incorporating the inorganic nanomaterials into specific polymeric materials to avoid problems induced by the size of the ultrafine nanomaterials, such as transition materials losses and extreme pressure drops that induced during their applications. Recently, the scientists verified that the

Fig. 5.5 Impregnation of inorganic nanomaterials onto the polymeric matrix for formation of nanocomposite materials

porous polymeric nanocomposite adsorbent materials demonstrated the perfect alternatives to fabricate comparable composite adsorbents considering their excellent mechanical strength and the adaptable surface chemistry of the polymeric chains. They also elucidated that the charged functional groups included in the polymeric matrix improve the permeation of the counter-charged pollutants at the composite nanomaterials. The latterly reported polymeric matrices are usually saturated carbon chain polymeric materials or fluoropolymers such as polyvinylpyrrolidone (PVP), poly(dimethylsiloxane) (PDMS), polyethylene (PE), polyaniline (PANI), polypropylene (PP), poly(3-hexylthiophene) (P_3HT), poly(tetrafluoroethylene), and Nafion (Zhang et al. 2016).

Recently, iron oxide polymeric-based nanocomposite was prepared as an advanced composite material. The magnetite nanomaterials were incorporated into decyltrimethoxysilane (DTMS) to functionalize its surface and change its hydrophilic properties to lipophilic (Huo et al. 2007). The adsorption properties of the prepared advanced composite nanomaterial were tested against hydrogen sulfide adsorption from polluted air.

Also, the microporous structure of polystyrene (PS)-based nanocomposite represents an interesting polymeric matrix for hosting various nanomaterials with specific functionalities (Zhang et al. 2007). Diverse approaches for nanomaterials incorporation into the polymeric PS matrix were applied. The first approach is based on the nanostructured material post-growth onto the cross-linked PS polymeric matrix. The more recent second approach is based upon the direct impregnation of the pre-synthetized nanomaterial during the pre-polymerization step that is followed by the hyper-crosslinking. Up to now, PS-based nanocomposites may be classified into three main categories: magnetic nanocomposites, carbon-based nanostructured fillers, and metal nanomaterials. Magnetic polymeric-based nanomaterials with high absorption properties can be usefully developed. This strategy is usefully applied for magnetic separation of fabricated magnetic composite nanomaterials at many industrial sectors for recovering the spent sorbents using an external magnetic field after finishing the process as indicated in Fig. 5.6 (Yavuz et al. 2009). So, the incorporation of magnetite nanomaterials into the polymeric-based materials is utilized to import the magnetic properties into the nanocomposite material. So, many scientists embedded the synthesized Fe_3O_4 nanoparticles into suspension polymerization of polystyrene divinylbenzene (DVB), followed by a hyper-crosslinking step as reported by Zhou et al. (2012).

This magnetic nanocomposite material showed promising adsorption characteristics compared with commercial resins for toxic gas pollutants such as chlorine gas. Furthermore, recently, Li et al. (2017) developed a magnetic polystyrene–ethylene diamine tetraacetic acid (PS/EDTA) nanomagnetic microsphere (Li et al. 2017). This material showed high VOCs adsorption capacity from polluted air.

On the other hand, the polymeric carbon-based nanomaterials such as graphene and carbon nanotubes (CNT) incorporated into polymeric matrices were considered extremely attractive for their remarkable electronic, adsorption, thermal, and mechanical properties. Recently, the interaction between carbon-based nanomaterials with polymeric matrix characterized by its high surface area such as HCL

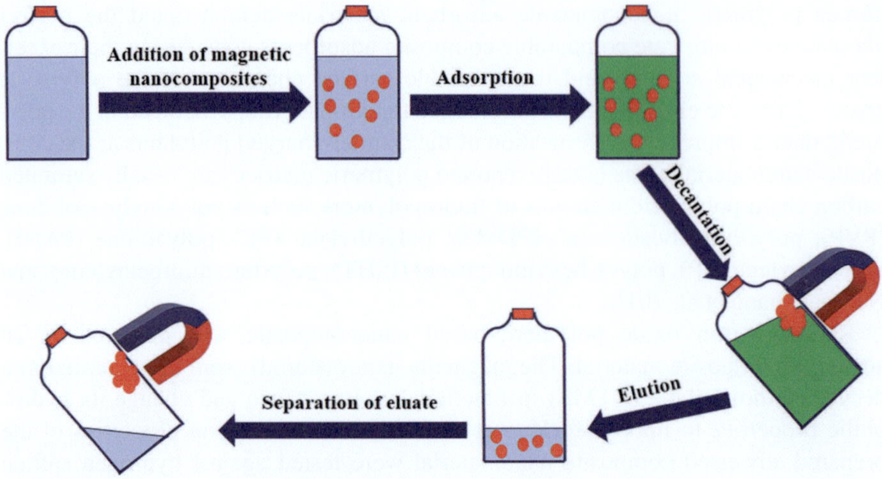

Fig. 5.6 Recovery of magnetic-based composite nanomaterial after finishing the adsorption process

styrene-based polymers was investigated. A new class of nanocomposites polymer base composed of divinylbenzene (DVB), styrene (ST), and vinylbenzyl chloride (VBC) that included surface-modified carbon nanotubes, graphene, and surface-modified graphene oxide were developed and tested as novel nanocomposites for air pollution decontamination (Castaldo et al. 2017).

Finally, the third category of the polymeric composite nanomaterial, which is concerned with metallic nanomaterials, supported onto polymeric matrices is widely used mainly as heterogeneous catalysts in various industrial sectors. The catalytic process depends mainly on the adsorption process as will be discussed later in the next section. The first investigation on the possible utilization of nanocavities characteristic of cross-linked polystyrenes as nanoreactors took place by Sidorov et al. (1999). By the same strategy, platinum nanomaterials were impregnated into the pores of the same microporous polymeric resin (Sidorov et al. 2001). These materials were proved to be highly selective for the catalytic oxidation of air pollutants with a high degree of reactivity tends to 99%.

5.4.1.2 Degradation Technique Using Nanocatalysis for Air Remediation

In the beginning of the 1990s, indoor air pollution received important consideration, human generally spend more than 80% of their time indoors subjected to indoor risks including inhalation of air pollutants that represent higher risk compared with outdoor pollution (Fujishima et al. 2020). Consequently, indoor air pollution can be controlled using nanotechnology in several manner. One of the promising techniques that implement nanotechnology is the catalytic application for air pollutant degradation.

So, nanocatalysts show promise for improving air quality and reducing air pollutants., whereas nanocatalytic process allows fast and selective transformations of polluted chemicals into safe products associated with the availability of catalyst recovery much better than traditional catalysts. So, nanoscale catalysis has more advantages compared with conventional catalysis, which are distinguished by high reactivity and selectivity and long-life time (Samy et al. 2020). Semiconducting nanomaterials represent the most utilized materials with photocatalytic properties and are used for air remediation. The exposure of the semiconducting nanomaterials to a light source with energy equal to the material band gap that forms a pair from electron–hole at the material structure and represents the active sides of the material (Samy et al. 2023). The active sides that are usually present at the material surface are considered the most significant catalyst properties that are responsible for reaction occurrence. As the catalyst's size decreased for nanocatalysis, its active surface amplified, leading to the enhancement of the catalytic efficiency of the material. So, most of the semiconductor metal oxides such as titanium dioxide, zinc oxide, tungsten oxide, and iron(III) oxide (Fe_2O_3) served as photocatalysts (El-Bestawy et al. 2023). Concerned with the application of these photocatalyst materials in environmental applications, especially the air remediation processes, the photocatalytic material can oxidize the air pollutants, especially the organic ones into nontoxic substances. Using the semiconductor principle, the organic pollutants can be oxidized using enough light level that is followed by the charge transfer process of electron transfer from the valence band to the conduction band, oxidizing the surrounding pollutants into harmless substances. So, the development in the field of semiconductor nanotechnology for their photocatalyst applications is developed. So, the most commonly used photocatalytic nanomaterials that are utilized for air remediation applications may be classified into three categories: metal and metal oxide nanomaterials, carbon-based nanomaterials, and aerogels nanomaterials (Elkady and Hassan 2021).

5.4.1.2.1 Metals and Metal Oxides-Based Nanocatalysts for Air Remediation

At the heterogeneous photolysis process, the semiconducting material is primarily excited through the absorption of the photons that have the energy equivalent to or greater than the material band gap (E_g), leading to electron transfer from the valence band (VB) to the semiconduction band (CB), with hole formation as indicated in Fig. 5.7 and according to the following equation:

$$\text{Material} + h\nu \rightarrow \text{negative electron} + \text{positive hole} \tag{5.1}$$

This semiconducting material may be de-energized in various strategies. The very expected but unfavorable consequence is the electron–hole recombination, which represents an unfavorable unproductive process. Another possible and promising strategy is the reactions between either the electrons or the holes and the pollutants adsorbed onto the material surface (Elkady et al. 2020). So, the semiconductor

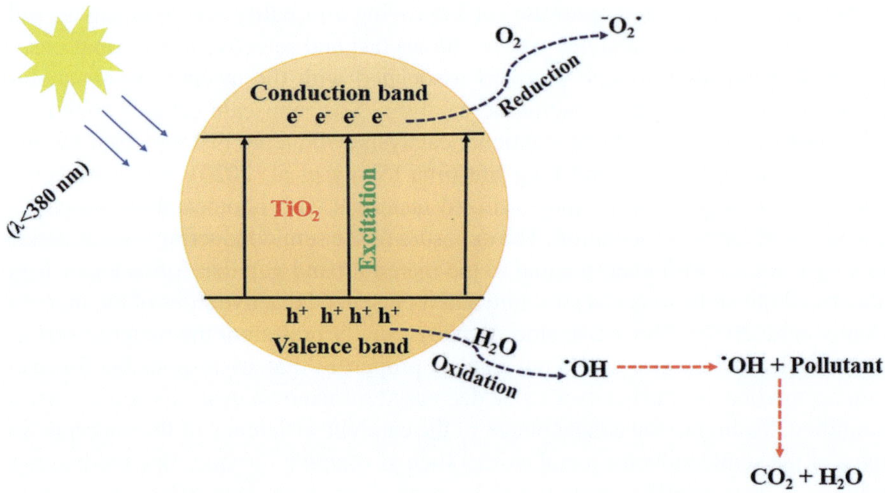

Fig. 5.7 Mechanism of photocatalytic degradation in the metal oxide nanomaterials

material can donate electrons and act as reducing electron acceptors (*A*), or the hole can collect electrons from donor molecules (*D*), which act as oxidizing agents. The dependence of the oxidation and reduction processes is based upon the relative redox sites of the conduction band and the valency band at the semiconductor material and their relations to the redox potential of the adsorbed pollutants (Mensah et al. 2022b). So, commonly, the photocatalytic reduction reaction can be summarized into four main stages, which are

1. Adsorption of pollutant,
2. Generation of electron–hole pair through absorbing of sufficient photon energy,
3. Separation of electron–hole pair and their migration to the surface of the photocatalyst material, and finally
4. Reduction of pollutant.

The most beneficial semiconductor material utilized for photocatalytic applications is TiO_2; this is due to its exceptional characteristics including optical, nontoxicity, electronic, stability, and cheapness (Gar Alalm et al. 2018). The energy band gaps of the TiO_2 phases utilized for photocatalytic applications, which are anatase and rutile, are 3.23 eV (corresponding to 384 nm) and 3.02 eV (411 nm), respectively (Murruni et al. 2007). The most popular photocatalytic commercial TiO_2 was initially developed commercially by Degussa from Germany under the product name P-25. Then the photocatalytic activity of other TiO_2 commercial products was evaluated. However, it was indicated through scientific research that TiO_2 exhibits some photocatalytic capability limitations; these limitations were overcome through the principles of material doping (Shokry Hassan 2019), and where the material is normally doped with transition metal ions to enhance its catalytic activity. Consequently, numerous studies have investigated metal-doped TiO_2 nanomaterials.

The fabrication of TiO_2 nanofibers and its silver doped nanofibers by the new trend of electrospinning technique were investigated by Park and Lee (2014). These advanced nanofiber materials were evaluated as photocatalyst matrices for the degradation of VOCs from polluted air using UV irradiation. It was indicated that the Ag-doped TiO_2 nanofibers exhibited higher photodegradation efficiency compared with TiO_2 nanofibers. This enhancement in the photocatalytic performance of silver-doped TiO_2 nanofibers was returned to the main expected four different factors which are the adequate utilized Ag amount onto the fiber surface which are successfully captured photoinduced electrons and holes. The second factor is the fast photoinduced electrons transfer to the oxygen adsorbed onto the nanofiber surface. The third parameter is the improvement of nanofiber surface hydroxyl groups that extend the material response range for light to include the visible region (Elkady et al. 2016).

Besides, the titanium oxide nanomaterials or titanates were examined for the remediation of air pollutants. The catalytic reduction of polluted NO and ammonia from contaminated air using various hydrothermally synthesized basic, acidic, and neutral titanate nanotubes (TNTs) was reported by Chen et al. (2013). These titanate nanotubes were further doped with manganese oxides for the formation of Mn-doped titanate nanosheets, titanate nanorods, and titanate nanotubes at basic, acidic, and neutral conditions, respectively. It was evident that the fabricated neutral manganese oxides doped titanate nanotubes showed the highest surface area and the most efficient redox performance of the series prepared doped titanates.

Additionally, titanium oxide nanomaterials were mixed with other oxide materials, and the photocatalytic activity of the composite oxides' nanomaterial was investigated. For instance, TiO_2–SiO_2 binary mixed oxide composite nanomaterials were prepared using bamboo as a silica source in the presence of titanium salt. The photocatalytic performance of the composite oxide was tested against the degradation of VOCs (Rasalingam et al. 2014). It was indicated that the TiO_2–SiO_2 composite nanomaterials showed higher catalytic performance compared with the pure TiO_2 nanomaterials. Latterly, the mesoporous CuO–TiO_2 nanotubes were developed and evaluated for photocatalytic oxidation of CO to be converted into CO_2 at low temperatures (Wang et al. 2016).

On the other hand, metals and metal oxide nanofibers including silver, iron, gold, and manganese oxide represent the recent research in the field of metals and metal oxide nanomaterials utilized for photocatalytic applications for air pollution control (Essam et al. 2020), where various volatile organic compounds were treated and degraded from industrial smokestacks using these advanced nanomaterials. Nanogold-based catalysts have photocatalytic activity for the oxidation of CO gas from indoor air at ambient temperature to be converted into carbon dioxide. Also, Au–Pt nanomaterial has the greatest catalytic activity for trichloroethylene (TCE) decomposition equivalent to 100 times more than that fabricated from conventional material (Yadav et al. 2017).

5.4.1.2.2 Carbon-Based Nanocatalysts for Air Remediation

The diversity of researchers investigated the suitability of carbon nanotubes and graphene as adsorbent materials for environmental air remediation applications. They elucidated that the material surface modification, activation, or functionalization of the original carbon chain enhances its performance (Ren et al. 2011). They are also utilized for air pollutant remediation through photocatalytic techniques. As indicated in Fig. 5.8, the mechanism of the photocatalytic process onto the carbon-based materials for air-contaminated remediation takes place in the presence of UV irradiation, where photons with energy larger than or equivalent to the material band gap induced the production of both the valence band holes (h^+) and the conduction band electrons (e^-). Then the positive holes are responsible for the hydroxyl radical's formation that participates in the oxidation process of air contaminants such as the oxidation of CO into CO_2. However, the produced electrons form superoxide radicals and participate in the reduction process of air contaminates such as the degradation of organic pollutants as indicated in Fig. 5.8.

Recently, numerous studies have designated the utilization of graphene-based nanomaterials to fabricate advanced photocatalytic nanocomposites (Yang et al. 2013). The most common and promising carbon-based nanocomposite photocatalytic material is that combined with semiconductor materials, especially metal oxides such as titanium dioxide and zinc oxide nanomaterials, where graphene–TiO_2 nanocomposites demonstrate high photocatalytic activity compared with TiO_2 nanomaterials due to the high conductivity of the composite material (Zhang et al. 2010). The effect of the mixing ratio of graphene to TiO_2 on the photocatalytic activity of the produced nanocomposite was studied and optimized. It was indicated that although the initial addition of more quantities from graphene improves the photocatalytic activity of the produced nanocomposite, the higher ratios of

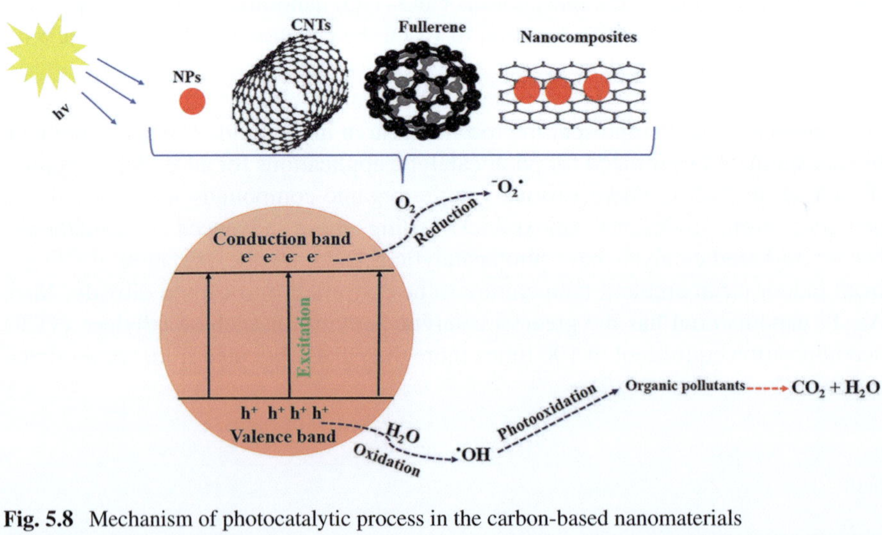

Fig. 5.8 Mechanism of photocatalytic process in the carbon-based nanomaterials

graphene beyond the optimum ratio tend to decrease the material photocatalytic activity. Accordingly, controlling the ratio of the material in the nanocomposite is critical for accomplishing an optimal photocatalytic performance of the produced nanocomposite. This optimum-prepared graphene–TiO_2 nanocomposite material showed promising photocatalytic activity for benzene remediation from contaminated air (Hassan et al. 2015).

Similar to the previous research concerned with the fabrication of graphene–TiO_2 nanocomposites, other research investigated the influence of the mixing ratio of graphene to either ZnO or CdS to fabricate graphene–ZnO and graphene–CdS nanocomposites that are tested as photocatalytic materials for air remediation. They also examined how the variation in weight percentages of graphene affects the photocatalytic activity of produced graphene-based nanocomposite materials (Liu et al. 2011; Zhang et al. 2013).

5.4.1.2.3 Aerogels Nanocatalysts for Air Remediation

Generally, the aerogel term is denoting to any material resulting from inorganic, organic, or composite precursors, which are regularly synthesized by sol–gel technique followed by suitable drying technology that conserved the three-dimensional with highly porous network materials. The continuous progress in aerogel research enables many categories of advanced aerogel materials including inorganic aerogels such as ZrO_2, Al_2O_3, TiO_2, and SiO_2; organic aerogels such as polystyrene, polyurethane, and polyimide; carbon-based aerogels such as carbon nanotubes (CNTs) and graphene; natural-based aerogels such as cellulose, proteins, and polysaccharides; and semiconductor aerogels such as PbTe, CdS, and CdSe (Aegerter and Prassas 2009).

The synthesis technique of most of the aerogels is almost similar. As outlined schematically in Fig. 5.9, the schematic diagram of the aerogel's synthesis procedure is summarized in three main steps, which are sol–gel processing followed by aging and finally the drying process, either this process is super- or subcritical drying. However, concerning carbon and carbide aerogels, further carbonization reduction stages were performed after the drying process (Zuo et al. 2015).

The aerogel materials can act efficiently as good adsorbents of polluted gas from contaminated air due to their porous structure. However, to combine the benefits of both adsorption and photocatalytic activity of the aerogel's materials, they can simply act as a catalyst substrate matrix for hosting different catalytic active materials at their nanostructured porous matrix that are selective for specific catalytic reactions of the air pollutants (Maldonado-Hódar 2011). The flexibility of the sol–gel synthesis process permits the introduction of various molecular precursors via direct material addition to the sol–gel mixture or at the post-gelation stage through incorporation of the metal salts in the wet gels. In this regard, many researchers investigated the support of noble metals or metal oxides onto the carbon or silica aerogel (Suh 2004). Moreover, the network of aerogel itself can act as photocatalytic active

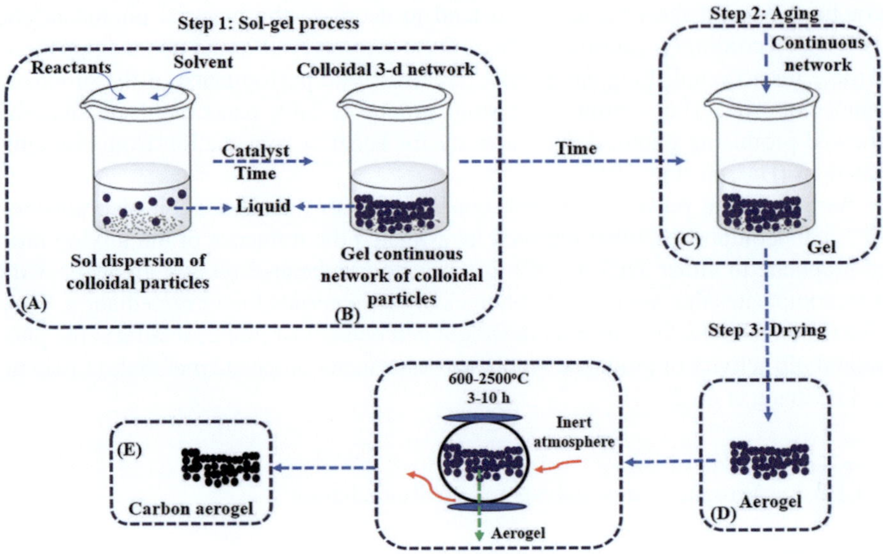

Fig. 5.9 Schematic diagram of the synthesis process of aerogels

material such as titania (TiO_2) or binary vanadium–titania (V_2O_5–TiO_2) and titania–silica (TiO_2–SiO_2) aerogels (Cao et al. 2006).

Regarding the application of aerogels in the air pollution control field as photocatalytic material, they have been examined for catalytic reduction of various hazardous gas pollutants such as NOx, oxidation of CO, and VOC burning (Rotter et al. 2004). In this respect, the binary V_2O_5–TiO_2 aerogels showed a higher catalytic activity for NO_x gases compared with conventional catalytic materials. Moreover, the application of aerogel as a supported matrix for both noble metals such as Pt, Pd, and transition metal oxides and their application for air cleaning from VOCs has been reported (Larsson and Andersson 2000). Also, the utilization of carbon aerogels as a supported matrix for metal and metal oxides and their applications for VOCs degradation is very attractive. This is due to the expansion properties of the carbon aerogels that are their tenable structures that allow the homogenous distribution of hosted metals or metal oxides beside their hydrophobicity and stability at elevated temperatures (Maldonado-Hódar 2011). Table 5.1 shows the different treatment techniques to eliminate the release of toxic gases and air pollution by different nanocatalysts (Saleem et al. 2022).

5.4.1.3 Filtration Technique Using Nanofilters for Air Remediation

Nowadays, membrane technologies represent the most efficient approach for air pollution control due to their availability for air removal contaminants without any hazardous side product production. The main principle of membrane filtration

Table 5.1 Different nanocatalysts used for eliminating the toxic gases and air contaminants from the air

Target contaminant	Nanomaterial used	Results	References
NO_x pollutants	Graphitic carbon nitride/titania composite	A 1:4 graphitic carbon nitride/ titanium dioxide composite showed increased photocatalytic capability in NO_x	Giannakopoulou et al. (2017)
Gaseous ammonia separation	Titanium dioxide	Achieved 60% gas separation	Wu et al. (2014)
Sulfur dioxide (SO_2)	Titanium dioxide	It was observed that the inherent adsorption capability of the small nanomaterials is higher adsorbed species	Baltrusaitis et al. (2011)
Elemental mercury	Titanium dioxide nanorods	Superior Hg^0 removal efficiencies	Jampaiah et al. (2019)
SO_2	Mn/copper slag	SO_2 removal was approximately 99% accomplished	Rabiee and Mahanpoor (2019)
CO_2	Titanium dioxide	CO_2 interacts strongly with TiO_2 nanomaterials disks	Tumuluri et al. (2017)
Rhodamine B dye	Titanium dioxide	Minute amounts of TiO_2 in cement mixtures improve the ability of the resulting concrete for remove contaminants as well as self-clean	Elia et al. (2018)

technology in air remediation processes is the separation process through the application of semipermeable membranes with small pores suitable for separating the different pollutants from the polluted air. Apart from the different available membrane categories such as microfiltration (MF), ultrafiltration (UF), and nanofiltration (NF), the final type (NF) represents the most promising membrane technology. It can remove the low molecular weight contaminants from polluted air. So, the nanofiltration process includes various advantages such as facile process, dependability, no additives, and facile to upscale. The adsorption process of air pollutants onto any membrane may be classified into (1) naturally physical adsorption, which is a fully reversible process, or (2) chemical adsorption, which may be either irreversible for hard chemical bonds or reversible for breakable secondary chemical bonds such as hydrogen bonds, or (3) include both physical and chemical adsorption (Lu et al. 2020).

Generally, there are four main types of NF membranes utilized for the air remediation process: carbon nanotubes-based, polymeric, inorganic, and composite membranes. The carbon nanotubes-based membranes are comprised of hollow cylinders associated with CNTs. This nanotube represents the pores at the nanostructure membrane. Also, this structure looks like the needle nanotube geometry that is characterized by their microscale pores associated with the low surface energy, and this structure induced the super hydrophobicity and super oleophilicity of the prepared membrane (Taheran et al. 2016). On the other hand, polymeric membranes are characterized mainly by their easy fabrication, selectivity, and relatively low cost (Fig. 5.10) (Hassan et al. 2021).

However, the polymeric membranes have their limitations of low chemical and chemical that enable the inorganic membrane to be compotators of the polymeric-based membranes. In contrast, the inorganic membranes have better chemical resistance and stability, good mechanical strength, and good tolerance at high pH and oxidation conditions compared with the organic membranes. However, they are not utilized widely due to their large costs and low control in their pore size distribution (Ibrahim et al. 2021). Thus, with the continuous progress of nanotechnology, scientists tend to incorporate inorganic nanomaterials into polymeric membranes to fabricate the composite membranes, which is considered to make polymeric membranes more commercially attractive. Respecting the application of nanofilter membranes for air remediation, the researchers were focused on how to optimize the fabrication of the nanostructured membranes to enhance their ability to capture many gas pollutants, where, at the industrial scale, the nanofiber-coated filter media was utilized for air filtration from contaminated dust and air filtration of inlet gas turbines (Ye et al. 2006). Also, it was proved that the nanostructured membranes are suitable for the separation of various VOCs gases (Devi et al. 2018). One of the most recent types of nanofiters membranes utilized for air purification is the nanofibers. These nanofibers membranes are characterized by their effective diameters, which are ranged between 1 and 200 nm. The fabrication technique for this membrane type for both organic and inorganic membranes is the spinning nanofiber from either soluble

Fig. 5.10 Air filter nanofiber membranes

or molten state of materials using the spinner technique (Shokry Hassan et al. 2014). The most famous and advanced technique utilized for nanofiber membrane fabrication is electrospinning. In the air pollution control field, the electrospun polyacrylonitrile (PAN)-based carbon nanofiber (CNF) membrane characterized by their high microporosity and their nitrogen-containing functional groups, which are considered as active adsorption sites, were evaluated for separation of the indoor formaldehyde (HCHO) even at very low concentrations from polluted air (Scholten et al. 2011). Also, these nanofiber membranes were utilized as biological filters for decontamination of the bioaerosols air pollutants including the biological pollutants such as viruses, bacteria, and fungi. The incorporation of silver and copper nanoparticles into the nanofibers is widely investigated in bioaerosol applications.

5.5 Conclusions

The availability of different types of nanomaterials and improved knowledge of the materials integration has opened new insights into the development of advanced functional composite nanomaterials that provide superior and promising characteristics for cleaning air pollution. Novel strategies using a wide range of nanomaterials and their composites have been explored for the manufacturing of efficient devices for cleaning air pollutants with significantly improved performance. Various carbon-based nanomaterials, inorganic nanomaterials such as metals and metal oxides, and composites either polymeric composites or inorganics including aerogels were investigated for air pollution remediation. Both metal, metal oxide nanomaterials, and carbon-based nanomaterials can be used for the removal of pollutants either by adsorption process as nanoadsorbents, oxidation, or reduction of pollutants due to their photocatalytic activity. The degree of air pollutant decontamination using these nanomaterial categories can be enhanced through either material functionalization or doping with chemical groups or conducting elements that can capture selectively target air pollutants. So, the development of advanced nanomaterials through their functionalization or elemental doping represents the challenge of the nano-engineered materials for enhancing the properties of the nanomaterials for a high degree of air pollution treatment. Accordingly, the research on nanocatalysts and adsorbents for sustainable air purification field is mostly still in progress to be ready for practical applications. Consequently, many science-based protocols were developed for the construction and control fabrication of composite or functionalized nanomaterials to simulate the scaling-up of their manufacturing that will be required to broadly allow penetration of nano-engineered materials into new commercial applications, especially in the field of the environmental remediation. On the other hand, nanomembranes have found applications in the fabrication of potable air filters for air pollutants decontamination. Continuous progress was undergone in developing nanofilter membranes with selective properties, greater stability for long operation time intervals, and cost optimization. Nanofibrous filter membranes are characterized by their high permeability with a small pore size that makes them suitable for a wide range of air filtration processes. Moreover, nanofibrous membranes offer unique properties such as good interconnectivity of the pores, high specific surface area, and the potential to incorporate active material or surface functionality of the nanofiber. Furthermore, new fabrication methods are currently under investigation for developing different fiber membranes at various nanoscale diameters and morphologies to enhance their filtration performance toward air pollutants.

References

Abdelraheem A (2019) Comparable investigation of polyaniline behavior towards gaseous ammonia and toluene adsorption. Environ Sci Pollut Res 26:3991–3999

Abozeid MA, Hassan HS, Morsi I, Kashyout AB (2019) Development of nano-WO_3 doped with NiO for wireless gas sensors. Arab J Sci Eng 44:647–654. https://doi.org/10.1007/s13369-018-3608-1

Aegerter MA, Prassas M (2009) Advances in sol-gel derived materials and technologies. Springer, Cham

Ali I, Kashyout AE-HB, Tayel M, Shokry Hassan H, Rizk M (2020) Ruthenium (Ru) doped zinc oxide nanostructure-based radio frequency identification (RFID) gas sensors for NH_3 detection. J Mater Res Technol 9:15693–15704. https://doi.org/10.1016/j.jmrt.2020.11.033

An H, He X, Li J, Zhao L, Chang C, Zhang S, Huang W (2015) Design, synthesis of uniform Au nanoparticles modified Fe2O3–TiO2 coaxial nanotubes and their enhanced thermal stability and photocatalytic activity. New J Chem 39:4611–4623. www.rsc.org/njc, https://doi.org/10.1039/C5NJ00239G

Arora AK, Jaswal VS, Singh K (2016) Metal/mixed metal oxides and their applications as adsorbents: a review. Int J Chem Sci 14:3215–3227

Asyadi M, Muiz F, Weng L (2017) Electronic properties and gas adsorption behaviour of pristine silicon-, and boron-doped (8, 0) single-walled carbon nanotube: a first principles study. J Mol Graph Model 75:85–93. https://doi.org/10.1016/j.jmgm.2017.05.003

Baltrusaitis J, Jayaweera PM, Grassian VH (2011) Sulfur dioxide adsorption on TiO_2 nanoparticles: influence of particle size, coadsorbates, sample pretreatment, and light on surface speciation and surface coverage. J Phys Chem C 115:492–500. https://doi.org/10.1021/jp108759b

Bergmann CP, Machado F (2015) Carbon nanomaterials as adsorbents for environmental and biological applications. Springer, Cham

Bhatnagar A, Hogland W, Marques M, Sillanpää M (2013) An overview of the modification methods of activated carbon for its water treatment applications. Chem Eng J 219:499–511. https://doi.org/10.1016/j.cej.2012.12.038

Bhawana P, Fulekar MH (2012) Nanotechnology: remediation technologies to clean up the environmental pollutants. Res J Chem Sci 2:90–96

Cao S, Lun K, Yue P (2006) Preparation of freestanding and crack-free titania–silica aerogels and their performance for gas phase, photocatalytic oxidation of VOCs. Appl Catal B Environ 68:99–108. https://doi.org/10.1016/j.apcatb.2006.07.022

Castaldo R, Avolio R, Cocca M et al (2017) Synthesis and adsorption study of hyper-crosslinked styrene-based nanocomposites containing multi-walled carbon nanotubes. RSC Adv 7:6865–6874. https://doi.org/10.1039/c6ra25481k

Chen I, Hegde U, Chang C, Yang S (2008) Methane and carbon dioxide emissions from closed landfill in Taiwan. Chemosphere 70:1484–1491. https://doi.org/10.1016/j.chemosphere.2007.08.024

Chen X, Cen C, Tang Z et al (2013) The key role of pH value in the synthesis of titanate nanotubes-loaded manganese oxides as a superior catalyst for the selective catalytic reduction of NO with NH3. J Nanomater 2013:3

Cheng D, Liu Y, Wang H et al (2015) Enhanced CO2 adsorptive performance of PEI/SBA-15 adsorbent using phosphate ester based surfactants as additives. J Environ Sci 38:1–7. https://doi.org/10.1016/j.jes.2015.04.025

Daly A, Zannetti P (2007) An introduction to air pollution – definitions, classifications, and history. In: Ambient air pollution. ASST and The EnviroComp Institute, pp 1–14

DeCoste JB, Demasky TJ, Katz MJ, Farha OK, Hupp JT (2015) A UiO-66 analogue with uncoordinated carboxylic acids for the broad-spectrum removal of toxic chemicals. New J Chem 39:2396–2399. www.rsc.org/njc, https://doi.org/10.1039/C4NJ02093F

Devi KS, Lakshmi VV, Alakanandana A (2018) Impacts of cement industry on environment – an overview. Asia Pac J Res I:156–161. https://doi.org/10.3945/ajcn.110.003483

Dong H, Lin B, Gilmore K et al (2015) B 40 fullerene: an efficient material for CO2 capture, storage and separation. Curr Appl Phys 15:1084–1089. https://doi.org/10.1016/j.cap.2015.06.008

El Fawal GF, Hassan HS, El-Aassar MR, Elkady MF (2019) Electrospun polyvinyl alcohol nanofibers containing titanium dioxide for gas sensor applications. Arab J Sci Eng 44:251–257. https://doi.org/10.1007/s13369-018-3529-z

El-Bestawy EA, Gaber M, Shokry H, Samy M (2023) Effective degradation of atrazine by spinach-derived biochar via persulfate activation system: process optimization, mechanism, degradation pathway and application in real wastewater. Environ Res 229:115987. https://doi.org/10.1016/j.envres.2023.115987

Elia HN, Ghosh A, Akhnoukh AK, Nima ZA (2018) Using nano- and micro-titanium dioxide (TiO_2) in concrete to reduce air pollution. J Nanomed Nanotechnol 9:505. https://doi.org/10.4172/2157-7439.1000505

Elkady MF, Hassan HS (2021) Photocatalytic degradation of malachite green dye from aqueous solution using environmentally compatible Ag/ZnO polymeric nanofibers. Polymers 13:2033. https://doi.org/10.3390/polym13132033

Elkady MF, Hassan HS, Salama E (2016) Sorption profile of phosphorus ions onto ZnO nanorods synthesized via sonic technique. J Eng 2016:2308560. https://doi.org/10.1155/2016/2308560

Elkady M, Salama E, Amer WA, Ebeid EM, Ayad MM, Shokry H (2020) Novel eco-friendly electrospun nanomagnetic zinc oxide hybridized PVA/alginate/chitosan nanofibers for enhanced phenol decontamination. Environ Sci Pollut Res 27:43077–43092. https://doi.org/10.1007/s11356-020-10247-8

Esrafili MD (2017) N_2O reduction over a fullerene-like boron nitride nanocage: a DFT study. Phys Lett A 381:2085–2091. https://doi.org/10.1016/j.physleta.2017.04.009

Essam A, El-Shazly A, Shokry H (2020) Piezoelectric PVDF nanofibrous membrane electrospinning optimization parameters. Key Eng Mater 897:71–76. https://doi.org/10.4028/www.scientific.net/KEM.897.71

Fereidoun H, Nourddin MS, Rreza NA, Mohsen A, Ahmad R, Pouria H (2007) The Effect of Long-Term Exposure to Particulate Pollution on the Lung Function of Teheranian and Zanjanian Students. 3(2):1–5. https://doi.org/10.69656/pjp.v3i2.637

Font X, Artola A (2011) Detection, composition and treatment of volatile organic compounds from waste treatment plants. Sensors 11:4043–4059. https://doi.org/10.3390/s110404043

Fujishima A, Zhang X, Tryk DA (2020) TiO_2 photocatalysis and related surface phenomena. Surf Sci Rep 63:515–582. https://doi.org/10.1016/j.surfrep.2008.10.001

Gar Alalm M, Samy M, Ookawara S, Ohno T (2018) Immobilization of S-TiO_2 on reusable aluminum plates by polysiloxane for photocatalytic degradation of 2,4-dichlorophenol in water. J Water Process Eng 26:329–335. https://doi.org/10.1016/j.jwpe.2018.11.001

Giannakopoulou T, Papailias I, Todorova N et al (2017) Tailoring the energy band gap and edges' potentials of g-C_3N_4/TiO_2 composite photocatalysts for NOx removal. Chem Eng J 310:571–580. https://doi.org/10.1016/j.cej.2015.12.102

Gupta VK, Saleh TA (2013) Sorption of pollutants by porous carbon, carbon nanotubes and fullerene – an overview. Environ Sci Pollut Res Int 20:2828–2843. https://doi.org/10.1007/s11356-013-1524-1

Hassan HS, Kashyout AB, Morsi I, Nasser AAA, Raafat A (2015) Fabrication and characterization of nano-gas sensor arrays. AIP Conf Proc 1653:020042. https://doi.org/10.1063/1.4914233

Hassan HS, Elkady MF, Serour NM (2021) Intelligent nanosensors (INS) for environmental applications. In: Handbook of nanomaterials for sensing applications. Elsevier, Amsterdam, pp 321–344. https://doi.org/10.1016/B978-0-12-820783-3.00017-8

Hilal N, Darwish NA, Mohammad AW, Arabi MA (2004) A comprehensive review of nanofiltration membranes: treatment, pretreatment, modelling, and atomic force microscopy. Desalination 170:281–308. https://doi.org/10.1016/j.desal.2004.01.007

Huang Q, Zhang Q, Cicek N, Mann D (2011) Biofilter: a promising tool for mitigating methane emission from manure storage. J Arid Land 3:61–70. https://doi.org/10.3724/SP.J.1227.2011.00061

Huo Z, Dai S, Wang K et al (2007) Nanocomposite gel electrolyte with large enhanced charge transport properties of an I3-/I- redox couple for quasi-solid-state dye-sensitized solar cells. Sol Energy Mater Sol Cells 91:1959–1965. https://doi.org/10.1016/j.solmat.2007.08.003

Huy NN, Thuy VTT, Thang NH et al (2019) Facile one-step synthesis of zinc oxide nanoparticles by ultrasonic-assisted precipitation method and its application for H2S adsorption in air. J Phys Chem Solids 132:99–103. https://doi.org/10.1016/j.jpcs.2019.04.018

Ibrahim Y, Kashyout AB, Morsi I, Shokry Hassan H (2021) Development of nano-SnO_2 and SnO_2:V_2O_5 thin films for selective gas sensor devices. Arab J Sci Eng 46:669–686. https://doi.org/10.1007/s13369-020-04735-9

Jampaiah D, Chalkidis A, Sabri M, Mayes E et al (2019) Low-temperature elemental mercury removal over TiO_2 nanorods-supported MnOx-FeOx-CrOx. Catal Today 324:174–182. https://doi.org/10.1016/j.cattod.2018.11.049

Karn B, Kuiken T, Otto M (2009) Nanotechnology and in situ remediation: a review of the benefits and potential risks. Environ Health Perspect 117:1823–1831. https://doi.org/10.1289/ehp.0900793

Kesselmeier J, Staudt M (1999) Biogenic volatile organic compounds (VOC): an overview on emission. Physiol Ecol 33:23–88

Lanjouw JO, Mody A (1996) Innovation and the international diffusion of environmentally responsive technology. Res Policy 25:549–571

Larsson P, Andersson A (2000) Oxides of copper, ceria promoted copper, manganese and copper manganese on Al2O3 for the combustion of CO, ethyl acetate and ethanol. Appl Catal B Environ 24:175–192

Li Z, Liao F, Jiang F et al (2016) Fluid phase equilibria capture of H2S and SO2 from trace sulfur containing gas mixture by functionalized UiO-66 (Zr) materials: a molecular simulation study. Fluid Phase Equilib 427:259–267. https://doi.org/10.1016/j.fluid.2016.07.020

Li B, Ma J, Zhou L, Qiu Y (2017) Magnetic microsphere to remove tetracycline from water: adsorption, H2O2 oxidation and regeneration. Chem Eng J 330:191–201. https://doi.org/10.1016/j.cej.2017.07.054

Liu X, Pan L, Lv T et al (2011) Microwave-assisted synthesis of ZnO – graphene composite for photocatalytic reduction of Cr (VI). Catal Sci Technol 1:1189–1193. https://doi.org/10.1039/c1cy00109d

Lu D, Sha S, Luo J et al (2020) Treatment train approaches for the remediation of per- and polyfluoroalkyl substances (PFAS): a critical review. J Hazard Mater 386:121963. https://doi.org/10.1016/j.jhazmat.2019.121963

Maldonado-Hódar FJ (2011) Removing aromatic and oxygenated VOCs from polluted air stream using Pt–carbon aerogels: assessment of their performance as adsorbents and combustion catalysts. J Hazard Mater 194:216–222. https://doi.org/10.1016/j.jhazmat.2011.07.082

Mensah K, Mahmoud H, Fujii M et al (2022a) Dye removal using novel adsorbents synthesized from plastic waste and eggshell: mechanism, isotherms, kinetics, thermodynamics, regeneration, and water matrices. Biomass Convers Biorefinery 14:12945. https://doi.org/10.1007/s13399-022-03304-4

Mensah K, Samy M, Ezz H et al (2022b) Utilization of iron waste from steel industries in persulfate activation for effective degradation of dye solutions. J Environ Manag 314:115108. https://doi.org/10.1016/j.jenvman.2022.115108

Moreno-Castilla C (2004) Adsorption of organic molecules from aqueous solutions on carbon materials. Carbon 42:83–94. https://doi.org/10.1016/j.carbon.2003.09.022

Murruni L, Leyva G, Litter MI (2007) Photocatalytic removal of Pb (II) over TiO2 and Pt – TiO2 powders. Catal Today 129:127–135. https://doi.org/10.1016/j.cattod.2007.06.058

Paltanea G, Nemoianu IV, Popovici D, Radu-mircea C (2018) Electrical characterization of titanium dioxide nanotubes using dielectric and electrochemical impedance spectroscopy. Rev Chim 69:3494–3499. https://doi.org/10.37358/RC.18.12.6777

Park J, Lee I (2014) Photocatalytic degradation of 2-chlorophenol using Ag-doped TiO2 nanofibers and a near-UV light-emitting diode system. J Nanomater 2014:250803

Portela R et al (2014) Nanostructured ZnO/sepiolite monolithic sorbents for H2S removal. J Mater Chem A Mater 3:1306–1316. https://doi.org/10.1039/C4TA04440A

Rabiee F, Mahanpoor K (2019) Photocatalytic oxidation of SO_2 from flue gas in the presence of Mn/copper slag as a novel nanocatalyst: optimizations by Box-Behnken design. Iran J Chem Chem Eng 38(3):69–85

Ranveer AC, Pollution C, Board C, Pawar P (2016) Odour pollution and its measurement. Int J Res Appl Sci Eng Technol 3:221–229

Rasalingam S, Peng R, Koodali RT (2014) Removal of hazardous pollutants from wastewaters: applications of TiO2-SiO2 mixed oxide materials. J Nanomater 2014:10. https://doi.org/10.1155/2014/617405

Ren X, Chen C, Nagatsu M, Wang X (2011) Carbon nanotubes as adsorbents in environmental pollution management: a review. Chem Eng J 170:395–410. https://doi.org/10.1016/j.cej.2010.08.045

Rotter H, Landau MV, Carrera M et al (2004) High surface area chromia aerogel efficient catalyst and catalyst support for ethylacetate combustion. Appl Catal B Environ 47:111–126. https://doi.org/10.1016/j.apcatb.2003.08.006

Saleem H, Zaidi SJ, Ismail AF et al (2022) Advances of nanomaterials for air pollution remediation and their impacts on the environment. Chemosphere 287:132083. https://doi.org/10.1016/j.chemosphere.2021.132083

Samy M, Ibrahim MG, Alalm MG, Fujii M (2020) Modeling and optimization of photocatalytic degradation of methylene blue using lanthanum vanadate. Mater Sci Forum 1008:97–103. https://doi.org/10.4028/www.scientific.net/MSF.1008.97

Samy M, Gar Alalm M, Khalil MN et al (2023) Treatment of hazardous landfill leachate containing 1,4 dioxane by biochar-based photocatalysts in a solar photo-oxidation reactor. J Environ Manag 332:117402. https://doi.org/10.1016/j.jenvman.2023.117402

Santhosh C, Velmurugan V, Jacob G et al (2016) Role of nanomaterials in water treatment applications: a review. Chem Eng J 306:1116–1137. https://doi.org/10.1016/j.cej.2016.08.053

Scholten E, Bromberg L, Rutledge GC, Hatton TA (2011) Electrospun polyurethane fibers for absorption of volatile organic compounds from air. ACS Appl Mater Interfaces 3:3902–3909

Shokry Hassan H (2019) Role of preparation technique in the morphological structures of innovative nano-cation exchange. J Mater Res Technol 8:2854–2864. https://doi.org/10.1016/j.jmrt.2019.04.023

Shokry Hassan H, Kashyout AB, Morsi I, Nasser AAA, Raafat A (2014) Fabrication and characterization of gas sensor micro-arrays. Sens Biosensing Res 1:34–40. https://doi.org/10.1016/j.sbsr.2014.04.001

Sidorov SN, Bronstein LM, Davankov VA et al (1999) Cobalt nanoparticle formation in the pores of hyper-cross-linked polystyrene: control of nanoparticle growth and morphology. Chem Mater 11:3210–3215

Sidorov SN, Volkov IV, Davankov VA et al (2001) Platinum-containing hyper-cross-linked polystyrene as a modifier-free selective catalyst for L-sorbose oxidation. J Am Chem Soc 123:10502–10510

Singh U, Sachan P (2016) Application of nanotechnology in monitoring and control of pollution. J Crit Rev 3:23–26
Su F, Lu C, Cnen W et al (2009) Capture of CO2 from flue gas via multiwalled carbon nanotubes. Sci Total Environ 407:3017–3023. https://doi.org/10.1016/j.scitotenv.2009.01.007
Suh DJ (2004) Catalytic applications of composite aerogels. J Non Cryst Solids 350:314–319. https://doi.org/10.1016/j.jnoncrysol.2004.08.230
Taheran M, Brar SK, Verma M et al (2016) Membrane processes for removal of pharmaceutically active compounds (PhACs) from water and wastewaters. Sci Total Environ 547:60–77
Tumuluri U, Howe JD, Mounfield WP III et al (2017) Effect of surface structure of TiO_2 nanoparticles on CO_2 adsorption and SO_2 resistance. ACS Sustain Chem Eng 5(10):9295–9306
Wang W, Liao S, Liu M et al (2014) Polymer composites reinforced by nanotubes as scaffolds for tissue engineering. Int J Polym Sci 2014:805634
Wang Y, Wen Z, Zhang H et al (2016) CuO nanorods-decorated reduced graphene oxide nanocatalysts for catalytic oxidation of CO. Catalysts 6:214. https://doi.org/10.3390/catal6120214
William D, Connell O, Birkinshaw C et al (2008) Heavy metal adsorbents prepared from the modification of cellulose: a review. Bioresour Technol 99:6709–6724. https://doi.org/10.1016/j.biortech.2008.01.036
Wu S, He Q, Tan C et al (2013) Graphene-based electrochemical sensors. Small 9:1160–1172. https://doi.org/10.1002/smll.201202896
Wu H, Ma J, Li Y et al (2014) Photocatalytic oxidation of gaseous ammonia over fluorinated TiO_2 with exposed (0 0 1) facets. Appl Catal B Environ 152:82–87. https://doi.org/10.1016/j.apcatb.2014.01.021
Yadav KK, Gupta N, Kumar V (2017) A review of nanobioremediation technologies for environmental cleanup: a novel biological approach. J Mater Environ Sci 8:740–757
Yang M, Zhang N, Xu Y (2013) Synthesis of fullerene–, carbon nanotube–, and graphene–TiO2 nanocomposite photocatalysts for selective oxidation: a comparative study. ACS Appl Mater Interfaces 5:1156–1164
Yavuz CT, Prakash A, Mayo JT, Colvin VL (2009) Magnetic separations: from steel plants to biotechnology. Chem Eng Sci 64:2510–2521. https://doi.org/10.1016/j.ces.2008.11.018
Ye P, Xu ZK, Wu J et al (2006) Nanofibrous membranes containing reactive groups: electrospinning from poly(acrylonitrile-co-maleic acid) for lipase immobilization. Macromolecules 39:1041–1045. https://doi.org/10.1021/ma0517998
Yunus IS, Kurniawan A, Adityawarman D (2012) Nanotechnologies in water and air pollution treatment. Environ Technol Rev 1:136–148. https://doi.org/10.1080/21622515.2012.733966
Zhang L, Fang M (2010) Nanomaterials in pollution trace detection and environmental improvement. Nano Today 5:128–142. https://doi.org/10.1016/j.nantod.2010.03.002
Zhang X, Shen ÆS, Fan ÆL (2007) Studies progress of preparation, properties and applications of hyper-cross-linked polystyrene networks. J Mater Sci 42:7621–7629. https://doi.org/10.1007/s10853-007-1763-y
Zhang Y et al (2010) TiO_2 graphene nanocomposites for gas-phase photocatalytic degradation of volatile aromatic pollutant: is TiO2-graphene truly different from other TiO2-carbon composite materials? ACS Nano 4:7303–7314
Zhang N, Yang M, Tang Z, Xu Y (2013) CdS – graphene nanocomposites as visible light photocatalyst for redox reactions in water: a green route for selective transformation and environmental remediation. J Catal 303:60–69. https://doi.org/10.1016/j.jcat.2013.02.026
Zhang Q, Teng J, Zou G et al (2016) Efficient phosphate sequestration for water purification by unique sandwich-like MXene/magnetic iron oxide nanocomposites. Nanoscale 8:7085–7093. https://doi.org/10.1039/c5nr09303a
Zhou Q, Li Z, Shuang C et al (2012) Efficient removal of tetracycline by reusable magnetic microspheres with a high surface area. Chem Eng J 210:350–356. https://doi.org/10.1016/j.cej.2012.08.081
Zuo L, Zhang Y, Zhang L et al (2015) Polymer/carbon-based hybrid aerogels: preparation, properties and applications. Materials 8:6806–6848

Chapter 6
Discovery and Application of Nanotechnology Toward Air Purification in Reducing Airborne Pathogens, Canine Allergies, Microflora in Ice Machines and Application of Silver Nanoparticles in Facial Mask

Nabarun Ghosh, Shaily Goyal, Jay Vitale, Aubrey D. Howard, and Prabir K. Banerjee

Abstract Innovations in technology continue to significantly impact both business and society. Emerging markets are increasingly becoming centers of innovation, particularly in efforts to cater to the expanding middle-class and low-income consumers worldwide. Nanotechnology has seen significant advancements in the past decade, particularly in the realm of indoor air quality (IAQ) control. Intellectual capital has fostered innovation in organizations, propelling industrial growth. Many scientific breakthroughs are driven by collaborations between industry and academia. A notable collaboration involves air purification industries in the United States and the United Kingdom working with West Texas A&M University. We have developed, modified, and adopted Advanced Hydrated Photocatalytic Oxidation (AHPCO) and plasma nanotechnology. The company Air For Life (UK) adopted this technology and renamed it AFLPCO nanotechnology. These nanotechnologies have been employed to create air purification systems that reduce indoor aeroallergens, improve air quality, and enhance food preservation. Plasma nanotechnology

---

N. Ghosh (✉) · S. Goyal
Department of Life, Earth & Environmental Sciences, West Texas A&M University, Canyon, Texas, USA

J. Vitale
Air for Life UK LTD, Milton Keynes Business Centre, Milton Keynes, UK

A. D. Howard
Science, Roswell High School, Roswell, NM, USA

P. K. Banerjee
Department of Gastroenterology, RN Tagore Hospital, Kolkata, India

N. Ghosh et al. (eds.), *Nanotechnology in Air Quality Management*, Sustainable Development and Biodiversity 39, https://doi.org/10.1007/978-981-96-8339-0_6

has also been used to sterilize icemakers. This novel system generates redundant oxidizers that efficiently clean the air, targeting and neutralizing particulate matter (PM) on surfaces. Microbial populations in the air and on equipment surfaces can be controlled using this technology. It has proven effective in reducing particulate matter and indoor volatile organic compounds (VOCs), as well as controlling methicillin-resistant *Staphylococcus aureus* (MRSA). We also described the use of silver nanoparticles in the new generation face masks to prevent against contagious diseases.

Keywords Indoor air quality · Air purification · COVID-19 pandemic · Aerosol · PM 2.5 · Advanced Hydrated Photocatalytic Oxidation nanotechnology · Air purifiers · Spore count · VOC · AFL Mask® · LUNA Wall Mount Sanifier®

6.1 Introduction

The rise of respiratory ailments among students in public schools is a growing concern, with allergy and asthma rates in children increasing (Chapman 2003). Air pollution remains a critical public health issue, contributing to an increasing array of adverse health effects that have been extensively documented through comprehensive research efforts across various regions globally. A substantial portion of the populations in both developed and developing nations are exposed to air that fails to meet the quality standards established by the World Health Organization Air Quality Guidelines (WHO-Air Quality Guidelines 2025). Over the past decade, air pollution has emerged as a significant concern, with numerous acute episodes reported in many urban centers worldwide.

Air pollution is associated with a wide range of acute and chronic illnesses, including lung cancer, chronic obstructive pulmonary disease (COPD), and cardiovascular diseases. Ambient air pollution was responsible for 4.2 million deaths as of 2016. Globally, ambient air pollution was estimated to cause approximately 16% of lung cancer deaths, 25% of COPD deaths, about 17% of ischemic heart disease and stroke deaths, and approximately 26% of deaths from respiratory infections. Overall, air pollution is projected to account for around 29% of lung cancer deaths, 43% of COPD deaths, 25% of ischemic heart disease deaths, and 24% of stroke deaths on a global scale (WHO 2025).

Air Quality Index (AQI) forecasting, also known as "Atmospheric Dispersion Modeling," simulates how pollutants like PM 2.5 or Ozone disperse in the atmosphere. The simulation results provide ambient concentrations of each pollutant to calculate the AQI. Air pollution spikes annually in northern India during winter due to crop residue burning, combined with colder temperatures that trap smoke, which is then blown into cities and mixed with auto emissions. On November 20, 2024, Delhi's air pollution reached an unhealthy peak of 225, reported by the New Delhi US Embassy. Severe air quality in Delhi, with an AQI of 978, equates to inhaling about 49 cigarettes daily (AP News, AQI, Delhi 2024).

We have tested the Outdoor and Indoor Air Quality (IAQ) in public schools across the Texas Panhandle area. This project assessed classroom air quality by analyzing mold composition in air ducts, ceiling tiles, and surfaces. Notably, children in Amarillo and Canyon, Texas, spend at least 1260 h per year in school. Indoor Air Quality testing can provide valuable data to recommend preventative measures against the rising rates of respiratory ailments, such as asthma, influenza, rhinovirus, and colds, which are associated with poor air quality. Ensuring a healthy environment in public schools is paramount (US-EPA). Numerous airborne pathogens and allergens can be quantitatively measured. Among them are aeroallergens such as pollens, fungal spores, dust, plant fibers, burnt residues, insect debris, and plant products such as gums and resins.

We used Digital Microscopy, Fluorescent Microscopy (FM), and Scanning Electron Microscopy (SEM) to identify major aeroallergens. The most significant findings were fungal spores including *Alternaria*, *Drechslera*, *Stachybotrys*, and *Cladosporium*. Pollen grains from grass (Poaceae), Short Ragweed (*Ambrosia artemisiifolia*), and Lamb's Quarters (*Chenopodium album*) were also found on ceilings and air vents. This IAQ survey led our research toward developing an efficient air purification system aimed at reducing indoor airborne allergens and volatile organic compounds (VOCs), ultimately improving IAQ. This system would enhance comfort measures and reduce ailments related to asthma and allergies (Ghosh et al. 2005).

6.2 Methodology

In recent years, nanotechnology has permeated various facets of life, including information technology, energy, medicine, food safety, and air quality. We have assessed the air purification systems in novel air purifiers that utilize a new generation of Advanced Hydrated Photocatalytic Oxidation (AFLPCO) nanotechnology, which was modified by Air For Life (UK) and named as AFLPCO. Traditional purifiers mostly rely on filters or the passage of air through the purifier. The novel AFLPCO nanotechnology produces continuous blankets of redundant oxidizers that cleanse the surrounding air and sanitize surfaces by targeting particulate matter in both the air and on surfaces.

6.2.1 Outdoor Air Sampling: Collection of Aeroallergens Using the Burkard Volumetric Spore Trap

Understanding aeroallergens helps clinicians determine the onset, duration, and severity of pollen seasons, guiding allergen selection for skin tests and treatments. This interdisciplinary field includes biology, chemistry, ecology, environmental science, toxicology, and physics. The air around us is dynamic and studying it is complex due to fluctuating factors. Airborne particles directly impact health, necessitating better understanding to prevent allergic reactions, tissue irritation, and disease spread. Knowledge about air-related diseases is crucial for developing preventive measures.

Aeroallergens can induce significant allergic and asthmatic reactions. Characterizing these aeroallergens provides vital information regarding the onset, duration, and severity of pollen seasons, which clinicians use to guide allergen selection for skin testing and treatment protocols. Digital microscopy offers valuable techniques to analyze the structure and function of microscopic objects. The analysis of airborne particles was conducted using the Burkard 7-day Volumetric Spore Trap (Burkard Scientific Sales Manufacturing Co. Ltd., Rickmansworth, Hertfordshire, UK). Typically, volumetric spore traps yield superior data by drawing a fixed volume of air into the trap, facilitating the calculation of pollen and spore concentrations.

During our study, the spore trap was installed on the flat roof of the university building and was standardized using the air flow meter. The sampling equipment was strategically placed in wind-protected locations on a flat surface away from walls and other obstacles. This location is advantageous due to its exposure to the prevailing winds of West Texas and its elevation above the surrounding trees, allowing for effective sampling of wind-borne pollen and spores while minimizing contamination from dirt or sand. Additionally, the absence of overhanging trees or vegetation ensures the integrity of the collected data. The spore trap operates on 25 W of electricity at 240 V, or 50 W at 110 V, with air being drawn through the trap at a rate of 10 L/min. A fan located on the ventral portion of the trap rotates at 2900 rpm to facilitate air intake. The drum, driven by a precision 7-day jewel clock located on the lid inside the canister, rotates at a consistent speed, completing one full rotation over a 7-day period. Hourly observations are possible due to the drum's standard rotation rate of 2 mm/h.

Initially, Melinex tape was used, but after an extensive literature review, the protocol was modified to employ regular transparent Scotch tape, which allows for better aeroallergen attachment and circumvents the emulsification issues associated with Melinex tape. A piece of Scotch tape was affixed to the drum using double-sided tape at the orifice start position. Once attached, the drum was placed inside the equipment and began collecting aerial particles, including fungal spores, pollens, plant debris, and carbonized materials from burnt elements. The distance from the tape's starting point indicates the spore collection time. A trimming block segmented the tape into 24-h periods. A wet mount was prepared for the slide examination by placing two to three drops of distilled water on a clean glass slide. The exposed tape was removed from the drum with fine forceps, lay flat on the trimming block, and aligned with the segment marks. The tape was cut along the segment lines to obtain daily samples. Each strip was positioned sample side up on a slide using forceps, ensuring the tape's long edges were parallel with the slide's long edges, aligning with the microscope's transverse. Excess water was removed with a paper towel to keep the strip in place (Fig. 6.1).

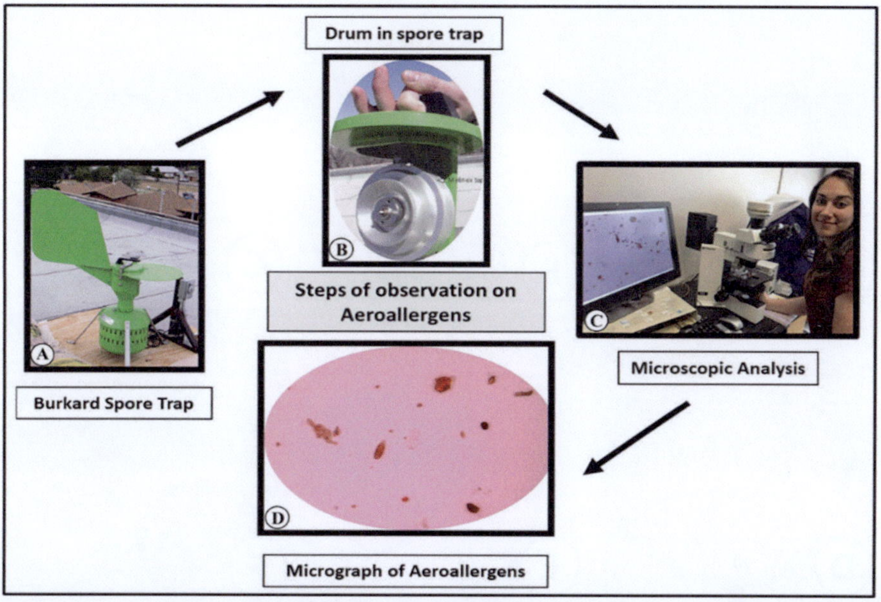

Fig. 6.1 (**a–d**) Steps of analysis of ambient aeroallergen levels using the Burkard Spore Trap. (**a**) Burkard Spore Trap, (**b**) collection of the sample from the inner drum, (**c**) slide preparation and microscopic analysis, (**d**) example micrograph of aeroallergens

The aeroallergens were stained with 1% Safranin O (Sigma Cat no. 84120, Fluka, Microscopy Grade) and mounted using Gelvatol (Burkard Agronomics Division of Burkard Scientific Sales Ltd., UK), a permanent mounting medium. Gelvatol consists of 35 g Gelvatol powder (Burkard Manufacturing Co Ltd., UK), 50 mL glycerol, 100 mL distilled water, and 2 g phenol. Gelvatol preparation involves mixing the Gelvatol powder and phenol in water, letting it sit overnight, then adding glycerol and distilled water while heating over a water bath (65 °C), with continuous stirring to produce the emulsion. Adding a few drops of 1% Safranin O to the Gelvatol enhances the visibility of pollen, highlighting their detailed morphology such as cell wall patterns, pore orientation, and colpi. The Gelvatol-Safranin-O emulsion was uniformly applied to the cover slip using a glass rod, which is essential for producing a high-quality slide. The cover slip was placed at a 45° angle on one side of the tape and slowly lowered across the remainder of the tape using a bent needle to ensure proper placement. The slide was then ready for examination under a compound microscope (Ghosh et al. 2003a, b, 2006). Samples were collected using Melinex tape placed in a Burkard Spore Trap. The tape was stained with safranin and mounted on slides. Micrographs of aeroallergens were captured utilizing an Olympus BX-40 microscope connected to a DP-70 digital camera with *ImagePro* 6.0 software (Fig. 6.2). Scanning Electron Microscopy (SEM) was employed to observe ultrastructural details such as pores and ornamentations on pollen. SEM measured pollen grains, revealing the size of colpi or sulci

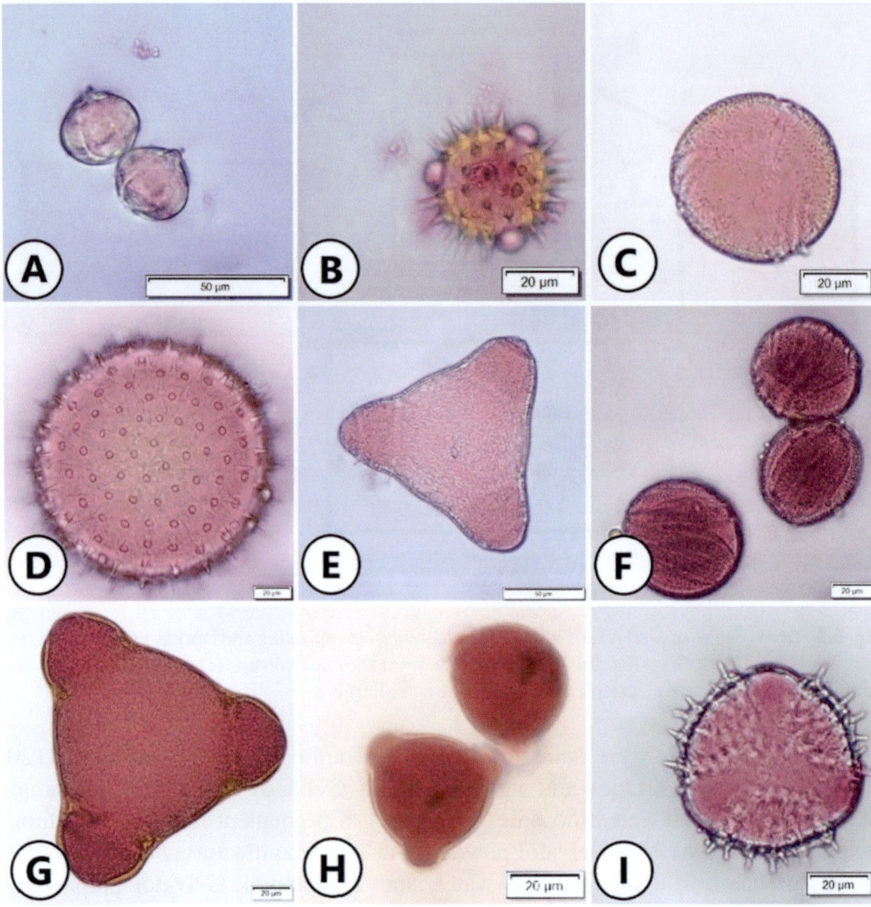

Fig. 6.2 (**a–i**) Most common pollen specimen from the locality. (**a**) Frog fruit (*Phyla nodiflora*), (**b**) Echinacea (*Echinacea angustifolia*), (**c**) Bractless Blazingstar (*Mentzelia nuda*), (**d**) Winecup (*Callirhoe involucrata*), (**e**) Scarlet Gaura (*Gaura coccinea*), (**f**) Golden Dalea Prairie Clover (*Dalea aurea*), (**g**) Hartweg Evening Primrose (*Calylophus hartwegii*), (**h**) Silverleaf nightshade (*Solanum elaeagnifolium*), (**i**) Ironweed (*Vernonia marginata*)

and distances between microstructures, crucial for pollen identification (Fig. 6.3). Aeroallergens were examined under an Olympus BX-40 microscope equipped with FITC and TRITC fluorescent filters, a mercury lamp. Aeroallergens were viewed, recorded, and analyzed. A high-pressure mercury lamp was employed to excite storage molecules or proteins exhibiting autofluorescence (Figs. 6.4, 6.5, 6.6, 6.7 and 6.8).

The aeroallergen analysis data that we obtained offer an extensive overview of allergy and asthma trends in the Texas Panhandle region over the past two decades. Since 2007, allergy and asthma cases have doubled in the Texas Panhandle area, with data showing a steady increase in aeroallergen concentrations (Ranaivo 2011).

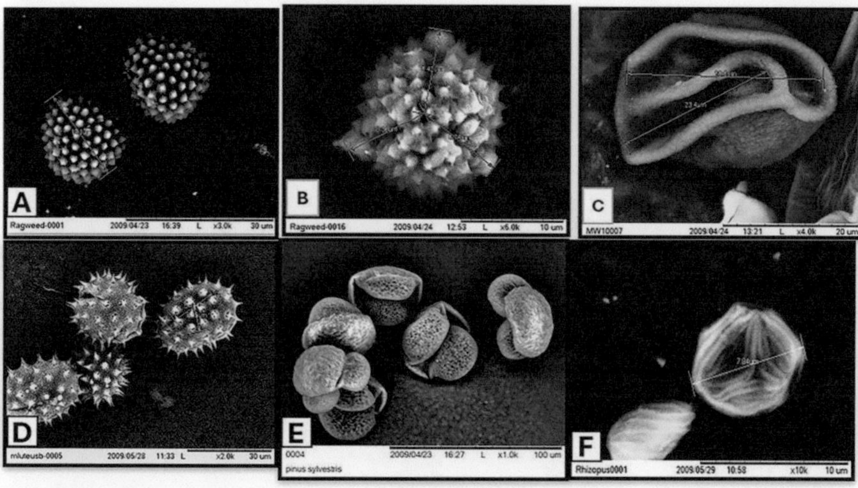

Fig. 6.3 (**a–f**) Most common Aeroallergen specimen from the locality. The micrographs were captured using an SEM TOP 3000 Scanning Electron Microscope attached to a computer. SEM micrographs showing: (**a**) Short Ragweed (*Ambrosia artemisiifolia*), (**b**) Measurement of Short Ragweed (*Ambrosia artemisiifolia*), (**c**) Measurement of Oak pollen (*Quercus robur*), (**d**) Malva (*Malva sylvestris*), (**e**) Pine (*Pinus sylvestris*), (**f**) Golden Dalea Prairie Clover (*Dalea aurea*), (**g**) Hartweg Evening Primrose (*Calylophus hartwegii*), (**h**) Silverleaf nightshade (*Solanum elaeagnifolium*), (**i**) Ironweed (*Vernonia marginata*)

This trend correlated with warmer climates and shifts in flowering seasons, contributing to the rise in allergy cases. Aeroallergen monitoring, utilizing a Burkard Spore Trap, revealed a consistent rise in pollen and mold spores. Temperature and precipitation directly influence mold spore concentration. Higher temperatures decrease mold, while increased rainfall raises mold levels.

The continuous aeroallergen monitoring and innovative technologies highlighted in our analysis played a vital role in addressing the rise of allergy and asthma cases in the Texas Panhandle region. The most significant aeroallergens recorded include pollens from grass (Poaceae), Short Ragweed (*Ambrosia artemisiifolia*), Scotch Pine (*Pinus sylvestris*), Common Sunflower (*Helianthus annuus*), Hairy Sunflower (*Helianthus hirsutus*), Buffalo Bur (*Solanum rostratum*), Purple Nightshade (*Solanum elaeagnifolium*), and Lamb's Quarters (*Chenopodium album*). Fungal spores such as *Alternaria*, ascospores from Pezizales, *Drechslera*, *Stachybotrys*, *Cladosporium*, and *Curvularia* were also significant (Fig. 6.8). Temperature was found to have an inverse relationship with mold spore concentration, while rainfall directly affected mold count, with increases in precipitation bringing higher mold spore concentrations. Ragweed emerged as the major culprit, alongside other important allergens such as sagebrush, red-root pigweed (*Amaranthus retroflexus*), lamb's quarters (*Chenopodium album*), Russian thistle or tumbleweed (*Salsola*

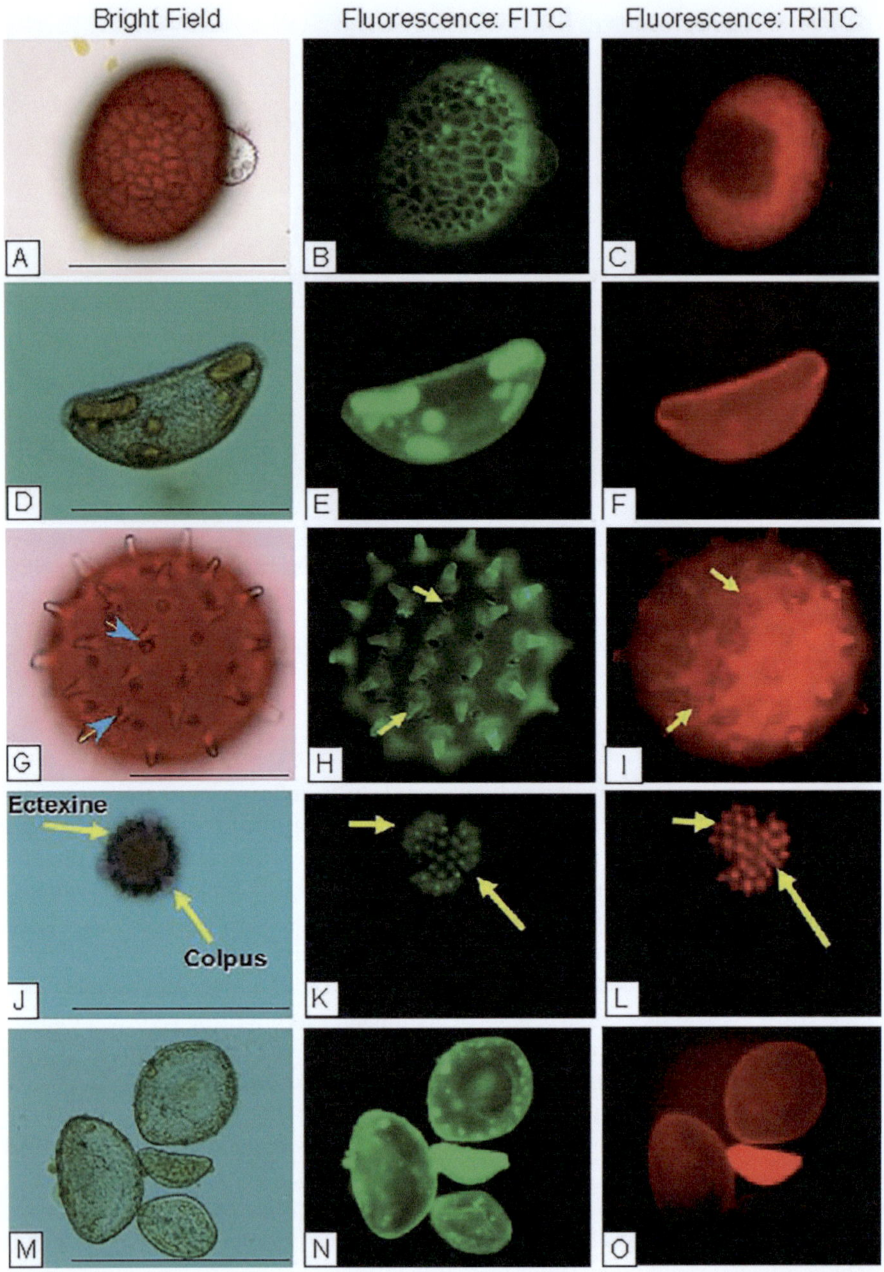

Fig. 6.4 (**a**) *Lilium longiflorum* (Easter lily) under bright field, (**b**) with FITC fluorescent filter, (**c**) with TRITC fluorescent filter. (**d**) *Alstroemeria* (Easter lily) under bright field, (**e**) with FITC filter, (**f**) with TRITC filter (side view), (**g**) shows *Hibiscus rosa-sinensis* (China rose) pollen under bright field, (**h**) with FITC filter, (**i**) with TRITC filter, (**j**) shows *Bellis perennis* (common daisy) under bright field, (**k**) with FITC filter, (**l**) with TRITC filter, (**m–o**) top views of *Alstroemeria*. Pollen grains were collected, spread uniformly, and stained with Fluorol Yellow 88, an anthraquinone-derived azo dye. Fluorescent stains enhance visualization of pollen ektexine architecture (Ghosh et al. 2010, 2017b)

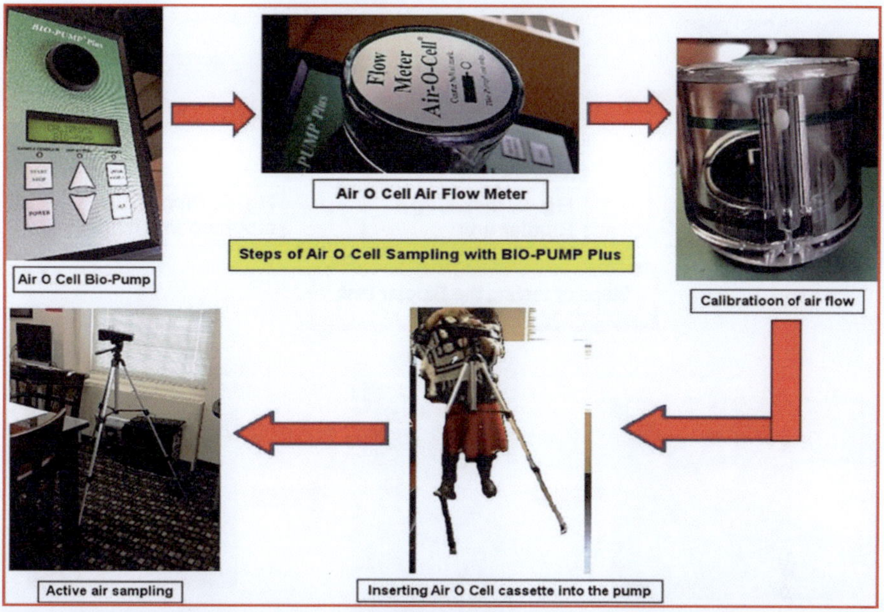

Fig. 6.5 Steps of sampling the indoor air using the Air-O-Cell technology

oppositifolia), and English plantain. Grass and trees, too, were important sources of allergenic pollen, including oak, ash, elm, hickory, pecan, box elder, and mountain cedar. Regional aerobiology plays a crucial role in the diagnosis and treatment of allergic rhinitis.

6.2.2 Indoor Air Sampling

Air sampling was conducted in a room using two techniques: Petri plates with Brain Heart Infusion Agar were placed 5, 10, and 20 feet from the air purifier and exposed for 24, 48, 72, 96, and 120 h. Data on microbial colonies formed were collected with and without the air purifier. Bacteria found included *Bacillus*, *Coryneform*, coagulase-negative *Staphylococcus*, *Micrococcus*, and gram-negative bacilli. Fungal colonies included *Alternaria alternata*, *Cladosporium*, *Drechslera*, *Stachybotrys*, and *Curvularia*. A gradual decrease in microbial colonies was observed with longer use of the air purifier.

In the second part, we used Air-O-Cell® sampling equipment, designed for rapid collection of a wide range of airborne aerosols, including mold spores, pollen, insect parts, skin cells, fibers, and inorganic particulate (Fig. 6.5). Optical microscopy analysis of samples using Air-O-Cell® cassettes reported the presence of *Aspergillus*/*Penicillium* at 200 count/m^3 in a Lab room.

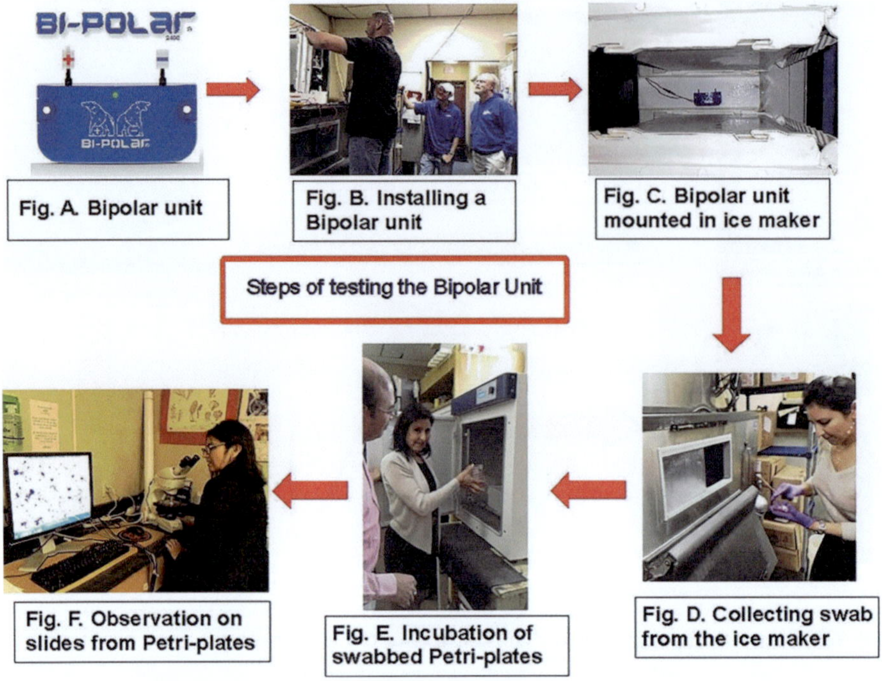

Fig. 6.6 Stepwise experimental procedures followed for testing the Bi-Polar® unit

Fig. 6.7 The Petri plates that were exposed to the icemaker without the Bi-Polar® unit

6.2.3 Improving Indoor Air Quality (IAQ) Using AFLPCO Nanotechnology

Breathing the clean air has an impact on individual health and daily life or work. Indoor air quality significantly impacts our well-being and productivity. Clean air enhances cognitive functions, sleep quality, and overall health. Research indicates that the air we breathe plays a role in energy production by aiding in the breakdown

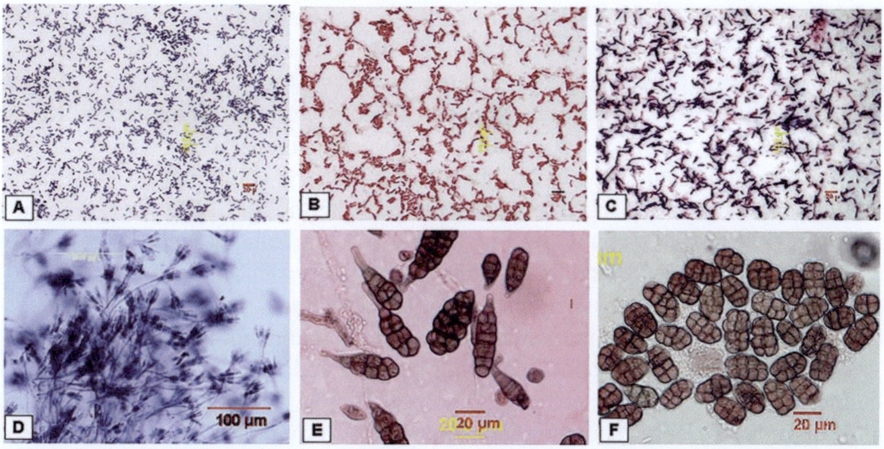

Fig. 6.8 Microscopic study of bacteria and fungi obtained from different part of icemaker prior to installation of air purifier. (**a**) Gram-positive bacilli from a swab sample of the upper wall of the icemaker. (**b**) Gram-negative bacilli from the icemaker floor. (**c**) Spore-forming bacilli from the icemaker floor. (**d**) *Penicillium* sp. from the wall of the icemaker. (**e**) *Alternaria alternata* conidia and (**f**) *Pithomyces* sp. spores from the corners. Swab cultures were incubated for 24 h at 37 °C

of food and drinks. Our evaluation focused on air purifiers utilizing AFLPCO nanotechnology, which operate without filters and do not require air to pass directly through the purifier. This technology generates a blanket of redundant oxidizers that clean the surrounding air and sanitize surfaces by targeting particulate matter both in the air and on surfaces. Technological advancements continue to profoundly influence business and society. Technology transfer, along with related concepts such as technology diffusion and dissemination, have become an integral part of research and development policy at both national and international levels (Hameri 1996). The continuous innovation and collaboration between corporate and academic worlds are crucial for advancing technology and addressing global challenges. It is truly remarkable how advancements in technology continue to shape our lives, improving health, safety, and efficiency.

We assessed air purifiers targeting particulate matter (PM) in the air and on surfaces, focusing on novel air purifiers developed by companies in the United States and the United Kingdom. These devices aim to reduce aeroallergens such as pollen, bacteria, fungal spores, dust particles, fibers, animal dander, and VOCs. Our goal is to alleviate allergy and asthma symptoms by minimizing indoor aeroallergens. AFLPCO nanotechnology has proven effective in reducing indoor aeroallergens, MRSA in hospitals, and microflora in food processing. Evaluations at BSA Hospital in Amarillo, Texas, showed a 68.5% average reduction in bacterial population using the induct air purification system. The technology demonstrated no Ozone production or side effects on human cell cultures.

These units were presented at the China Clean Expo (2013) and the Trade Fair in Chicago (2015) and are marketed in China, Hong Kong, Singapore, Bangladesh,

Dubai, the United States, and the United Kingdom. Our analysis of indoor aeroallergen data clearly indicated a significant correlation between the concentration of pollen from trees, grass, and weeds and the number of patients suffering from allergies and asthma. The application of nanotechnology shows potential for improving quality of life by adding new features to the original functions of products. We recorded a significant reduction in the spore count per cubic meter (*Aspergillus/Penicillium*) with increased intervals while running the AFLPCO air purification unit using Air-O-Cell technology (Fig. 6.5). After continuously running the unit for 2 weeks, the count decreased to 40 count/m^3 from 200 count/m^3, and after 4 weeks, it reached 0 count/m^3. This demonstrated that using a negative ion purification system effectively eradicates microbial aeroallergens in indoor air (Revanna et al. 2016).

A major group of microbes, including bacteria and fungi, can cause food contamination during processing. This technology, if used, will prove to be an efficient way of reducing food contaminants, especially during meat processing, which affects thousands of lives worldwide. AFLPCO nanotechnology has ushered in a new era of air purification, advanced contaminant-free food processing, and a mobile phone sanitizer system, which are being marketed in the United States, the United Kingdom, China, Hong Kong, Singapore, Bangladesh, and Dubai (Ghosh et al. 2015a, b).

6.3 Nanotechnology in Purification Equipment Reducing Microflora in Ice Machines

Scientific tests revealed that ice served in six out of ten Britain's most popular high street restaurants had higher levels of bacteria than samples of water taken from their lavatory bowls (The Daily Mail 2013). Dirty ice machines causing contamination via ice cubes are also a major health problem in different countries (Triggiano et al. 2024). In 2013, in the United States, an outbreak of norovirus caused by dirty ice at a golf course in Phoenix sickened 80 people. In 2018, in North Carolina, "dirty ice machines" was often cited in the public reports. In China, in 2013, ice cubes at fast-food chains were found to be riddled with 13 times more bacteria than toilet water. In 2017, in Hong Kong, edible ice samples collected under the routine Food Surveillance Program of the Centre for Food Safety (CFS) were found to contain excessive coliform bacteria, organisms that are present in the environment and in the feces of all warm-blooded animals and humans. European countries have shown dirty ice cases as well. The BBC investigated iced beverages in the United Kingdom with an undercover study in 2017 and found that out of 30 samples taken from three different fast-food restaurants, more than half were contaminated with fecal coliform bacteria (BioSure Professional 2024). To prevent any potential contamination, the interior surfaces of the ice machine must be cleaned and sanitized regularly. This report justifies the demand for and use of novel equipment that will sanitize the surface of an ice machine continuously. Bacteria can form biofilms on equipment surfaces, creating a slimy layer that is difficult to clean. Bacteria that adhere to equipment can encase themselves in a hydrated matrix of polysaccharide and protein, forming a slimy layer known as a biofilm. The presence of moisture and organic matter aids in biofilms formation. They can form on the walls of icemakers and refrigerators, leading to contamination through ice cubes. Cleaning and sanitizing biofilms are challenging, so regular sanitation of ice machine interiors is crucial to prevent contamination.

6.3.1 Principles and Methodology of Air Purifier in Sanitizing the Ice Machines

As the applications of nanotechnology expand across virtually all industry sectors, it is essential to remember that only a few in the business world have experienced its implementation (Nanowerk Spotlight 2024). There is a significant demand for an energy-efficient air purifier that effectively cleans the air. Air For Life (UK) has modified plasma nanotechnology in Bi-Polar units to create cold plasma discharge with positive and negative ions from water vapor. Plasma ionization nanotechnology is a safe and effective air purification method. Experiments showed a significant reduction in microbial entities after running the Bi-Polar attached in the ice making machine for 120 h.

We evaluated the efficiency of the Bi-Polar® unit in sanitizing the ice machine surface. The Bi-Polar® unit creates a cold plasma discharge consisting of positive and negative ions from water vapor in the air. These ions cluster around airborne particles such as mold, viruses, bacteria, and allergens like pollen. The ions not only help to make these particles larger, making them easier to capture by air filter, but they also physically impact viruses and other pathogens. In the case of viruses, the ions affect the molecular structure of the protein coat, which is a key part of the infection process. With the proteins altered, viruses cannot infect our cells and thus cannot make us ill (Bi-Polar Ionization 2024) (Figs. 6.9 and 6.10).

The Petri plates exposed in the icemaker without the Bi-Polar® unit showed extensive microbial growth (Fig. 6.7).

Chances of contamination in the ice cubes reduced considerably on reduction of microbial population inside the icemaker. Restaurants, universities, public schools, hospitals, and hotels are the predominant users of the Bi-Polar 2400, but this technology has been proved very beneficial in ice machines due to some of the advantages like it reduces ice machine cleaning time, increases the life span of ice machines, reduces service calls relating to ice machines, eliminates the threat of health code violations for unsanitary ice machines, protects the brand and reputation.

For the best result, we recommend continuous use of the Bi-Polar® unit and clean the icemaker regularly (at least twice a month). This is necessary to remove the inactivated particulate matter from the icemaker. The use of multiple Bi-Polar® units is recommended based on the size of the icemaker for more efficient performance

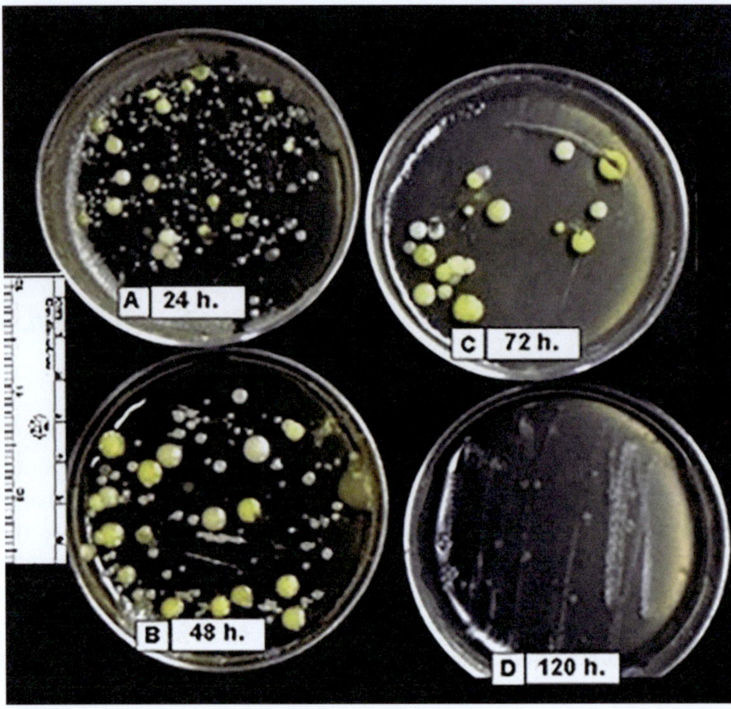

Fig. 6.9 Petri plates labeled **a** through **d**, illustrating swab cultures while operating the Bi-Polar® unit. Observations were made as follows: microbial colonies after 24 h (**a**), 48 h (**b**), 72 h (**c**), and 120 h (**d**). There was significant decrease in bacteria, fungi, slime molds, and cyanobacteria observed after 120 h or more

and cleaning. A dirty icemaker in restaurants or hospitals could be hazardous to the public's health. After running the Bi-Polar® for 120 h or more, there was a significant reduction in microbial entities including bacteria, fungi, slime molds, and cyanobacteria that has left blue-green patches on the icemaker's wall.

6.4 Photocatalytic Oxidation Nanotechnology in Reducing Canine Allergies

The COVID-19 pandemic has highlighted the importance of indoor air quality. Air pollution remains a major issue, causing millions of deaths annually. There is a need for efficient air purifiers that clean air effectively. The AFL Mini Sanifier II® uses filter-less technology to purify rooms up to 330 square feet, including various living spaces and vehicles. It features a UVC lamp that destroys germs with UV light and a nano metal catalyst that generates negative ions to eliminate bacteria, viruses, and VOCs. Its ionizers also reduce allergens such as dust, pollen, and pet dander. A

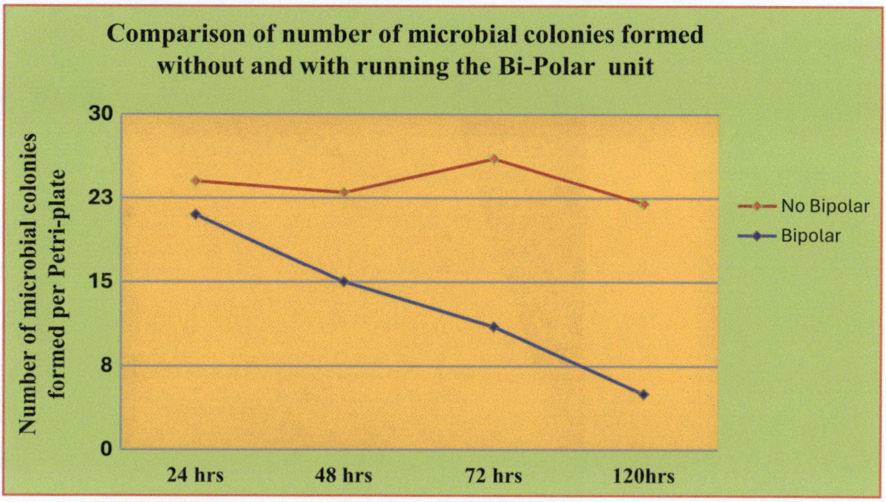

Fig. 6.10 The graph shows reduction in number of colonies formed with the use of air purifier

study with dogs assessed the AFL Mini Sanifier II®'s efficiency in reducing indoor aeroallergens. Data from dogs treated for allergies were compared with aeroallergen indices. Prepared slides exposed to air before and after using the purifier showed a significant correlation between aeroallergen levels and allergy incidence in dogs. Nanotechnology allows manipulation of atoms and molecules at the nanometer scale. Photocatalytic oxidation (PCO) is a powerful air purification technology that destroys particles as small as 0.001 microns. It eliminates carbon-based impurities such as bad odors, volatile organic compounds (VOCs), allergens such as dust mites, mold, pollen, and fungal spores. PCO can also neutralize harmful pathogens, including bacteria, viruses, and fungi responsible for contagious diseases (Miyazaki and Islam 2007).

6.4.1 Assessing AFL Mini Sanifier II® Nanotechnology in Reducing Aeroallergens

We utilized a fiberglass chamber to evaluate the efficacy of the AFL Mini Sanifier II® air purifier (Fig. 6.11) in mitigating airborne pathogens and Particulate Matter PM 2.5. Our methodology consisted of several steps: Initially, we cleaned the surfaces using Clorox wipes and allowed them to dry for a duration of 24 h. Subsequently, we dispersed ISO 12103-1 Ultrafine Dust Particles, with an average size of 2.75 microns (PTI Powder Tech., Minnesota), employing four small blower fans positioned at the four corners of the chamber. The aerosol was permitted to saturate for 24 h and settle. We measured the initial PM 2.5 concentration utilizing a Garosa Air Quality Monitor Formaldehyde Detector and Accurate PM 2.5 Micron

Fig. 6.11 AFL Mini Sanifier II®

Particulate Matter Dust Pollution Multi Tester. The AFL Mini Sanifier II® was operated for intervals of 24 h, 48 h, 72 h, 120 h, and weekly. PM 2.5 concentrations were recorded at each interval. Testing in the AFL fiberglass chamber showed a gradual reduction in mold spore count and PM 2.5 concentration over time. Total VOC (TVOC) and formaldehyde readings significantly decreased after running the AFL Mini Sanifier II® for 1 week. Specifically, the formaldehyde count dropped from 563 to 15, TVOC count decreased from 2134 to 185, and PM 2.5 count nearly reached 0. Simultaneously, the colony count reduced to 0 after 1 week of using the air purifier (Fig. 6.12). This proved the efficiency of this air purifier to test against aeroallergens.

6.4.2 Comparing the Incidence of Allergy Among Dogs with Aeroallergen Data

Dogs can suffer from allergies caused by pollen and molds, just like humans. According to the SPCA (Society for the Prevention of Cruelty to Animals), about 20% of pets suffer from allergies. As one of the most common types of companion animals, the health and well-being of dogs are of utmost importance. No one wants to see their pet scratch themselves raw or end up hairless and covered with sores.

The most common symptom is canine atopic dermatitis, often leading to secondary issues. This study examined the link between aeroallergens and allergies in dogs and evaluated the effectiveness of air purifiers in a clinic in Amarillo, Texas. The

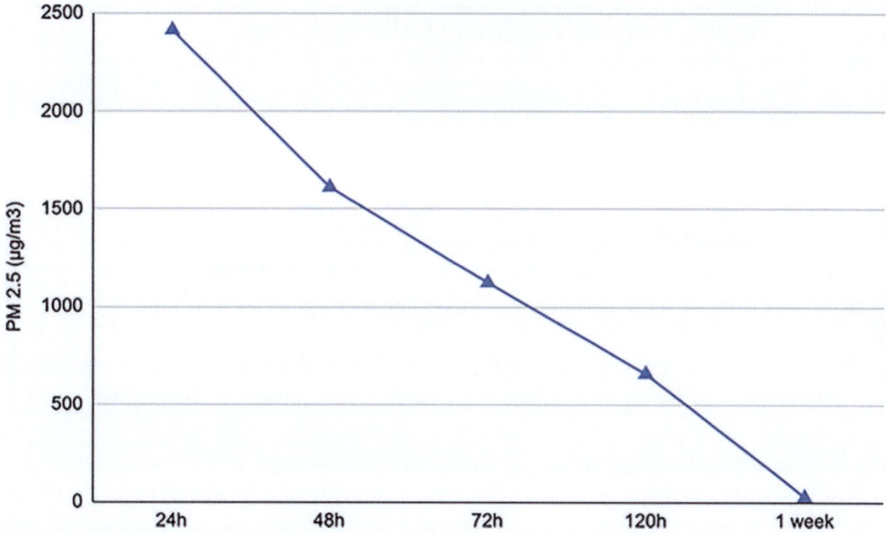

Fig. 6.12 Decrease in PM 2.5 concentration in glass chamber over specified intervals with the use of AFL Mini Sanifier II®

AFL Mini Sanifier II® air purifier, fitted with a UVC lamp, destroys germs as they pass through UV light. The light rays react with the nano metal AFLPCO® catalyst, producing an abundance of negative ions that seek out and destroy carbon-based molecules, including microbial organisms and VOCs. Hydrocarbon contaminants are broken into water and carbon dioxide. Safety evaluations showed no side effects on human cell cultures. Indoor aeroallergens, such as airborne fungal spores, bacteria, and animal dander, significantly reduce when using these air purification units, leading to improved indoor air quality and decreased allergy episodes in dogs.

Treating allergies is not simple, with treatments ranging from oral medications, topical treatments, and hyposensitization to complete avoidance. One of the purposes of this study was to determine if there is a correlation between aeroallergens and allergies in dogs. The next part of the study aimed to assess whether the concentration of aeroallergens in the air is related to inhalant allergies in dogs.

The data were collected from dogs admitted to an animal hospital in Amarillo, Texas, and compared to the aeroallergen indices of respective years. The impact of air purifiers on aeroallergens in the indoor air of the clinic was also analyzed by setting up slides and culture plates around the clinic and observing them under a microscope with a digital camera.

The analysis begins with descriptive statistics, including time plots that examine potential trends in the data, such as aeroallergen concentrations over time (seasonally, daily, and hourly). Daily weather data, including temperature, wind speed, precipitation, humidity, and average soil temperatures, were recorded. The characteristics of the pollen season were assessed using the Principal Pollen Period (PPP), defined as the first day with at least 1 pollen grain/m^3, with subsequent days

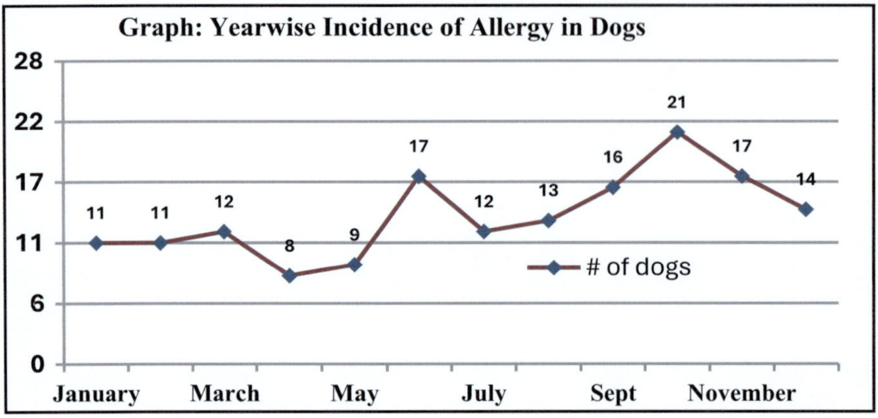

Fig. 6.13 Incidence of allergy in dogs, an average of 3 years of data

also having at least 1 pollen grain/m^3. The end of the season was marked by the first of 5 consecutive days without pollen grains. The correlation between pollen concentrations during the PPP and primary meteorological parameters was analyzed using Pearson's correlation coefficient and Spearman's correlation test, with significance levels set at 1% and 5%. These data assisted in identifying the most prevalent pollen and spores. Atmospheric pollen counts were indexed using a contemporary application of Thommen's postulates, allowing estimation of the clinical significance of various pollen types by combining data on in vivo allergenicity and terminal velocity. At the conclusion of each year, pollen and spore data were charted to identify trends in peaks throughout the year. Daily recordings of pollen and spore counts, along with meteorological data, reflect the weather patterns. The data were utilized to determine pollen peaks and basic patterns for the pollination periods of local vegetation (Ghosh et al. 2004, 2017a, b; Creip 1967).

The graph (Fig. 6.13) shows peaks in dog allergy cases in June and September–October, coinciding with the highest yearly aeroallergen counts (Ghosh et al. 2004).

6.4.3 Result and Discussion: Nanotechnology in Reducing Canine Allergies

The analysis of dog allergies, comparing cases in the control group (with no air purifier) to those in the treated group (using the AFL Mini Sanifier II®), revealed a gradual reduction in allergy symptoms with the use of the air purifier. The geographical location of the Texas Panhandle contributes to a varied composition and concentration of aeroallergens (Fig. 6.14). Dogs display various types of hypersensitive reactions, including redness on the skin and paws, continuous scratching, and respiratory discomfort (Fig. 6.15) (Muller and Krik 1969; Ralf et al. 2002). The

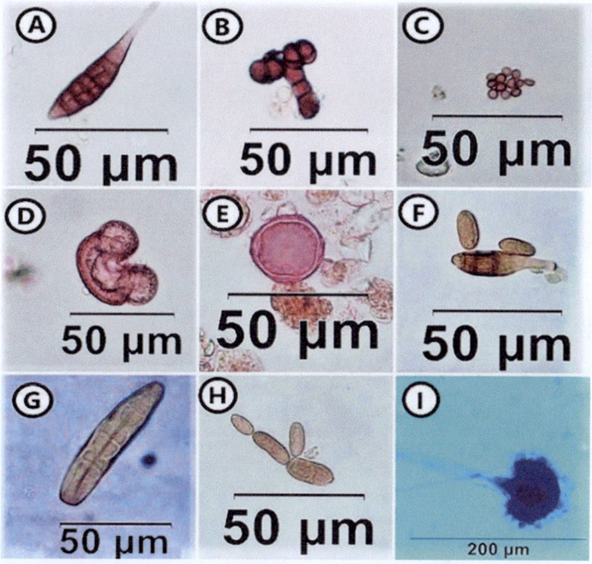

Fig. 6.14 (**a–i**) The most frequent aeroallergens of Texas Panhandle: Fungal spores: (**a**) *Alternaria alternata*, (**b**) *Ustilago*, (**c**) *Cladosporium*, (**f**) *Alternaria*, (**g**) *Drechslera*, (**h**) *Stachybotrys*, (**i**) *Aspergillus* spp.; Pollens: (**d**) *Pinus sylvestris*, (**e**) *Solanum elaeagnifolium*. These aeroallergens were trapped using a Burkard Volumetric Spore Trap, stained and mounted with Safranin-Gelvatol. The images were captured using a BX-40 Olympus Microscope with a DP-74 Digital camera

analysis of 3 years of data revealed a constant variation in the presence of different aeroallergens. A seasonal variation was observed in the Texas Panhandle area. The combinations of outdoor and indoor aeroallergens also varied significantly, as shown in Figs. 6.2, 6.3, 6.4, and 6.14.

6.4.4 Conclusion: AFL Mini Sanifier II® in Reducing Canine Allergies

Nanotechnology can be employed for air purification by developing air purifiers or equipment with catalytic materials that break down pollutants into harmless substances through chemical reactions. This technology can trap and remove contaminants such as particulate matter, VOCs, and microbes from the air. Additionally, nanotechnology can enhance the efficiency and effectiveness of filtration systems by increasing surface area and improving the capture of pollutants (Answers.com: Nanotechnology 2024). Dogs suffer from similar inhalant allergies like humans, which is caused by pollen, mold, and other allergens (Creip 1967; Muller and Krik 1969; Ralf et al. 2002). Data on dog allergies were collected and analyzed from dogs admitted to an animal hospital in Amarillo. AFLPCO® (Air For Life Photocatalytic Oxidation) is a highly advanced air purification technology capable of destroying particles as small as 0.001 microns (nanometers), including carbon-based impurities such as bad odors, volatile organic compounds (VOCs), allergens such as household dust mites and their droppings, mold, pollen, and fungal spores. This AFLPCO® nanotechnology was applied to build the AFL Mini Sanifier II®.

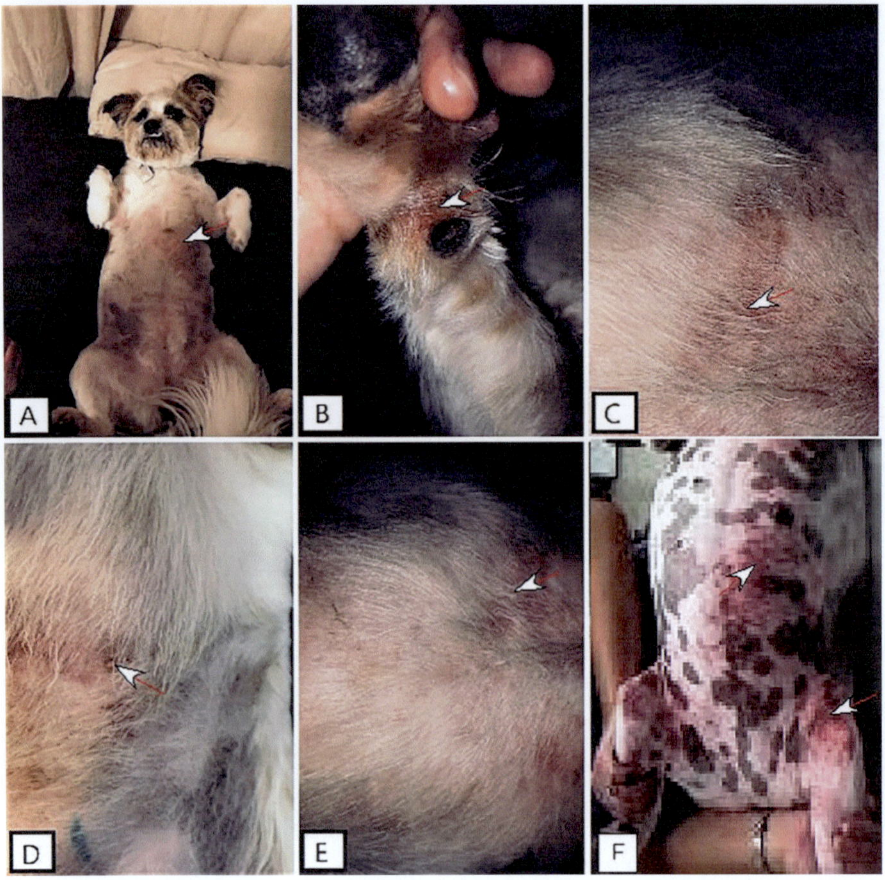

Fig. 6.15 **(a–f)** Arrows display hypersensitive skin reactions and scratch wounds on the body and paws

Using the AFL Mini Sanifier II® resulted in a gradual reduction of aeroallergens, including PM 2.5 and mold spore concentrations, total VOC (TVOC), and formaldehyde concentration in room air (Ghosh et al. 2015a, 2021). Consequently, this reduced the symptoms of allergy among sick dogs. The use of the AFL Mini Sanifier II® can improve indoor air quality (IAQ) and help people combat infections, allergic rhinitis, and other respiratory ailments.

6.5 Nanotechnology Applied to the Air Purifiers in Eradication and Prevention of Indoor Insects

Observing the link between weather, local aeroallergens, and respiratory issues is vital in allergy and asthma studies. In addition to natural allergens like pollen and fungi, insect parts also contribute to airborne particles and allergies such as asthma and allergic rhinitis. Insects can become airborne and attach to other allergens, impacting health by contaminating food, household items, and pets. Symptoms can include sneezing, itchy eyes, skin rashes, and breathing difficulties. Various types of micro-insects have been recorded using our Burkard Volumetric Spore Trap (Fig. 6.16a–d).

The employment of Scanning Electron Microscopy (SEM) in analyzing the micro-arthropods (Figs. 6.17 and 6.18) has revolutionized the field of Entomology. The remarkable microstructures found in the insects' exoskeleton have facilitated detailed examinations that were previously unattainable. This advancement has not only enriched our comprehension of insect morphology but also fostered the development of extensive databases on airborne microarthropods, aiding in the accurate identification and classification of insect species. Such databases have proven invaluable for aerobiological and studies on micro-arthropods, as well as for

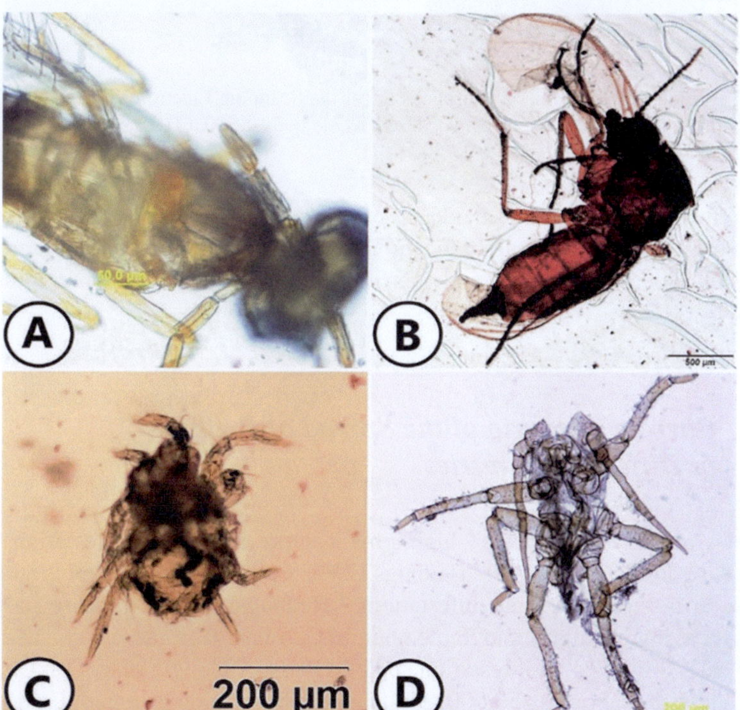

Fig. 6.16 (**a–d**) Micro-arthropods trapped in the Burkard Spore Trap

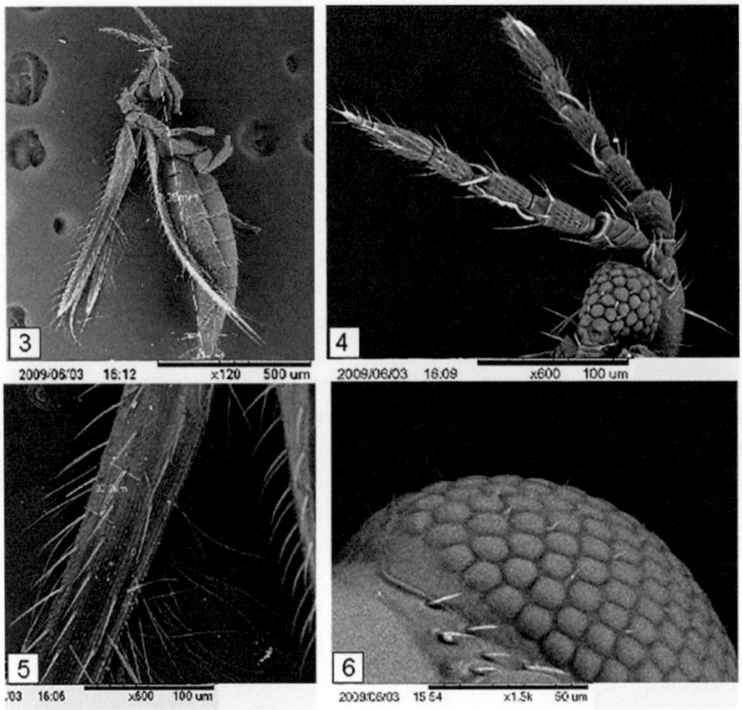

Fig. 6.17 Various airborne micro-arthropods under the Scanning Electron Microscope, Tabletop. (3) Pear Thrip (*Taeniothrips inconsequens* Uzel) whole body X500, (4) Antennae X600, (5) Hairs on wings X600, (6). Compound eyes X1500

forensic investigations where evidence of micro-insects can be used to trace geographic origins. The practical applications of SEM in entomological research offer insights into local biodiversity and contribute to broader environmental monitoring efforts.

6.5.1 Current Scenario of the Spread of Bedbugs in Different Countries

The news article on "Express" highlights a surge in bedbug infestations in the United Kingdom with the title: "Bedbugs LIVE: Full list of UK hotspots and symptoms as Brits report surge in infestations." Bedbugs tend to increase in warmer months, peaking in August and September. In October 2023, there was a rise in pest control calls following an infestation spike in Paris. A Google search on December 4, 2023, revealed about 8,150,000 results on bedbugs' global spread. France saw a 65% increase in pest control visits due to rising temperatures and humidity. Bedbugs

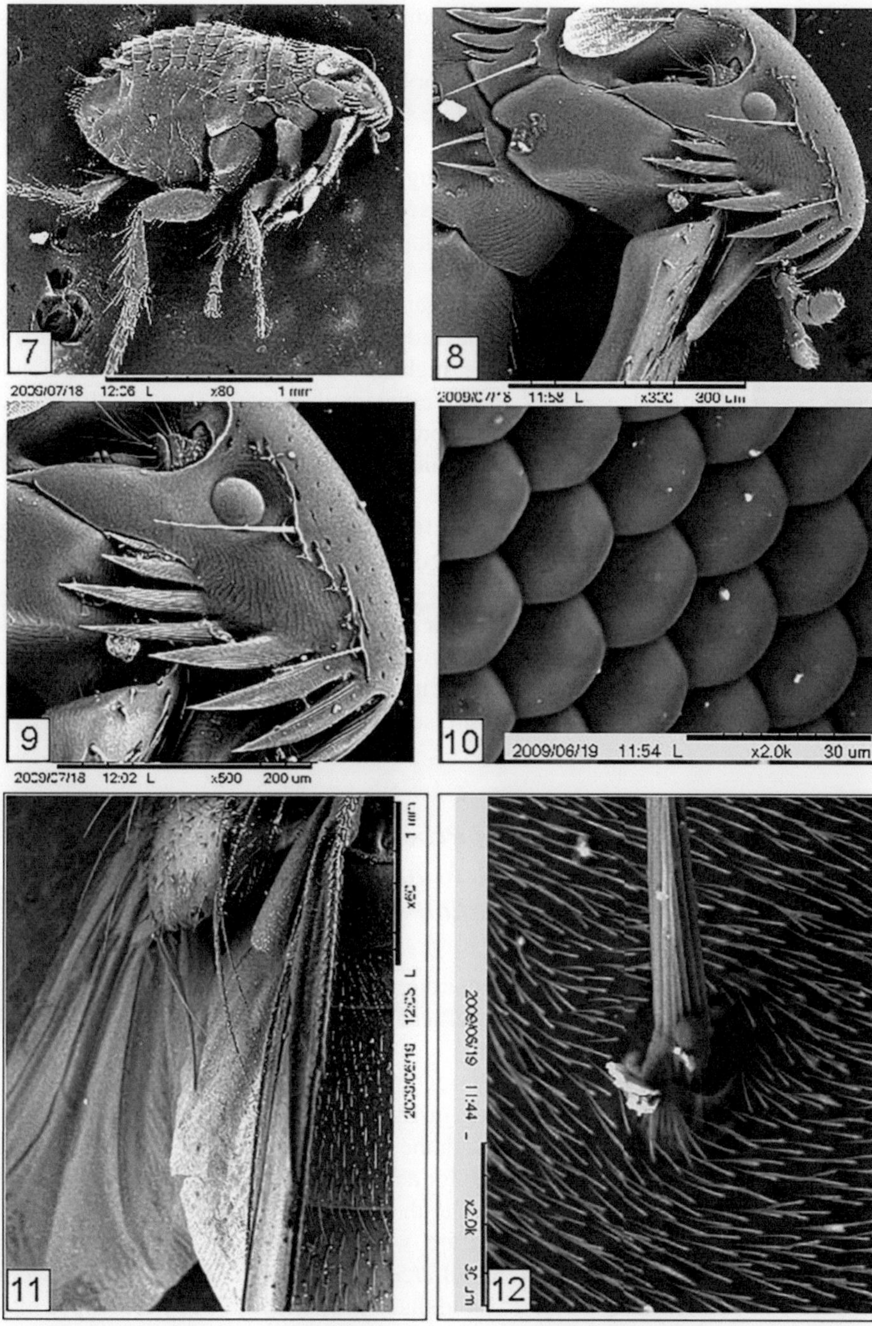

Fig. 6.18 (7–12) Scanning Electron Micrographs of Cat flea (*Ctenocephalides felis*) and its parts

are small insects living on furniture or bedding, causing itchy bites without serious health issues.

Control methods include reducing indoor humidity below 50%, regular cleaning, and encasing mattresses and pillows. While some chemicals are effective, they pose risks of toxicity. Physical strategies like UV irradiation offer residue-free advantages over chemicals. Bedbugs spread through individual transmission, contaminating bedding in hotels and homes. They cause red, itchy bites and skin swelling. The United Kingdom has identified three areas in London with high bedbug search rates, particularly Croydon.

Research on insects and other aeroallergens is essential for improving air quality. Advanced air purification techniques can reduce allergenic reactions. The spread of bedbugs, especially in Asia and Europe, highlights the need for effective solutions. There is a substantial demand for energy-efficient air purifiers that can effectively target insect pests on-site. Common indoor pests include cockroaches, booklice, dust mites, and various beetles, all of which are linked to health issues. Outdoor insects, such as moths, crickets, and midges, exacerbate seasonal allergies and occasionally invade homes such as *Drosophila* fruit flies and Asian lady beetles, causing allergic reactions in the United States. Airborne insect parts also contribute to respiratory diseases.

We conducted an analysis of outdoor air samples at the Aerobiology Lab using a Burkard Volumetric Spore Trap, identifying insect parts among other particles. Additionally, we performed an experiment utilizing the AFL Mini Air Sanifier II® to evaluate its effectiveness against indoor insects.

6.5.2 *Experimental Steps and Procedures to Assess the Capacity of the AFL Mini Sanifier II® in Reducing and a Complete Eradication of the Bedbugs*

Equipment was set up in a simulated environment of a fiberglass chamber. An insect test system was utilized to assess the efficacy of this air purifier in reducing and eradicating bedbugs.

Drosophila melanogaster (fruit fly)—an insect, a model lab organism. The similar size of the fruit fly and the bedbug justifies the usage of the fruit flies as the model organism for this testing. Although the size of the bedbugs varies widely, the most common one has a size of 3.5 mm for adults (Fig. 6.19). The adult form of the fruit fly is 3.22 mm, which is very close to the size of the bedbugs (Fig. 6.20).

The *Drosophila melanogaster* culture was sourced from Wards Biological Supply. *Drosophila* (fruit flies) were cultured on *Drosophila* media. Stocks were maintained by periodically transferring adults to fresh media. Flies were tapped into a new bottle, or some back into the old one, with minimal escapes due to efficient techniques. Cultures were kept at room temperature and refreshed every 3–4 weeks. To begin the experiment, we set up the fiberglass chamber with the AFL Mini

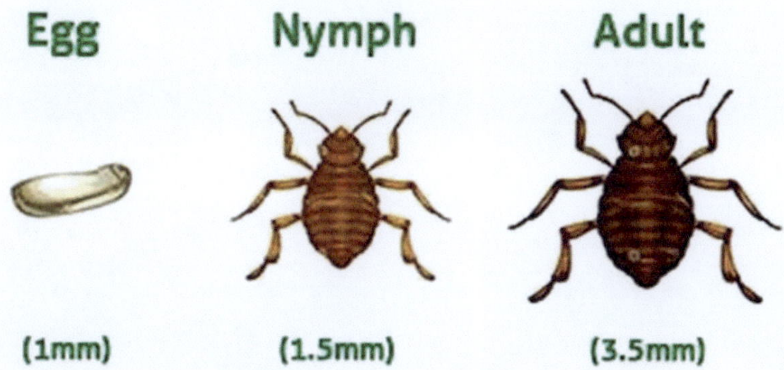

Fig. 6.19 Size of the bedbug at different life forms

Fig. 6.20 Size of an adult form of fruit fly

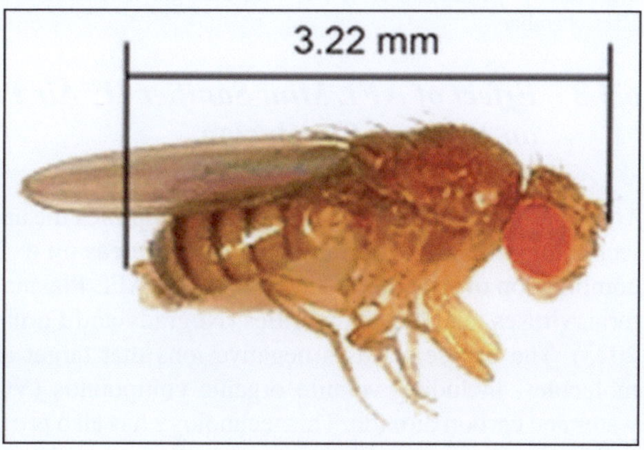

Sanifier II®, an air purifier, food for the flies, and two fans. We briefly refrigerated fruit flies for 5–10 min to immobilize them. After counting the live flies under a dissecting scope, we placed 50 flies on each of two Petri dishes in the center of the chamber. With the fans running at low speed to aerate the flies, we operated the air purifier for 24, 48, 72, and 120 h (Fig. 6.21). At these intervals, we counted the living and dead flies to measure lethality (Fig. 6.22). We repeated this procedure three times to average the data, captured images at various steps, and analyzed the results statistically (Fig. 6.23).

Fig. 6.21 Experimental setup showing flies got repelled and moved to the wall of the fiberglass chamber

6.5.3 *Effect of AFL Mini Sanifier II® Air Purifier on Bedbug: Conclusion*

The AFL Mini Sanifier II® air purifier exemplifies the application of advanced nanotechnology in air and surface cleaning for areas up to 330 square feet. It utilizes a combination of AFLPCO® technology and AFL Plasma Technology to combat bacteria, viruses, and other impurities (Nogrady and Furnass 1983; Ghosh et al. 2020, 2023). The device releases negative ions that target and neutralize carbon-based molecules, including volatile organic compounds (VOCs), converting them into water and carbon dioxide. This technology has also proven highly effective in eliminating insects comparable in size to bedbugs, making it an invaluable tool for their eradication. Continuous operation and monitoring are recommended to achieve complete eradication.

6.6 Application of Silver Nanoparticles to Develop the AFL Mask® and Nanotechnology Applied to the LUNA Wall Mount®, the New Developments in the Air Purifier Industry to Prevent

With population growth and industrial expansion, many cities are experiencing poor air quality. Many articles have drawn the attention of the scientific community toward the elevated risks of airborne transmission of diseases and the associated risks of epidemics or pandemics. The complexity of the problem and the need for multidisciplinary research is highlighted. Bolashikov and Melikov (2009) reviewed the airborne transmission of pathogens, including the generation of droplets from

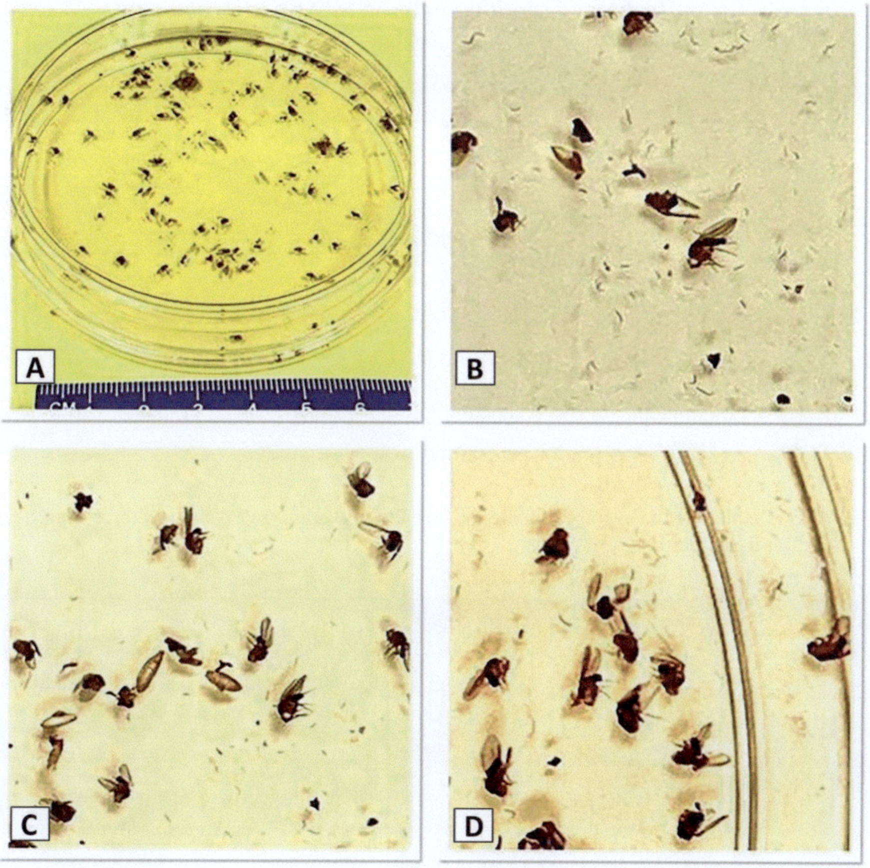

Fig. 6.22 (**a–d**) Showing the death of the fruit flies at different intervals of exposure with the AFL Mini Sanifier II®

infected individuals, pathogen survivability, indoor dispersal, and transfer to healthy individuals.

There is an urgent need for an advanced air purifier capable of addressing airborne pathogens and significantly improving air quality beyond what most of the current market offerings provide. In the context of the COVID-19 pandemic, facemask usage has become widespread in both healthcare settings and the general community. However, the public's indiscriminate use of surgical or filtering face piece (FFP) masks has led to a global shortage, adversely affecting healthcare workers who are most at risk. This situation underscores the necessity to clarify the indications for different mask categories and to rationalize their use (Lepelletier et al. 2020).

Numerous types of facemasks, composed of various materials, are available on the global market (Fig. 6.24). Facemasks are a significant component of personal

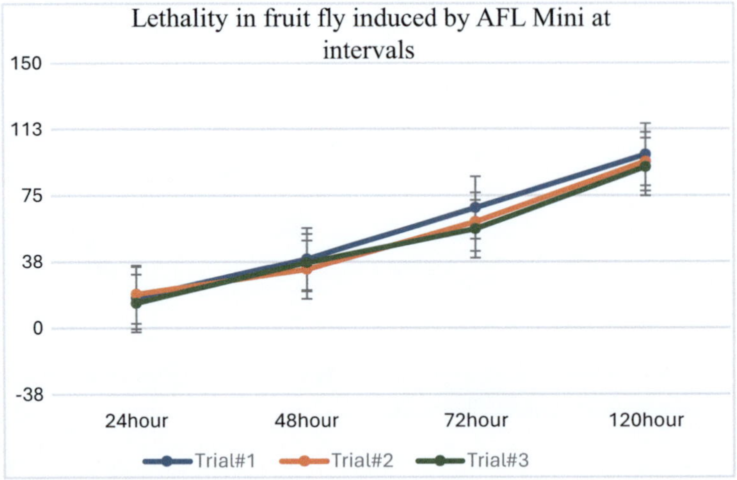

Fig. 6.23 Displaying the average lethality of fruit flies at various intervals over three trials using the AFL Mini Sanifier II®

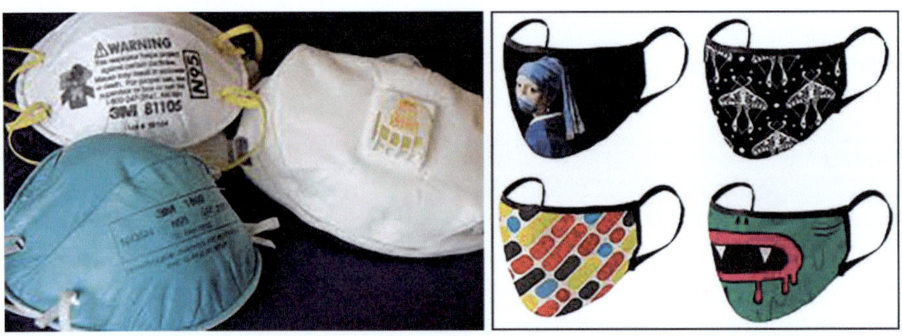

Fig. 6.24 Different types of face masks are available on the market

protective equipment (PPE) for healthcare workers and the public. Depending on their filters, they can provide protection against contagious diseases and severe pollution. They filter fine airborne particles from reaching the respiratory system and prevent infection.

Some studies show a relation between air pollution and increasing heart rate and blood pressure (Langrish et al. 2009), indicating that wearing a facemask may help maintain healthy blood pressure and heart rates in susceptible individuals. However, wearing a mask may be challenging for some people with heart or lung disease, as it can increase breathing difficulty. Some masks, such as N95, may create exhaled moisture and breathing resistance if worn for more than 4 h (Roberge et al. 2010).

Wearing masks could affect the wearer's overall thermal sensation. Extended use of N95 respirators may induce physiological stress on the wearer. Healthcare providers may experience headaches from using the N95 mask, and shorter usage durations may reduce the frequency and severity of these headaches (Lim et al. 2006; Bolashikov and Melikov 2009). The mask recently developed at our laboratory incorporates a four-level filtration system supplemented by an internal air supply facilitated by a fan. This device is designed to mitigate various airborne pathogens, including bacteria, viruses, mold spores, and harmful volatile organic compounds (VOCs).

6.6.1 Methods and Operation: AFL Masks: Filters

The AFL Mask was created collaboratively to solve specific issues. It is comfortable and user-friendly, featuring an electric fan for continuous airflow. The air supply system ensures easy breathing with positive pressure and prevents condensation with a unique vent design. The fan operates quietly with a durable rechargeable lithium-ion battery. The mask has a one-touch control for two fan speeds: press once to start, twice to increase speed, and three times to turn off (red arrow pointing in Fig. 6.25).

AFL filters consist of high-efficiency four-stage nano silver ion filters for maximum protection against 99.9% of all indoor and outdoor impurities such as bacteria, dust, pollen, mites, viruses, pet dander, PM 2.5, smoke, formaldehyde, benzene and other gaseous pollutants, and other volatile organic compounds (VOCs). Wearing the Air for Life Mask can eliminate the potential health risk caused by air pollution and airborne diseases.

6.6.2 Application of Silver Nanoparticles

Silver nanoparticles (AgNPs) have been recognized as effective antimicrobial agents capable of combating bacterial infections in vitro and in vivo. The antibacterial efficacy of AgNPs encompasses Gram-negative and Gram-positive bacteria, including multidrug-resistant strains. AgNPs employ multiple simultaneous mechanisms of action and demonstrate synergistic effects when combined with organic compounds or antibiotics, effectively targeting pathogens such as *Escherichia coli* and *Staphylococcus aureus*. The attributes of silver nanoparticles make them suitable for application in medical and healthcare products, efficiently treating or preventing infections. Given the urgent need for new antibacterial agents, recent reports highlight factors influencing the antibacterial and cytotoxic effects of silver nanoparticles and emphasize their advantages as new antibacterial agents (Bruna et al. 2021). Nanotechnology has emerged as a rapidly growing field with numerous applications in biomedical science. Silver has been adopted as an antimicrobial material and disinfectant with relatively few adverse effects. Silver nanoparticles can penetrate bacterial cell walls, altering cell membrane structures and potentially causing cell death. This efficacy is attributed not only to their nanoscale size but also

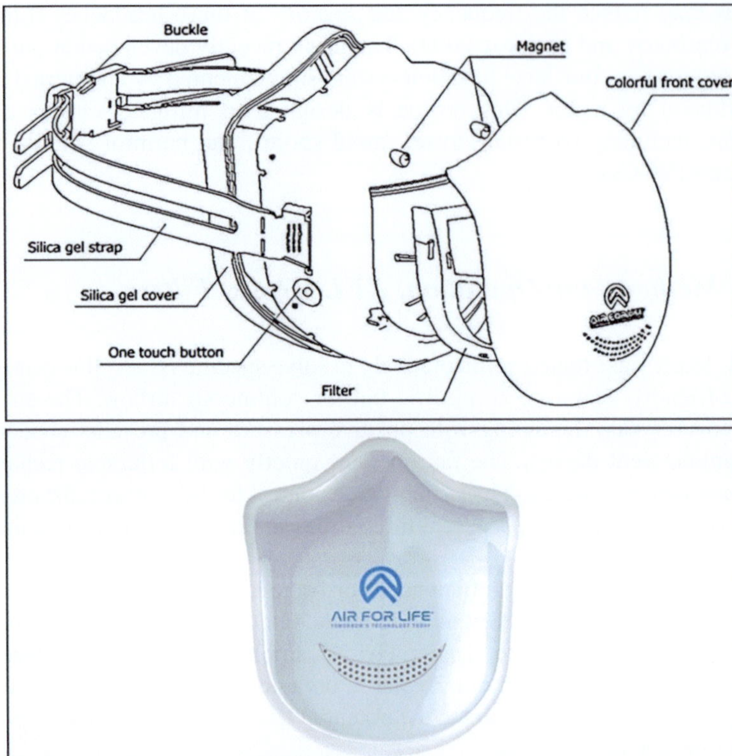

Electrical specification	
Voltage	5 VDC
Watts	0.75W

Mechanical specification			
Material	ABS+Silica gel+Polymer Lithium Ion Batteries	weight	78g
Colour	White/Pink/Blue/Grey	Shipping weight	340g
Fan speed	2 speed	Sound level	40dB-50dB
1st speed air flow	4.0m³/h	2nd speed air flow	8.0m³/h

Fig. 6.25 Diagram of AFL Mask with feature details

to their large surface area-to-volume ratio. AgNPs can increase cell membrane permeability, produce reactive oxygen species, and disrupt DNA replication by releasing silver ions. Recent research has demonstrated that silver nanoparticles possess the capability to neutralize or inactivate bacteria, bacterial endospores, viruses, mold spores, and harmful VOCs present in the air or on surfaces.

The antibacterial actions of silver nanoparticles (AgNPs) include: (1) disrupting cell walls and membranes as Ag+ ions adhere or pass through; (2) denaturing ribosomes, inhibiting protein synthesis; (3) halting ATP production by deactivating respiratory enzymes; (4) disrupting membranes via reactive oxygen species; (5) interfering with DNA replication and cell multiplication; (6) causing membrane denaturation by accumulating in cell wall pits; and (7) perforating membranes and releasing organelles (Yin et al. 2020). In the AFL Mask, a four-stage filtration system incorporates a third-stage filter impregnated with fine silver particles (Fig. 6.26).

AFL Masks have replaceable filters. Clean and maintain the masks as follows: *Surface*: The surface of the mask can be cleaned by wiping with any alcohol-based solution. The silicone cover can be removed and washed with alcohol or sterilize in boiling water for 5 min. Once cleaned, these parts can be cooled down, dry, and reassembled. The internal components cannot be cleaned with water as they are not waterproof.

Effortlessly, the filter can be changed with one button for easy operation. The frequency of changing the filters depends on the environmental conditions. In a lightly polluted environment, change the filter every 2–3 months; in a non-tolerable highly polluted environment, change it every 1–2 weeks. The following are the steps to follow:

1. Remove the magnetic front cover of the mask.
2. Replace the old filter with a new one.
3. Reattach the magnetic front cover of the mask.

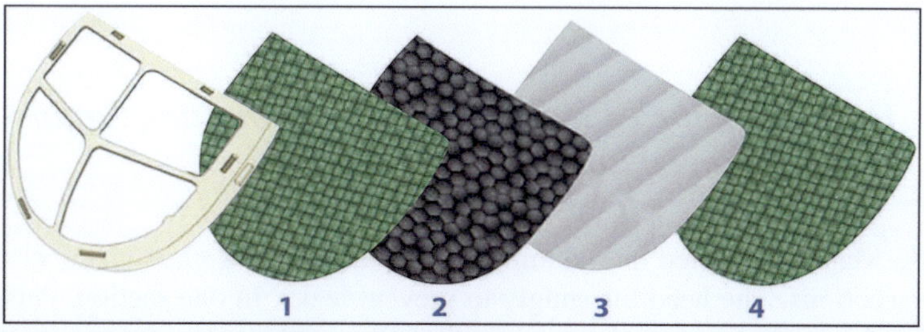

Fig. 6.26 Four-stage filtration system used in the AFL Mask. 1. The first stage is an antibacterial cotton filter designed to block dust and larger particles. 2. The second stage is a carbon filter that blocks bad odors and harmful gases. 3. The third stage is a nano silver ion layer sprayed on the fourth stage HEPA layer. 4. The fourth stage includes a HEPA filter featuring 3M's H11 Filtrate® technology to efficiently remove particles up to 0.3 microns. The PM 2.5 filtration efficiency is assessed to be over 99%

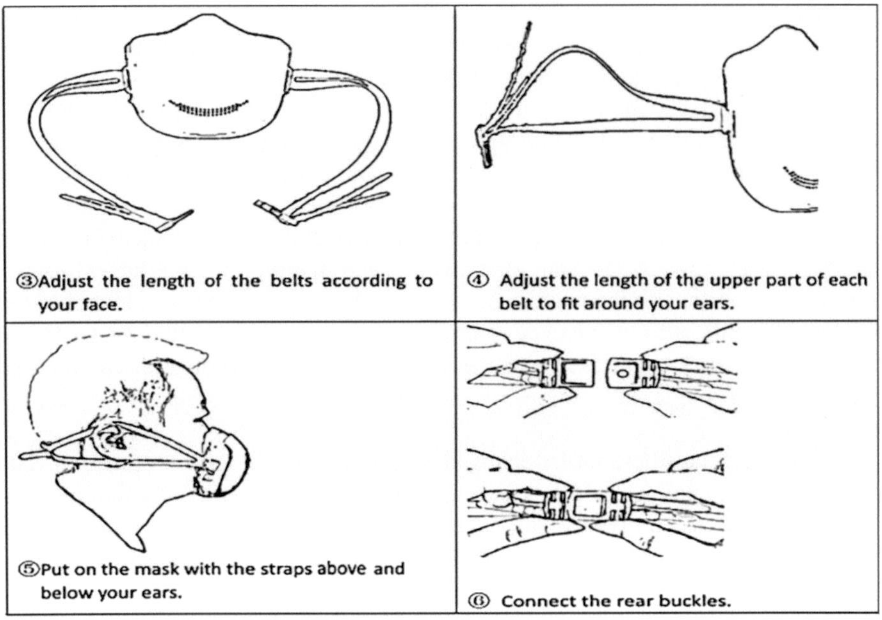

Fig. 6.27 Steps to wear the AFL Mask

6.6.3 *AFL Mask Is Connected to an Adjustable Comfortable Belt, Filters Can Be Changed*

During the COVID-19 pandemic, efficient PPE was scarce, and healthcare professionals struggled with keeping masks on continuously. Many masks have elastic ear loops that cause pain or pressure injuries on the ears (Jiang et al. 2020). AFL Masks solve this issue with an adjustable soft belt that goes above and below the ears, connecting at the back of the head to minimize ear discomfort (Fig. 6.27).

6.6.4 *Evaluating the Efficacy of Mask in Protecting Against Fungal Spores, Bacteria, and PM 2.5*

The AFL Mask was tested on a dummy head placed into a glass chamber divided into two sections. The head fits snugly to prevent leaks. In one section, Petri plates with microbial colonies were kept, while the other contained an air quality monitor and sterilized media plates. Spore concentration was recorded, and the presence of fungal or bacterial colonies in the second chamber indicated mask permeability. A similar setup was used for testing PM 2.5. This procedure assessed the mask's effectiveness.

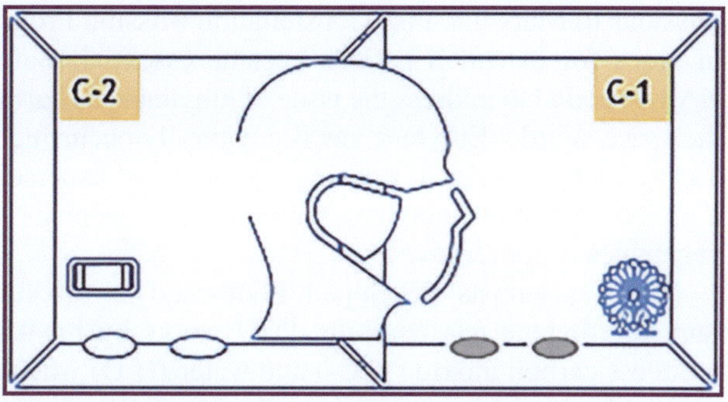

Fig. 6.28 Experimental setup of two glass chambers: C-1 with a fan and Petri plates holding fungal colonies and C-2 with media-containing Petri plates and an air quality monitor

The fiberglass chamber was divided into two sections, C-1 and C-2, with an airtight slot for the AFL Mask. Initially, both the sides were sterilized. The unfiltered air (C-1) was compared with the filtered air passing through the mask (C-2). Variable colony counts were observed at each interval (Fig. 6.28). Subsequently, PM 2.5 levels were measured in both chambers after 24, 48, 72, 96, and 120-h using ISO 12103-1 Ultrafine Dust Particle and a Temtop M2000C Air Quality Monitor. The particulate matter was uniformly distributed and allowed to reach equilibrium within 72 h. Airtight seals ensured containment, and PM 2.5 counts were recorded. Chamber-1 (C-1) shows particulate count with unfiltered air. Chamber-2 (C-2) uses AFL Mask to filter air, blocking particles from C-1 to C-2, indicating protection from particulates. With the observations after 120 h it was noted that there was non-significant amount of PM 2.5 found in C-2, and only one or two bacterial colonies were formed on the Petri plates.

6.7 Nanotechnology Applied to the LUNA Wall Mount® Air Purifier to Prevent the Airborne Pathogens

Airborne allergens such as dust, smoke, pollen, and animal dander can trigger allergies, asthma, and COPD. Symptoms include sneezing, coughing, wheezing, and shortness of breath, affecting the eyes, throat, nose, and lungs. Around 300 million people globally suffer from asthma, with 250,000 deaths annually. By 2025, asthma cases are expected to increase by over 100 million. Workplace exposure to fumes, gases, or dust accounts for 11% of global asthma cases, contributing to 15% of adult asthma. Nanotechnology advancements have the potential to significantly improve materials and manufacturing processes. One such innovation, the Air For Life Photocatalytic Oxidation (AFLPCO®) Technology, helps reduce asthma symptoms by eliminating allergens such as dust mites, mold, pollen, and fungal spores. AFLPCO® emits negative ions, which sterilize air and surfaces, making it valuable in medical settings for preventing infection spread. Photocatalytic Oxidation (PCO) technology was originated with NASA in 1994 to purify the air in spacecraft. When NASA began planning for deep space exploration, including proposed manned missions to Mars like the "Constellation Mission Program" in 2013, sustaining life in space for extended periods became essential. Before deep space exploration, NASA needed to address the issue of ethylene gas harming plants and vegetables in the space shuttle. Ethylene gas is a naturally occurring hormone released by plants and vegetables to signal ripening. On Earth, it disperses into the atmosphere, but in a sealed spacecraft, ethylene builds up and becomes toxic, killing the plants and vegetables if not removed.

NASA scientists developed Photocatalytic Oxidation (PCO) technology to remove ethylene gas build-up. PCO works by breaking down ethylene gas into harmless carbon dioxide (CO_2) and water (H_2O) when exposed to ultraviolet (UV) light in the presence of a titanium dioxide (TiO_2) catalyst. Further tests revealed that PCO technology not only eliminated ethylene gas buildup but also destroyed all carbon-based impurities in the air, including bad odors, volatile organic compounds (VOCs), fungi, bacteria, and viruses (Maria et al. 2007).

Air For Life Photocatalytic Oxidation (AFLPCO®) is applied in the LUNA Green AFLPCO® Cell™, which is coated inside with the AFL Nano five metal catalyst and contains two high-quality, long-lasting AFL germicidal UVC bulbs. Photons of UV light react with the AFL Nano five metal catalyst, leading to the formation and release of millions of human-friendly negative ions. This abundance of negative ions is released into the air by the machine with the Air For Life AFLPCO® catalyst, flooding the room. The negative ions then collide and react with pathogens and allergens in the air and on surfaces, destroying or demobilizing them on contact.

6.7.1 AFL Filtration System vs HEPA Filter

Nearly 99% of the particles inhaled are under 1 micron in size (Weisburd and Smith 1962). Most HEPA filters capture particles up to 0.3 microns, with a 99.97% efficiency for approximately 0.2 micrometer particles. However, viruses smaller than 0.3 microns can pass through these filters, potentially exceeding minimum infection doses (MID50) (Heimbuch et al. 2007). In contrast, AFL Technology targets particles from 0.1 to 0.001 microns, destroying up to 99.997% of viruses, bacteria, fungi, VOCs, and contaminants.

The Green AFLPCO® Cell uses a germicidal UVC lamp to deactivate the DNA of bacteria, viruses, and other pathogens, preventing them from multiplying. The UVC lamp's photons react with our proprietary nano Nickel HCT catalyst (containing five reactive metals) to create oxidizers that clean air and sanitize surfaces. LUNA LNT2 employs this superior catalyst with nano-sized particles, enhancing reaction rates to produce millions of oxidizers. The 254 nm UV lamp provides the same benefits as natural sunlight by oxidizing and ionizing. By combining UV light with our nano catalyst and AFL Plasma® Technology, we achieve exceptional results (Fig. 6.29).

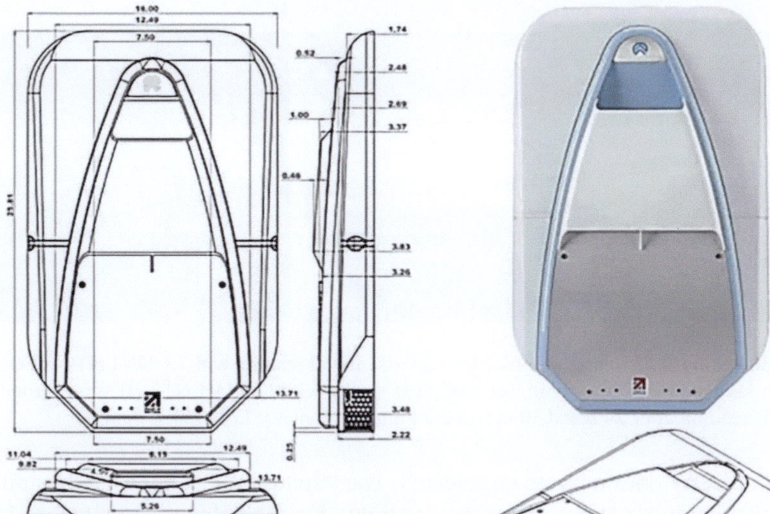

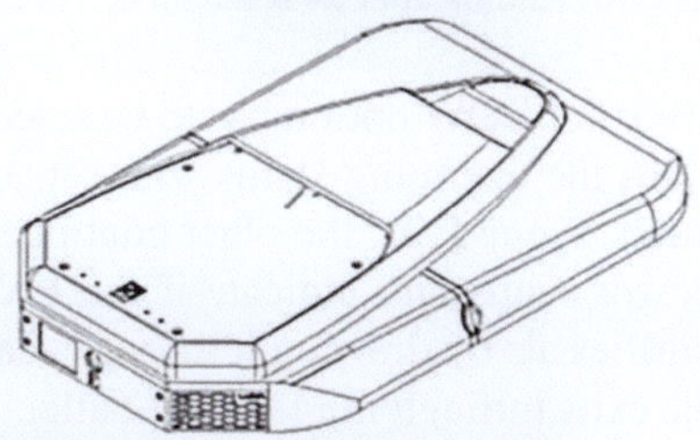

Electrical Specifications	
Input Voltage	230V AC
Max Watts	35W

Mechanical Specifications			
Material:	V0 rated PC and Aluminium	Weight	4.7Kg
Coverage:	3000 Sq ft (commercial use)	Shipping Weight	5.7Kg

Electrical Specifications			
Power Supply:	110V - 220V AC 50/60Hz Max Current 3A UK Plug to C7 CE, UKCA, RoHS listed	Ballast:	2 x 34V DC Ballast Lamp status indicator Lamp faulty indicator Lamp replace indicator
AFL Plasma:	Negative & Positive ions: 360 million PCS/cm^2		
Blower Fan:	2 x 24V DC - 0.20A Low noise - 86.4CFM	PCO Cell:	AFL PCO Technology

Fig. 6.29 The design and details of the LUNA LNT2 Green AFLPCO unit (www.airforlife.co.uk). Printed with permission from AFL, UK

6.7.2 LUNA LNT2 Green AFLPCO® Wall Mount Unit and Its Operation

The LUNA LNT2 uses AFLPCO technology and is controlled via a High/Low rocker switch on the bottom panel (Fig. 6.30). Based on contamination levels, either

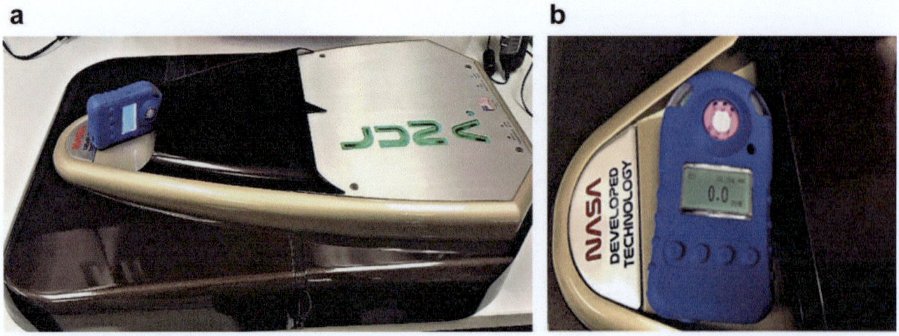

Fig. 6.30 (**a–d**) Setup in the Aerobiology Lab, (**b**). initial reading with LUNA LNT2-6000 at 1:54 p.m. on January 16, 2020: 0 ppm, (**c**) continued testing with LUNA LNT2-6000 for Ozone emission, (**d**) reading after 24 h and 30 min at 2:24 p.m. on January 17, 2020: 0 ppm

HIGH or LOW option has to be selected. The "Status Panel" opposite the unit indicates the operating status with separate lights for each electrical ballast and UVC lamp. If one fails, the other continues to work independently. The lights also show power status and indicate if the UVC lamp needs replacement. The Luna LNT2 purifies air by drawing it through rear dust guards that trap large particles. Purified air exits through the forward outlet. Install the unit away from objects obstructing airflow. AFL Plasma® Developed by AFL R&D, AFL Plasma® uses carbon brush electrodes to generate ions from natural water vapor, removing ferrous metals, reducing static charge, and destroying airborne impurities like PM 2.5. Ozone, a strong oxidizer, is safe at low concentrations but lethal to microorganisms. Tests conducted in an ultra-clean Aerobiology Lab showed Air For Life (UK) air purifiers effectively emit Ozone within safe levels.

1. LUNA WALL MOUNT SANIFIER®, LNT2-6000
2. AFL MINI SANIFIER®AFL M2
3. AFL CAR SANIFIER®
4. AFL MASK®

We have used the *FD-90 Ozone Detector* (Forensics Detectors, USA) to check the Ozone level emitted by the above-stated units.

6.7.3 *Results and Discussion for Experimentation with LUNA Wall Mount*

The results and the photographs taken during the tests are summarized in Figs. 6.30 and 6.31.

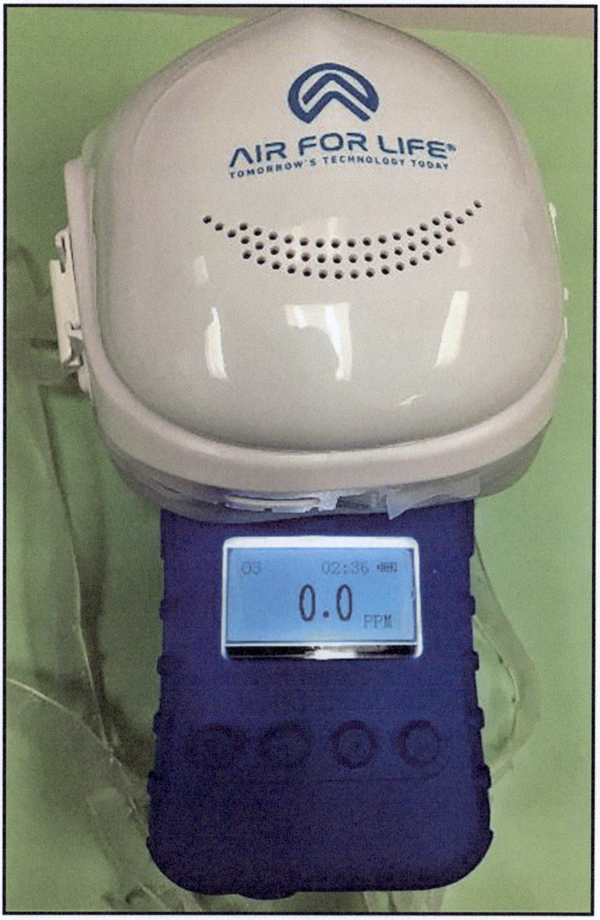

Fig. 6.31 8E-F: E shows no Ozone emission from the AFL Mask®

6.8 Conclusions

Indoor air quality is a significant concern as people spend approximately 90% of their time indoors at home, school, work, and other public spaces, with only about 7.6% spent outdoors. Individuals more vulnerable to pollution, such as children, older adults, and those with cardiovascular or respiratory diseases, tend to spend even more time indoors. Indoor air quality can be compromised due to higher concentrations of aeroallergens, VOCs, and particles compared to outdoor levels. Pollutants can vary substantially between and within buildings. Several factors contribute to indoor air pollution, including environmental tobacco smoke: a major contributor wherever smoking occurs; cooking: increases VOCs, various chemicals, and particle concentrations indoors. To improve indoor air quality, it is recommended to use more indoor plants and install a whole-home air purification system

(EPA 2024). The filter-less air purifiers have been proving to be very effective for cleaning large indoor areas with negligible ozone emissions.

Acknowledgment The authors are thankful to Dr. Jeff Bennert, PhD, CTN, former President and CEO of Air Oasis, Dr. Constantine Saadeh, MD, FICP of Allergy ARTS, USA, and Neil Vitale of Air For Life, UK, for their invaluable contributions to the research and development in nanotechnology and air purification.

References

AFL Mask® built by Air For Life (UK). Web link: https://www.airforlife.net/

Air For Life UK (2024) web site: Air for life – experience the purest air indoors

Answers.com (2024) How can air purification be done using nanotechnology? https://www.answers.com/chemistry/. Updated: 5/23/2024

AP News (2024) Severe Delhi AQI: GRAP-4 curbs imposed as air pollution chokes capital | AP News

BioSure Professional (2024) How dirty is restaurant ice? Website visited Dec. 2024 Link: How Dirty is Restaurant Ice? | BioSure Professional

Bi-Polar Ionization (2024) web site: Bipolar ionization may be the next big step in the fight against Coronavirus – ISO-Aire Clean Air

Bolashikov ZD, Melikov AK (2009) Methods for air cleaning and protection of building occupants from airborne pathogens. Build Environ 44(7):1378–1385

Bruna T, Maldonado-Bravo F, Jara P, Caro N (2021) Silver nanoparticles and their antibacterial applications. Int J Mol Sci 22(13):7202. https://doi.org/10.3390/ijms22137202

Chapman JA (2003) Toxic mold phantom risk vs. science. Ann Allergy Asthma Immunol:222–223

Creip (1967) Dermatologic allergy: immunology, diagnosis, management. W.B. Saunders Co., Philadelphia/London

Custovic A, Simpson A (2006) Dust mite. Encycl Respir Med 4:54–59. https://doi.org/10.19044/esj.2023.v19n12p1

EPA (2024) Improving your indoor environment. Web site visited on Dec 15, 2024. Link: Improving Your Indoor Environment | US EPA

Express (UK) (2023) Web link: https://www.express.co.uk/news/uk/1826893/bed-bugs-live-pests-surging

Fontes B, Cattani Heimbecker AM et al (2012) Effect of low-dose gaseous Ozone on pathogenic bacteria. BMC Infect Dis. 12:358. https://doi.org/10.1186/1471-2334-12-358. (PubMed). https://www.ncbi.nlm.nih.gov/pubmed/23249441

Geha R, Szefler SJ (eds) (2010) Pediatric allergy: principles and practice, 2nd edn. Elsevier, Philadelphia, pp 274–282

Ghosh N, Camacho R, Schneiderjan E, Saadeh C, Gaylor M (2003a) Correlation between the meteorological conditions with the aeroallergen concentration in the Texas Panhandle. Texas J Microsc 34(1):12–13

Ghosh N, Patten B, Lewellen G, Saadeh C, Gaylor M (2003b) Aeroallergen survey of the Texas Panhandle using a Burkard Volumetric Spore Trap. J Allergy Clin Immunol 111(2):S91

Ghosh N, Camacho R, Saadeh C, Gaylor M (2004) Assessment of pollen concentration in the atmosphere of the Texas Panhandle through the use of a Burkard Volumetric Spore Trap. Texas J Microsc 35(1):22–27

Ghosh N, Mahapatra D, Caraway E, Saadeh C, Gaylor MC, Smith D (2005) An investigation on the aeroallergens in the public schools of the Texas Panhandle. J Allergy Clin Immunol 115(2):S22:88

Ghosh N, Saadeh C, Gaylor M (2006) Quantification and characterizing the Aeroallergen by scanning and analyzing the tapes from the Burkard Spore-trap. J Scan Microsc 28(2):127–128

Ghosh N, Whiteside M, Ridner C, Celik Y, Saadeh C, Bennert J (2010) Characterization of aeroallergen of Texas Panhandle using Scanning and Fluorescence Microscopy. In: Postek MT, Newbury DE, Platek SF, Joy DC (eds) Scanning Microscopy 2010, vol 7729 Sec.1X. SPIE, pp 1–9

Ghosh N, Aguaiza C, Guzman A, Chudasama J, Bennert J, Das AB (2012) Abstract #174 reduction in allergic rhinitis index by decreasing aeroallergens and malodor causing volatile organic compounds by Luna Air Purifiers Using Photo-Catalytic Oxidation (PCO) Technology. World Allergy Organ J. 5(Suppl 2):S75. https://doi.org/10.1097/01.WOX.0000411931.03944.6f

Ghosh N, Bennert J, Bennert J, Saadeh C et al (2015a) Microbiome in ice machines and assessing the plasma nanotechnology in breaking the biofilm and improving air quality. In: Abstract #1018, WAO symposium on food allergy & the microbiome 2015, Miami

Ghosh N, Estrada C, Caraway E, Bennert J, Bennert J, Saadeh C (2015b) New world of business with plasma and AFLPCO nano-technology in marketing air purifier, cell phone and ice-maker sterilizer. Int J Comput Internet Manag (IJCIM):9–14. ISSN: 0858-7027

Ghosh N, Howard A, Sherali N, Revanna C, Pratt C, Saadeh C, Bennert J, Bennert J, Mullan K, Rogers J (2017a) Reduction of MRSA populations and aeroallergens on using AFLPCO® and plasma nanotechnology for air purification. Int J Adv Sci Eng Technol 5(3–1):10–12. ISSN:2321-9009

Ghosh N, Estrada G, Veloz M, Bouyi D, Bennert J, Bennert J, Saadeh C, Revanna C (2017b) Meteorological and clinical analysis of aeroallergen data: increase in allergy and asthma cases in Texas Panhandle. In: Allergy and allergen immunotherapy: new mechanisms and strategies. Book Chapter, Apple, CRC Press, New York, pp 101–124

Ghosh N, Vitale J, Bell J, Ray A, Goyal S, Howard H, Banerjee PB (2020) An assessment of the AFL Mask® and LUNA wall-mount, the new developments in the air- purifier industry for preventing the airborne pathogens. Eur Sci J 16(27) ISSN:1857-7881 (Print) e – ISSN:1857-7431

Ghosh N, Vitale J, Bell J, Goyal S, Howard A, Banerjee P, Mitra S (2021) An Assessment of the AFL Mini Sanifier II® in terms of reducing the incidence of canine allergy. In: June 2021 art, humanities, social sciences, STEAM and education conference proceedings. Hawaii University Place, Hawaii. https://huichawaii.org/ssec/proceedings-programs/proceedings-steam-2021/

Ghosh N, Goyal S, Howard A, Banerjee P, Vitale J (2023) Application of nanotechnology in a novel air purifier for remediation of airborne pathogen and to prevent the spread of COVID-19. Eur Sci J ESJ 19(12):1. https://doi.org/10.19044/esj.2023.v19n12p1

Goodman N (2006) The use of ionizers to destroy allergens: past, present and future research. Int Arch Allergy Immunol 141:311–312

Gorai AK, Tuluri F, Tchounwou PB (2014) A GIS based approach for assessing the association between air pollution and asthma in New York State, USA. Int J Environ Res Public Health 11(5):4845–4869. https://doi.org/10.3390/ijerph110504845

GRAP-4 curbs imposed as air pollution chokes capital | AP News

Hameri AP (1996) Technology transfer between basic research and industry. Technovation 16(2):51–57. Elsevier Science Ltd.

Heimbuch BK, Hodge JK, Wander JD (2007) Viral penetration of HEPA filters. J Mol Sci 22(13):7202. https://doi.org/10.3390/ijms22137202

Jeffrey DM, Lisa N, Shama S, Thomas AE (2007) Nonwoven in contrast to woven mattress encasing accumulate mite and cat allergen. J Allergy Clin Immunol 120(4):977–979

Jeong KY, Lee IY, Lee J, Ree HI, Hong CS, Yong TS (2006) Effectiveness of reduction for control of house dust mites and cockroaches in Seoul, Korea. Korean J Parasitol 44(1):73–79

Jiang W, Cao W, Liu (2020) Wear N95 mask with plastic handle reduce pressure injury. J Am Acad Dermatol. https://doi.org/10.1016/j.jaad.2020.04.001

Langrish JP, Mills NL, Chan JK et al (2009) Beneficial cardiovascular effects of reducing exposure to particulate air pollution with a simple facemask. Part Fibre Toxicol 6:8. https://doi.org/10.1186/1743-8977-6-8

Lepelletier D, Grandbastien B, Romano-Bertrand S, Aho S, Chidiac C, Géhanno JF, Chauvin F (2020) What face mask for what use in the context of COVID-19 pandemic? The French guidelines. J Hosp Infect 105(3):414–418

Lim ECH, Seet RCS, Lee K-H, Wilder-Smith EPV, Chuah BYS, Ong BKC (2006) Headaches and the N95 facemask amongst healthcare providers. Acta Neurol Scand 2006(113):199–202. https://doi.org/10.1111/j.1600-0404.2005.00560.x

Maciag MC, Phipatanakul W (2022) Update on indoor allergens and their impact on pediatric asthma. Ann Allergy Asthma Immunol. 128(6):652–658. https://doi.org/10.1016/j.anai.2022.02.009

Maria G, Antoniou MG, Dionysios D, Dionysiou (2007) Application of immobilized titanium dioxide photocatalysts for the degradation of creatinine and phenol, model organic contaminants found in NASA's spacecrafts wastewater streams. Catal Today 124(2007):215–223

Miyazaki K, Islam N (2007) Nanotechnology systems of innovation – an analysis of industry and academia

Muller GH, Kirk RW (1969) Small animal dermatology. W.B. Saunders Co., Philadelphia/London/Toronto

Nanowerk Spotlight (2024) How to talk to business leaders about nanotechnology. Website visited: Dec 2024: Business Leaders and Nanotechnology | MIT Professional Education

Nogrady SG, Furnass SB (1983) Ionizers in management of bronchial asthma. Thorax 38(12):919–922

Ozone Safety by OSHA. https://www.Ozonesolutions.com/knowledge-center/Ozone-safety.html

Publication at: https://www.researchgate.net/publication/235151297

Ralf S, Mueller et al (2002) Evaluation of cross-reactivity of allergens by use of intradermal testing in atopic dogs. AJVR 63:874–879

Ranaivo Y (2011) An increasing trend. Study: more Amarilloans Battle Asthma's effects. Amarillo Globe-News. August 28, 2011. Web link: http://amarillo.com/news/localnews/ 2011-08-28/increasing trend. US EPA: Creating Healthy Indoor Air Quality in Schools Web link http://www.epa.gov/iaq/schools

Ren H, Koshy P, Chen W-F, Qi S, Sorrell CC (2017) Photocatalytic materials and technologies for air purification. J Hazard Mater 325(2017):340–366

Revanna C, Sherali N, Veloz M, Bennert J, Rogers J, Ghosh N (2016) Microscopic evaluation of the AFLPCO Nano-technology in the an air purification company air purifiers using Ai-O-Cell samplers. Texas J Microsc 47(1)

Roberge RJ, Emily B, Powell JB, Coca A, Roberge, SM (2010) Effect of exhaled moisture on breathing resistance of N95 filtering face-piece respirators. Browsed on 7.7.2020 at https://netec.org/

Robert AW (2010) Environmental control. In: Leung DYM, Sampson HA, Geha R, Szefler SJ (eds) Pediatric allergy: principles and practice, 2nd edn. Elsevier, Philadelphia, pp 274–282

Suhaili ZA, Ho TM (2008) Residual activity of benzyl benzoate against *Dermatophagoides pteronyssinus* (Acari: Pyroglyphidae). Southeast Asian J Trop Med Public Health 39(3):507–510

The Daily Mail: Ice cubes have more bacteria (2013) Ice in six out of ten restaurants has more bacteria than water from toilets. Daily Mail Online

Uehara S, Franzolin MR, Chiesa S, Moreira D, Gambale W, Paula CR (2006) Effectiveness of house dust mite acaricide tri-n-butyl tin maleate on carpets, fabrics and mattress foam: a standard of methodology. Rev Inst Med Trop S Paulo 48(3):171–174

Weisburd MI, Smith GS (eds) (1962) Air pollution control field operations manual. A guide for inspection and enforcement. US Department of Health, Education and Welfare. Web link: https://play.google.com/books/reader?id=yFi6Sp0WBbQC&hl=en&pg=GBS.PA8

Wharton DRA (1971) Ultraviolet repellent and lethal action on the American cockroach. J Econ Entomol. 64:252–255

WHO- Air Quality Guidelines (2025) Link: What are the WHO Air quality guidelines?
WHO-The Global Health Observatory (2025) Link: Ambient air pollution
Yin IX, Zhang J, Zhao IS, Mei ML, Li Q, Chu CH (2020) The antibacterial mechanism of silver nanoparticles and its application in dentistry. Int J Nanomed 2020(15):2555–2562. https://doi.org/10.2147/IJN.S246764

Chapter 7
Engineered Carbon Nanoparticles: Exposure Risks, Respiratory Toxicity, and Underlying Mechanisms

Debajyoti Ghosh, Brijesh Yadav, Julia Eilers, Jayanta Gupta, and Jagjit S. Yadav

Abstract Carbon nanomaterials (CNMs) particularly carbon nanotubes (CNTs) have found widespread applications in diversified fields including biomedical, industrial, and agriculture sectors, because of their unique physicochemical attributes. However, these high-volume nanomaterials have the propensity to become airborne as carbon nanoparticles (CNPs) during manufacturing, handling, and usage and pose exposure hazards in occupational and nonoccupational settings. Consequently, CNPs are deemed as emerging members of the toxic particulate matter in air pollution. Animal model studies have shown that inhaled CNPs have the potential to cause respiratory (tissue injury, inflammation, fibrosis) and systemic health effects including various organ toxicities and gut dysbiosis; the respiratory

D. Ghosh (✉)
Division of Rheumatology, Allergy, and Immunology, Department of Internal Medicine, University of Cincinnati College of Medicine, Cincinnati, OH, USA
e-mail: ghoshdi@ucmail.uc.edu

B. Yadav
Pulmonary Pathogenesis and Immunotoxicology Laboratory, Department of Environmental and Public Health Sciences, Division of Environmental Genetics and Molecular Toxicology, University of Cincinnati College of Medicine, Cincinnati, OH, USA

Department of Nephrology and Renal Transplantation, Sanjay Gandhi Postgraduate Institute of Medical Sciences, Lucknow, Uttar Pradesh, India

J. Eilers
University of Cincinnati College of Medicine, Cincinnati, OH, USA

J. Gupta
Department of Health Sciences, Marieb College of Health & Human Services, Florida Gulf Coast University, Fort Myers, FL, USA

J. S. Yadav (✉)
Pulmonary Pathogenesis and Immunotoxicology Laboratory, Department of Environmental and Public Health Sciences, Division of Environmental Genetics and Molecular Toxicology, University of Cincinnati College of Medicine, Cincinnati, OH, USA
e-mail: yadavjs@ucmail.uc.edu

N. Ghosh et al. (eds.), *Nanotechnology in Air Quality Management*, Sustainable Development and Biodiversity 39, https://doi.org/10.1007/978-981-96-8339-0_7

effects have been corroborated in emerging human epidemiological studies in CNM-exposed individuals. Inhaled CNT nanoparticles have been shown to cause immunotoxicity, oxidative stress, DNA damage, and other cellular changes, though the specific underlying molecular mechanisms continue to emerge. In this direction, we have compared the available Omics-based studies and presented an account of adverse outcome pathways (AOPs) impacted by CNT exposures. In particular, activation of specific immune signaling pathways varies with the CNT particle type/shape (rods versus tangles), as well as exposure duration. In conclusion, while CNMs offer significant benefits across various fields, their propensity to become airborne posing exposure risks via air pollution and health consequences must not be overlooked. Continued research is crucial to fully understand the underlying molecular mechanisms of their health effects and to develop effective strategies for safe handling and use.

Keywords Carbon nanomaterials · Carbon nanotubes · Immunotoxicity · Respiratory toxicity

7.1 Introduction

7.1.1 Background

Nanotechnology is a branch of science that studies matters of size range 1–100 nanometer (nm) in at least one dimension (Shamim and Sharma 2013). A nanometer being one millionth of a millimeter, nanotechnology manipulates matters at the atomic and molecular levels, imparting them unique and distinct physical, chemical, optical, and electrical properties from their larger counterparts (Shamim and Sharma 2013; Chen et al. 2019). Typically, after manipulation, matters at nanoscales become chemically more reactive and have enhanced optical property, greater strength, lighter weight, and larger surface area. These properties make them suitable for wide ranging applications in diverse sectors such as medicine, agriculture, transport, energy, and communications among others (Ahmed and Hussain 2018).

Nanomaterials (NMs) can be classified based on methods and sources of the starting matter used. Nanomaterials may be generated via engineering or natural methods. Engineered nanomaterials are intentionally designed and synthesized for specific applications with technical interventions (Song et al. 2015). Natural nanomaterials, on the other hand, are generated spontaneously without human intervention in nature such as those derived from biological systems including plants, bacteria, and other living organisms (Alshammari et al. 2023). In terms of the starting material-based classification, engineered nanomaterials are synthesized from carbon or metals by chemical process to achieve the desired properties; natural (green) nanomaterials are synthesized using environmentally friendly and sustainable methods, involving plant extracts, microorganisms, or other natural polymers (Gong et al. 2024), as exemplified by silk, chitosan, cellulose, lipids, and so on.

Engineered nanomaterials (ENMs) are composed of nanoparticles, which are particles less than 100 nm in size in at least one of the dimensions. These may be carbon-based or metal-based. Carbon nanomaterials (CNMs) are a class of nanomaterials that are available in different forms: nanotubes with cylindrical structures having diameters in the nm range and length up to several micrometers; nanofibers, the fiber-like structures with a diameter in the nm range; nanowires having wire-like structures with a diameter in nm range; nanoplates/nanosheets having flat, plate-like structures with a thickness of a few nm and lateral dimensions of several micrometers; dendrimers having many branched, tree-like structures; fullerenes having hollow, cage-like structures composed entirely of carbon atoms; nanocomposites composed of a matrix embedded with carbon-based nanoparticles, nanotubes or other forms (Alshammari et al. 2023). On the other hand, engineered metal-based nanoparticles (MNPs) are derived from metal components by using either constructive or destructive methods of nanoparticle synthesis. The most common metals used for MNPs synthesis are aluminum, lead, silver, gold, iron, cobalt, zinc, cadmium, and copper. A few commonly used MNPs along with their applications are gold nanoparticles in medical imaging, cancer treatment, and diagnostics, silver nanoparticles used in wound dressing and preventing bacterial growth, titanium oxide used in cosmetics, quantum dots used in electronics displays, solar cells, and biological imaging.

Natural or organic nanomaterials are composed of biologically derived substances, exhibiting their high biocompatibility, biodegradability, and enhanced mechanical strength (Liu et al. 2021). Organic compounds have specific electrical charges, hydrophilic/hydrophobic surfaces, and adsorption-like property through which they form nanogels, micelles, and liposomes, referred to as nanospheres or nanocapsules (Alshammari et al. 2023). They can be conjugated with other nanoparticles, resulting in conjugated organic nanomaterials. They are highly photo- and heat-sensitive and get degraded easily, exhibiting nano-capsulation property for pharmaceutics transport. Nanocelluloses are examples of such nanomaterials that are widely used in medical devices, packaging materials, and wound healing dressing sheets (Liu et al. 2021).

7.1.2 Classification and Characteristics of Carbon Nanomaterials (CNMs)

Carbon-based nanomaterials (CNMs), the focus of this review, are entirely formed of carbon atoms arranged in different patterns (Georgakilas et al. 2015). Carbon atoms have unique bonding property with neighboring atoms, making them suitable to be engineered into different molecular structures such as linear, cylindrical, or hexagonal buckyball (Alshammari et al. 2023), a property technically described as allotropy. The type of chemical bonding and structural geometry of these atoms determine the physical and chemical properties of the CNMs. These NMs are synthesized by using high pressure and temperature from various starting base sources of carbon such as acetylene gas, graphite rods, carbon monoxide, glucose, and sucrose. The common methods for CNM production are chemical vapor deposition (CVD), arc discharge method (Patel et al. 2020), laser ablation methods, ball milling, high-pressure carbon monoxide (hipco), and hydrothermal and solvothermal methods. Various commercial forms of CNMs include carbon nanotubes (CNTs), graphene, fullerene, carbon nanofibers, and carbon black (Mekuye and Abera 2023).

Carbon Nanotubes (CNTs): CNTs are high volume nanomaterials with unique attributes such as small size, high surface area, high thermal and electrical conductivity, low thermal expansion coefficient, high tensile strength, among others. CNTs are essentially graphene sheets containing a series of carbon rings that are rolled into cylindrical structures. The rolled-up sheets of graphene may be in a single (single wall) or multiple (multiwalled) layers arranged in a hexagonal lattice (Georgakilas et al. 2015; Song et al. 2015). Accordingly, CNTs are classified into single-walled carbon nanotubes (SWCNT) and multi-walled carbon nanotubes (MWCNT). Because of significant electrical, thermal, and mechanical properties, CNTs are used in electronic devices, space vehicles, antimicrobial applications, phototherapy, water purification, drug delivery (Oh et al. 2019), and field-deployable devices (De Volder et al. 2013). Thus, CNTs have found applications in diversified industries including biomedical, electronics, automobile, aerospace, energy, agriculture, among others. In 2022, 8200 metric tons of CNTs were produced globally, which is expected to increase to 13,000 metric tons by 2025 with a projected cumulative annual growth rate of 13.3% between 2024 and 2032 (*Carbon Nanotubes Market Size, Share & Industry Analysis, By Product (Multi Walled Carbon Nanotubes (MWCNT) and Single Walled Carbon Nanotubes (SWCNT)), By Application (Plastics & Composites, Electrical & Electronics, Energy, and Others), and Regional Forecast, 2024-2032 Source* n.d.). Despite significant demands and production of CNTs for numerous applications, there are serious concerns associated with potential exposure risks, necessitating their judicious use.

Carbon Nanofibers: CNFs are cylindrical nanostructures with diameter ranging 10–100 nm and lengths up to several micrometers. These can be solid, hollow, or more complex herringbone-like structures. These have high tensile strength, thermal and electrical conductivity and are chemically inert. CNFs are used in various applications including composite nanomaterial preparation, supercapacitors for energy storage devices, and filter membrane preparation.

Graphene: It is made of a single layer of carbon atoms arranged in a two-dimensional honeycomb lattice (Alshammari et al. 2023). It is the building block of other based materials like graphite and CNTs. It has higher electrical conductivity, mechanical strength, and flexibility, widely used in electronics devices, sensors, and energy storage devices.

Fullerene/Buckyball: It is entirely made of carbon atoms and exists in the form of a hollow sphere, ellipsoid, or tube. The well-known fullerenes are C_{60} and $C_{70.}$ C_{60} fullerenes are made of 20 carbon–carbon atoms bonding together to form 12 pentagon and 16 hexagons, resulting in a total diameter of a fullerene molecule up to 7.0×10^{10} m. Being hollow in design, fullerenes can potentially be used in drug delivery systems as well as in materials science (Alshammari et al. 2023).

Carbon Dots (C-Dots): Carbon dots are small, quasi-spherical nanoparticles with sizes below 10 nm, providing high surface area. They often exhibit photoluminescence. C-dots are often used in bioimaging, biosensing, and as catalysts.

Carbon Nanocones: These are cone-shaped nanostructures with a sharp tip and a broad base. Their base size ranges in micrometers and tip size in nanometers, giving it high surface area in the bottom part and charge density at the tip. These NMs are used in preparing field emission displays, electron microscope, X-ray sources, catalysts, and so on.

CNM-Based Conjugates: These are hybrids of carbon-based nanomaterials, chemically or physically associated with other molecules, such as drugs, proteins, DNA, RNA, or polymers. These conjugates leverage the unique properties of CNMs, such as high surface area, biocompatibility, and ability to be functionalized with targeting molecules, enabling precise delivery to target tissue and reducing off-target effects.

7.1.3 Applications of CNMs

Biomedical Applications: The exceptional physical, chemical, and biological properties of CNTs such as high surface area, mechanical strength, and ability to be functionalized make these nanomaterials suitable for various biomedical applications such as targeted drug delivery in specific cells/tissues and biosensing. The desirable mechanical properties and electrical conductivity also allow use of CNTs to help in growth and differentiation of various cell types, such as neurons, muscles, and bone cells. CNTs are useful in various therapies including but not limited to photothermal therapy to kill cancer cells (as CNTs absorb near-infrared light and convert it into heat to kill cancer cell more precisely), chemotherapy (as attaching drugs to CNTs results in higher solubility, stability, and target cell killing), gene therapy, and antimicrobial therapy (Jeong et al. 2024; Murjani et al. 2022).

Agriculture Applications: Unique physical and chemical properties of CNTs have enabled their multifaceted use in agriculture sector for various applications such as in nano-fertilizers (because of efficient attachment and slow release, reducing fertilizer wash off and environmental impact), nano-herbicides (enabling targeted herbicide delivery), and microbial control agents (fungal and bacterial killing). CNTs have also been used in agricultural environment applications such as treating manure for odor reduction and remediation processes to adsorb heavy metals and other pollutants to clean the contaminated soils (Patel et al. 2020; Krishna et al. 2023).

7.2 Human Exposure to CNMs as Carbon Nanoparticles (CNPs)

While the wide-ranging applications of CNMs have revolutionized the technological advancements, these have posed concerns about potential human exposure risks and associated adverse health outcomes. The exposures may occur via various routes and have been reported to cause respiratory distress, cardiovascular malfunction, and even cancers, among other effects (Yadav and Yadav 2024; Yadav et al. 2024; Bhattacharya et al. 2022b). The focus in this review on CNMs is to understand the human exposure risks, respiratory toxicity/immunotoxicity, and the underlying mechanisms.

Humans may get exposed to engineered CNMs as CNPs in occupational and nonoccupational settings. The common routes of human exposure to CNMs are inhalation, dermal exposure/skin penetration, ingestion through contaminated hands and foods, and parenteral via injection or implantation of medical devices (Kobayashi et al. 2017; Shvedova et al. 2008).

7.2.1 Occupational Exposure in CNM Workers

Occupational exposure to carbon nanotubes/nanofibers (CNT/Fs) can occur in workers in manufacturing units and quality control units or research personnel in industry or academia. Given their fibrous asbestos-like structure, occupational exposure to CNT/Fs presents potential health hazards for individuals involved in their production, handling, and use. The primary route of human occupational exposure to CNT/Fs is through inhalation of airborne particles during manufacturing, grinding, or spraying, skin contact during handling, or/and accidental ingestion of nanoparticles in contact with uncleaned hands (*Current intelligence bulletin 65*, 2013). In an epidemiological study from 12 CNT/F production sites in the United States, mean exposure of 0.1275 CNT/F structures/cm^3 was reported. About 18% workers were found to be sputum positive for CNT and 70% were dermal exposure positive (Dahm et al. 2018).

7.2.2 Nonoccupational Exposure to CNPs

Environmental Exposure: Due to widespread utilization of CNMs, particularly CNTs in various industries, there is a higher chance of environmental release and human exposure to CNPs, posing a significant threat to human health and ecology. There are several routes of environmental release and human exposure to CNT nanoparticles. These nanoparticles can be released into the environment during the synthesis and processing stages, through by-products of manufacturing waste, during consumer product degradation and leaching, and from improper disposal and recycling, leaching from landfill, stormwater runoff, industrial effluents, and during washing of tires, surface textiles, cosmetics, sunscreens, and clothing (Köhler et al. 2008; Petersen et al. 2008; Nowack et al. 2012). Environmental exposure through inhalation can occur due to the release of CNT nanoparticles in the atmosphere during the manufacturing process as well as during the use, disposal, or recycling of commercial products containing these nanomaterials. Also, exposure via inhalation can occur when CNT-containing products degrade due to normal wear and tear, releasing CNT nanoparticles into the atmosphere (Guseva Canu et al. 2020).

Exposure via Food Chain: Human exposure to CNPs via food chain is a major concern. Environmental release of CNPs potentially contaminates soil, water, and air, leading to their incorporation into the food chain. CNMs used in biofertilizers, pesticides, and soil reclamation get deposited and absorbed by plants including bioaccumulation in edible parts such as fruits, thereby resulting in human exposure via food chain. Furthermore, certain CNMs are used in food packaging and food additives leading to direct human exposure. Additionally, industrial runoff of CNMs reaching surface water bodies such as ponds, rivers, and gutters results in their bioaccumulation in exposed aquatic life including fish and other sea food animals, ultimately reaching human consumers via food chain causing various health effects (Peters et al. 2016).

Exposure via Medical Applications: Carbon nanomaterials are used in various medical applications like drug delivery system in chemotherapy, medical imaging like MRI, scaffolds and implants for tissue engineering for cell growth, biosensors for physiological activity measurement, and in inactivation of drug-resistant microbes. These applications may lead to involuntary human exposure and may cause adverse health effects (Oh et al. 2019; Jeong et al. 2024; Murjani et al. 2022).

7.3 Adverse Health Effects of CNP Exposure

7.3.1 Respiratory Exposure Risks and Human Health Effects of CNMs

Carbon nanomaterials, being prone to aerosolization as CNPs during manufacturing, handling, use, and disposal, are inhalable nanoparticle hazards to human population via occupational and environmental exposures. Various in vitro (cellular) and in vivo (animal) toxicity studies including our own have revealed the CNT inhalation toxicity paradigm (Bhattacharya et al. 2013; Frank et al. 2016; Funahashi et al. 2015) indicative of potential respiratory risks to the individuals and professionals exposed to these nanoparticles. This has been corroborated in recent human exposure and epidemiological studies (Kolosnjaj-Tabi et al. 2015). Briefly, animal studies have highlighted that inhaled CNTs have the potential to cause lung inflammation, granuloma formation, and fibrosis (Bergamaschi et al. 2021). CNTs, by virtue of their fibrous nature like asbestos, also have the potential to be carcinogenic when inhaled. Unlike large number of published laboratory rodent model studies, there are limited human epidemiological studies examining the health effects of inhaled fibrous CNMs including carbon nanotubes/nanofibers (CNT/Fs), either in an occupational or an environmental setting. In fact, only a handful of research studies have examined the effects of exposure to CNT/Fs on human health in an occupational setting. Also, most of these studies are of a cross-sectional nature, with longitudinal data lacking. Important findings from these human studies are briefly described below.

Human occupational exposure and risk assessment studies on CNT/Fs have shown that these nanoparticles can induce various respiratory health effects. Prolonged exposure may even increase the risk of lung cancer (Oberdörster et al. 2005). Studies have shown that exposure to CNTs can lead to an increase in inflammatory markers in the lungs, which could predispose workers to chronic respiratory conditions. Although there is no clear consensus on extrapulmonary translocation and biodistribution of inhaled CNTs, a few studies have reported their systemic effects (Zheng et al. 2018; Principi et al. 2016; Mitchell et al. 2009).

In a cross-sectional study that compared workers handling nanomaterials ($n = 227$) including CNTs with nonexposed controls ($n = 137$), lung inflammation markers were not found to be higher among workers handling nanomaterials (Liou et al. 2012). A longitudinal assessment of workers handling nanomaterials that included CNTs (exposed = 124, controls = 77) showed that the reduction in some lung function parameters including maximal mid-expiratory flow, peak expiratory flow rate, and forced expiratory flow at 25% after 6 months when compared to baseline values was significantly higher in exposed workers compared to the nonexposed (Liao et al. 2014). A small health surveillance pilot study ($n = 13$) found normal values of all lung function parameters in MWCNT manufacturing workers and office controls (Lee et al. 2015). A study that measured biomarkers in sputum and serum to investigate the association between MWCNT exposure and health effects (exposed = 2, controls = 10) found elevated levels of cytokine biomarkers of inflammation, IL-1β, IL-6, and TNF-α, as well as KL-6, a biomarker for interstitial lung disease in exposed workers compared to the control group (Fatkhutdinova et al. 2016). A study compared the global coding RNA (mRNA) and noncoding RNA (ncRNA) expression profiles of exposed workers (n = 8) and nonexposed workers (n = 7) in an MWCNT manufacturing facility. Differential expression of 977 long noncoding RNAs (lncRNAs), 785 coding/messenger RNAs (mRNAs), and 17 microRNAs (miRNAs) were observed between MWCNT exposed workers and nonexposed controls, suggesting a possible role of MWCNTs in causing adverse pulmonary and cardiovascular outcomes, as well as their carcinogenic potential (Shvedova et al. 2016). Another cross-sectional study that investigated association between exposure to MWCNTs and biomarkers of lung health and the immune system found that levels of some immune markers (C–C motif ligand 20, basic fibroblast growth factor, and soluble IL-1 receptor II), FENO and the CBC parameters, neutrophils, monocytes, and mean platelet volume were significantly different in exposed workers (n = 22) compared to nonexposed workers (n = 39); however, lung function tests were not different between the groups (Vlaanderen et al. 2017). A cross-sectional study that examined the association of CNT/Fs with adverse health outcomes (n = 108) found a positive association between CNT/F exposure, assessed through inhalable elemental carbon in air samples and number of years of exposure, and respiratory allergies in workers; however, no association was observed for chest symptoms and lung function parameters (Schubauer-Berigan et al. 2018). Another study that examined sputum and blood biomarkers for CNT/F exposure in humans (n = 108) found that inhalable elemental carbon and inhalable CNT/F were associated with majority of the 36 sputum and 37 blood biomarkers that were measured in the study (Beard et al. 2018).

Taken together, the results from these small number of studies suggest possible deleterious lung effects in humans from exposure to CNTs. However, many of the studies were limited to small sample sizes, and issues remain with exposure assessment methods. Also, given the cross-sectional nature of all these studies except one (Liao et al. 2014), a causal relationship between CNT exposure and lung outcomes cannot be definitively established at this time. Longitudinal studies of sufficient duration to account for the long latency period of some of the conditions studied (e.g., cancer) in well-characterized cohorts are warranted. It has been suggested that until the health effects from exposure to CNTs are further clarified, these materials should be treated as respiratory hazards, and occupational exposure should be minimized as far as practicable (Bergamaschi et al. 2021).

7.3.2 *Experimental Model Studies on Respiratory and Systemic Effects of CNTs*

7.3.2.1 Respiratory Effects

7.3.2.1.1 Tissue Injury

Inhalation of CNTs has been shown to cause various forms of lung injuries in animal model studies. CNTs can penetrate deep into the lungs and reach the alveoli, leading to pulmonary toxicity. Studies indicate that exposure to CNTs can result in epithelial cell damage, alveolar collapse, and even granuloma formation, which are indicators of lung injury (Frank et al. 2016, 2017). The high aspect ratio, biopersistence, and fibrous nature of CNTs can cause them to behave similarly to asbestos fibers, leading to similar pathogenic features (Frank et al. 2016). The physical interaction between CNTs and lung tissue can cause direct mechanical damage, while their surface properties can induce cellular stress, leading to cellular injury and death (Warheit et al. 2004).

7.3.2.1.2 Inflammation

CNT exposure in rodent models can lead to both acute and chronic inflammation in the lungs. Acute exposure often results in neutrophil influx, increased levels of pro-inflammatory cytokines (TNF-α, IL-1β), and the activation of inflammatory pathways (Frank et al. 2016). Chronic exposure can lead to sustained inflammation, contributing to tissue damage and the development of fibrosis. Studies have shown that exposure to CNTs increases the expression of inflammatory markers in lung tissue, such as IL-6, IL-8, and CCL2 (Bhattacharya et al. 2022a; Yadav et al. 2024). The persistence of these markers is indicative of ongoing inflammatory responses, which may eventually lead to more severe lung damage (Donaldson et al. 2010).

7.3.2.1.3 Fibrogenic Potential

CNTs have been shown to induce pulmonary fibrosis in rodent toxicity models, a condition characterized by excessive deposition of extracellular matrix components such as collagen, α-smooth muscle actin protein, and fibronectin, leading to stiffening and scarring of lung tissue (Shvedova et al. 2008; Mercer et al. 2011). The fibrotic response is typically a result of chronic inflammation and the activation of fibroblasts in response to CNT exposure. The fibrogenic effects of CNTs involve several biological pathways, including the TGF-β/Smad signaling pathway, which play a crucial role in the differentiation of fibroblasts into myofibroblasts and the subsequent production of collagen (Wang et al. 2012). The oxidative stress induced by CNTs also contributes to the fibrotic process.

7.3.2.1.4 Granuloma Formation

Inhalation of CNTs has been associated with the formation of granulomas, small nodules of inflammation in the lungs, caused by the immune system's attempt to confine the foreign invader. Granulomas, often considered as a biomarker of persistent lung inflammation (Shvedova et al. 2008), can compromise lung function (Bhattacharya et al. 2022a).

7.3.2.2 Cardiovascular Effects

CNTs induce systemic inflammation and oxidative stress releasing reactive oxygen species (ROS) and inflammatory mediators, which may induce atherosclerosis. Interaction of CNTs with endothelial cells (ECs), which form the inner beds of blood vessels, leads to oxidative stress and reduced nitric oxide production. Nitric oxide controls the vasodilation of blood vessels; thus, CNT exposure leads to dysfunction of ECs that are required for blood pressure control, immune cell adherence, and diapedesis. A dysfunction of ECs can contribute to cardiovascular diseases such as atherosclerosis and hypertension (Li et al. 2007).

7.3.2.3 Kidney and Liver Toxicity effects

Various animal studies have suggested potential toxicity of CNTs to kidney and liver cells (Florek et al. 2023; Kim and Cho 2023; Ji et al. 2009). CNT exposure leads to systemic oxidative stress in kidney cells, resulting in tubular necrosis, glomerular sclerosis, interstitial fibrosis, proteinuria, and reduced glomerular filtration rate. Similarly, in liver, CNT exposure leads to hepatocyte necrosis, apoptosis, fibrosis, disruption of cholesterol metabolic pathways, and changes in cytochrome P450 gene expression.

7.3.2.4 Carcinogenic Effects

Various in vivo (animal) and in vitro (cell culture) experimental model studies have demonstrated the genotoxic and carcinogenic potential of CNTs (Frank et al. 2016; Yadav and Yadav 2024). CNTs have been shown to induce DNA damage, oxidative stress, chronic inflammation, and hyperplasia, all of which are pathways that may lead to carcinogenesis (Migliore et al. 2010; Yadav et al. 2024). As the carcinogenic potential of CNTs largely depends on dose, duration, and type of cell exposed (Palomäki et al. 2010), higher dose and duration often exhibit carcinogenic potential (Kasai et al. 2016). Also, studies on the carcinogenic potential of CNTs have focused only on certain CNT brands, particularly Mitsui. In 2014, after reviewing the available information, primarily animal studies, the International Agency for Research on Cancer (IARC) identified the Mitsui brand (MWCNT-7, produced by Mitsui

Chemicals, Tokyo, Japan) as an agent that could be carcinogenic in humans. Other MWCNTs and SWCNTs were not classifiable as to their carcinogenic potential in humans due to lack of data (Grosse et al. 2014).

7.3.2.5 Interindividual Susceptibility to CNM Exposure Effects

There exists interindividual variability for the susceptibility to CNMs. Genetic polymorphisms, especially in oxidative stress-related genes, may predispose individuals to different levels of toxicity. Our animal studies have shown that different strains of mice have variable susceptibility to CNTs (Frank et al. 2017). Similarly, younger individuals with more active immune system may respond differently from elderly people. Sex also influences the susceptibility toward CNMs toxicity (Gochfeld 2017).

7.4 Mechanisms of CNT Toxicity

7.4.1 Immunotoxicity

CNT exposure can alter immune responses depending on factors such as the CNT type, dosage, exposure route, and type of responding cells. CNTs have been shown to activate various immune cells, including macrophages, dendritic cells, and T-cells, leading to either suppression or hyperactivation (Mohammed et al. 2024; Palomäki et al. 2010). The activation leads to the release of pro-inflammatory cytokines and chemokines, contributing to acute and chronic inflammation and tissue damage (Shvedova et al. 2008). CNTs have also been shown to alter production of cytokines that influence both pro-inflammatory and anti-inflammatory pathways. This dysregulation can increase the risk of infections, allergic reactions, asthma, rheumatoid arthritis, and other autoimmune diseases (Muller et al. 2005; Mohammed et al. 2024).

Pulmonary immunotoxicity studies using CNT-exposure mouse models including our own have generated experimental knowledgebase on lung immune cells and CNT interactions and the induction of inflammatory and other immunopathological changes, although molecular mechanisms continue to evolve (Frank et al. 2015, 2016; Dong and Ma 2018; Palomäki et al. 2011). Our recent study has also reported an extrapulmonary effect of CNTs on gut microbiota dysbiosis after CNT instillation in mouse lungs, besides local immunotoxic effects in terms of differential modulation of cytokines and chemokines and granulomatous inflammation in mouse lungs (Bhattacharya et al. 2022a). Gut microbiota is an important systemic modulator of immune cells. Our acute exposure studies have shown that phagocytosed CNTs result in MyD88-dependent expression of proinflammatory soluble mediators (IL-1β and TNF-α) in the alveolar macrophage cells (Frank et al. 2015); these mediators may bind to their receptors on alveolar epithelial type II cells and induce the expression of inflammatory and proliferative markers (IL-1β and Ki67) in alveolar epithelial type II cells (AECII) (Yadav and Yadav 2024), as shown in our latest air–liquid interface (ALI) coculture studies. Such studies aimed at understanding mechanisms underlying the systemic effects without direct interaction between inhaled CNTs and impacted organs are in their infancy.

7.4.2 Oxidative Stress

CNT exposure has been linked to oxidative stress and systemic inflammation. CNTs promote the release of reactive oxygen species (ROS) and inflammatory mediators from immune and nonimmune cells. ROS can cause damage to various organs and systems, including the lungs, skin, and the cardiovascular system. Oxidative stress is a key mechanism through which CNTs exert their toxic effects, contributing to inflammation and fibrosis (Kagan et al. 2010). In cardiovascular dysfunction, ROS damages endothelial cells, leading to endothelial dysfunction and increased blood pressure (Li et al. 2007).

7.4.3 Genotoxicity and Carcinogenicity

CNTs have been shown to cause DNA damage in both in vitro and in vivo studies (Migliore et al. 2010). This genotoxicity may result from direct physical interactions between CNTs and DNA, as well as from oxidative stress-induced DNA damage (Migliore et al. 2010). DNA strand breaks and chromosomal aberrations have been observed following CNT exposure. While definitive evidence in humans is lacking, some animal studies have suggested that CNTs could have carcinogenic potential, particularly with long-term exposure. Particularly, MWCNTs with asbestos-like fibrous characteristics have been shown to induce mesothelioma, a type of cancer, in some animal studies (Takagi et al. 2008; Nagai et al. 2011).

7.5 Omics Studies and Adverse Outcome Pathways (AOPs) Underlying Carbon Nanoparticle (CNPs) Toxicity

Multiple groups around the world have used omics-based approaches to elucidate the effects of nanoparticle exposure at the molecular level. Genome-wide epigenetic (Pogribna and Hammons 2021; Moreira et al. 2021; Wong et al. 2017; Sierra et al. 2016) and proteomic (Matysiak et al. 2016; Nath Roy et al. 2017) changes caused by exposure to nanoparticles have been described in several studies. Additionally, recent investigations have extensively used transcriptomics, which has been regarded as a powerful approach to interpret the changes in and consequences of genome-wide gene expression patterns upon environmental exposures. Genome-wide gene expression data generated using exposed and unexposed subjects can be accessed from publicly accessible databases such as the NCBI Gene Expression Omnibus (GEO) (Clough and Barrett 2016). Genes that are differentially regulated (upregulated or downregulated) can be identified using appropriate computational packages such as deseq2 and used for Gene Set Enrichment Analysis (GSEA) (Love et al. 2014; Subramanian et al. 2005). Enrichment analysis enables the identification of specific biological pathways and processes significantly associated with the differentially expressed genes (DEGs) by Gene Ontology based annotations (The Gene Ontology Consortium et al. 2023). Recent studies in this area have focused on identifying DEGs from multiple tissue types to identify tissue-specific signatures of systemic diseases (Ghosh et al. 2020). In the following sections, we have outlined the available resources on nanoparticle exposure, the exposure models used, and the references, with focus on CNP exposures.

Our interrogation of the NCBI GEO database retrieved (date: August 20, 2024) multiple transcriptomic datasets (Table 7.1) generated by individual research groups. The data were generated using various tissues/cell types (e.g., lung/airway tissues and cells, placental tissue, cerebral cortex, and liver tissue) and organisms (e.g., mouse, rat, fish), providing insights into responses to nanoparticle exposure and associated molecular mechanisms underlying disease development.

In the cerebral cortex and placental tissues of mice (*Mus musculus*; GSE250286) exposed to carbon black nanoparticles (CB-NPs), significant changes in the expression of genes related to neural differentiation, synaptic functions, and placental development were observed, indicating the roles of these differentially expressed genes in brain maturation and maternal–fetal nutrient exchange (Onoda et al. 2017, 2018). This dataset further revealed that maternal exposure to CB-NPs led to dysregulation of genes involved in brain development stages, some of which were partially ameliorated by ascorbic acid pretreatment (Onoda et al. 2017).

Several research groups have utilized animal models and human cell lines to elucidate the response of lung cells to various types of inhaled metal nanoparticles at different concentrations and for varying durations of exposure (Boyadzhiev et al. 2021, 2023). Interestingly, nanoparticles elicited divergent cellular and molecular responses based on the type of the particle, and the responses were additionally influenced by multiple factors such as duration of exposure, time elapsed after exposure, and concentration of the nanoparticle. Carbon nanoparticles (CNPs) are one of the most well-studied nonmetal exposure agents that showed divergent responses based on the shape of the and duration of exposure nanoparticle (Rydman et al. 2014; Frank et al. 2017). Overall, experiments on lung epithelial cells (GSE246159, GSE161017, and GSE50176) identified DEGs involved in cell differentiation, oxidative stress, inflammation, and epithelial–mesenchymal transition (EMT), shedding light on the molecular basis of lung diseases, such as chronic obstructive pulmonary disease (COPD) and pulmonary fibrosis, associated with exposures to metal and nonmetal nanoparticles (Boyadzhiev et al. 2021, 2023; Rydman et al. 2014; Frank et al. 2017).

Table 7.1 Gene expression datasets related to nanoparticle (NP) exposure available from Gene Expression Omnibus (GEO)

	Species	Sample type	Nanoparticles	GEO ID	Submitters [References]
1	Mus musculus	Cerebral cortex	CB nanoparticle	GSE250286	Onoda A, et al; 2023 [PMID 28148272]
2	Mus musculus	Placental tissue	CB nanoparticle	GSE250286	Onoda A, et al; 2023 [PMID 29660869]
3	Mus musculus	lung epithelial cell line (FE1)	Multiple metal NPs	GSE246159	Boyadzhiev A, et al.; 2024 [PMID 38203705]
4	Mus musculus	Pulmonary tissue	CNT rod and tangled	GSE50176	Rydman E, et al.; 2014 [PMID 25318534]
5	Mus musculus	lung epithelial cells	CuONPs	GSE161017	Boyadzhiev A, et al.; 2022 [PMID 33507836]
6	Mus musculus	Lung tissue	DWCNT	GSE223818	Han L, et al.; 2023 [PMID 37856828]
7	Mus musculus	Lung tissue	Tangled CNP	GSE86339	Frank E.A., et al.; 2017 [PMID 28433707]
8	Mus musculus	Lung tissue	MWCNT	GSE140676	Khaliullin TO., et al 2020 [PMID: 31978390]
9	Mus musculus	Peripheral blood	MWCNT	GSE140676	Khaliullin TO., et al 2020 [PMID: 31978390]
10	Rattus norgicus	Lung tissue	MWCNT-7	GSE148869	Sager TM, et al.; 2022 [PMID 35648795]
11	Rattus norgicus	Lung tissue	Carbon nanotubes (Mitsui)	GSE160175	Valentina SA, et al.; 2023 [PMID 200160175]
12	Anguilla Anguilla	Liver tissue	Gold nanoparticles	GSE110011	Perrier F, et al.; 2019 [PMID 32833228]
13	Anguilla Anguilla	Brain tissue	Gold nanoparticles	GSE110011	Perrier F, et al.; 2019 [PMID 32833228]
14	Homo sapiens	Bronchial epithelium	Copper nanoparticles	GSE127773	Kooter I, et al.; 2019 [PMID 31188557]
15	Homo sapiens	Lung fibroblasts 16iu cell line	MWCNT	GSE43515	Martin SL, et al.; 2013 [N/A]
16	Homo sapiens	Lung tissue	Carbon NM	GSE92899	Kinaret P, et al.; 2017 [PMID 28380293]
17	Homo sapiens	Placental tissue	Copper, Polystyrene	GSE220756	Chortarea S, et al.; 2023 [N/A]

The table shows that the datasets are derived from different experimental systems (mouse, rat, European eel, or human-derived cell lines), different target tissues, and different types of NP types. However, most of the studies were designed to capture the effects of carbon nanoparticles

A rat model (*Rattus norvegicus*; GSE148869) provided insights into the effects of inhalation exposure to multi-walled carbon nanotube (MWCNT-7/Mitsui) on lung development and disease. The study revealed that MWCNT-7 exposure led to mild to moderate lung pathological changes, including increased cellularity, alveolitis, fibrosis, and granuloma formation. Global gene expression profiling identified significant changes in genes associated with cancer, leukocyte migration, inflammation, and mitosis, among others, illustrating a dose–response relationship to MWCNT exposure (Sager et al. 2022).

A *Mus musculus* macrophage cell line (GSE223818) model used for investigation of immune responses during murine gamma herpesvirus 68 (MHV-68) infection led to identification of key genes involved in antiviral responses and immune signaling pathways, offering potential therapeutic targets for managing infections from viral nanoparticles. This study also showed that repeated exposure to nanoparticles, such as soot-like carbonaceous NPs and double-walled carbon nanotubes (DWCNTs), can reactivate latent herpesvirus in the lungs, leading to progressive lung inflammation and emphysema-like injury (Han et al. 2023).

European eel (*Anguilla anguilla*, GSE110011) model was used to identify DEGs in brain and liver tissues (Perrier et al. 2020). The study revealed how these tissues adapt to environmental stressors by modulating energy metabolism and stress response genes. The study focused on exposure to gold nanoparticles (AuNPs) and found that, even at lower concentrations, AuNPs led to significant changes in immune-related gene expression in both organs.

Together, the above experimental model studies provide a comprehensive understanding of NP-modulated gene expression across different tissues and species, contributing to a broader knowledge of immunity, disease, and environmental adaptation.

Human studies have focused on placental and bronchial epithelium tissues, as well as on lung cell lines. The gene expression profiling of human placenta samples (GSE220756) investigated how exposure to copper oxide nanoparticles (CuO NPs) and polystyrene nano plastics (PS NPs) affect placental gene expression. This study showed that CuO NPs and PS NPs induced distinct cellular responses: CuO NPs were linked to pathways related to angiogenesis and protein misfolding, while PS NPs impacted inflammation and iron homeostasis (Chortarea et al. 2023). The bronchial epithelium dataset (GSE127773) was used to investigate the effects of CuO nanoparticle aerosols on gene expression in healthy and asthmatic airway epithelia (Kooter et al. 2019). The investigators found that asthmatic cells exhibited higher cytotoxicity and pro-inflammatory responses compared to healthy cells when exposed to nanoparticles. Transcriptome analysis revealed distinct changes in response to different nanoparticle types, with a set of genes related to mucociliary clearance showing significant differences between asthmatic and healthy cells. Next, the analysis of human lung cell line 12lu (GSE43515) revealed gene expression changes related to cell proliferation, differentiation, and apoptosis. This study examined how MWCNT affects gene expression in human lung fibroblasts, particularly in the presence of Interleukin 13 (IL-13), a key cytokine in asthma. MWCNT and IL-13 each induced distinct gene expression changes, with their combination resulting in a unique profile. Specifically, IL-13 and MWCNT together increased the expression of IL-17A and collagen (Col1A1), suggesting a shift in gene expression, while MWCNT alone increased interferon-inducible protein-27 (IFI27). This study is consistent with previous results showing pro-fibrotic effects of MWCNT exposure (Hussain et al. 2014). Additionally, Kinaretthe et al (GSE92899) investigated transcriptomic responses in the THP-1 macrophage cell line and mouse lung tissues exposed to various carbon nanomaterials (Kinaret et al. 2017). The study identified networks of genes with coordinately altered expression in response to specific CNM properties. Results highlighted the relevance of DNA repair and innate immunity. Although there was minimal overlap in gene sets across different exposure scenarios, significant commonalities emerged in altered molecular functions.

CNTs have been increasingly recognized as one of the most significant inhalant nanoparticles that can cause a spectrum of health hazards. Therefore, several studies have focused on understanding the effects of CNT exposure using human-derived cells or animal models. Our studies demonstrated that mice strains differ in their responses to CNT exposures. C57Bl/6 mice developed intense airway reactions compared to DBA/2 mice. Interestingly, B6D2F1 (C57Bl/6 × DBA/2 hybrid) mice retained sensitivity to CNT exposures, indicating the genetic basis of CNT sensitivity (Frank et al. 2017). The study also demonstrated the rationality of choosing inbred mice for CNT exposure studies.

Finally, as a specific example, we will focus on a study involving lung tissue from *Mus musculus* (GSE50176) to investigate the effects of CNTs. The study compared rod-shaped/rigid CNTs (rCNT) and tangled CNTs (tCNT) against untreated controls (UT). The experiments included 4 h of exposure and two different post-exposure time points, 0 and 24 h, to capture both immediate and delayed biological responses, respectively. The primary goal was to assess the impact of CNT inhalation on the pulmonary system, with particular focus on potential allergic-like airway inflammation. The results revealed that an inhalation period of 4 h can produce both immediate (identified at 0 h, i.e., immediately after the exposure period) and delayed/persistent (identified 24 h after the exposure period) gene expression changes. There are unique DEGs associated either with immediate or with delayed/persistent effects in addition to DEGs shared between these conditions. Moreover, these changes are also dependent on the shape of the nanoparticles, rod versus tangled. Rod-shaped CNTs induced a significant innate immunity-mediated allergic-like airway inflammation in healthy mice. This was characterized by marked eosinophilia, mucin hypersecretion, airway hyperresponsiveness, and the expression of Th2-type cytokines. Transcriptomics analysis demonstrated that a single 4-h exposure to rod-shaped CNTs caused a significant upregulation of genes involved in innate immunity and cytokine/chemokine pathways. In contrast, tangled CNTs did not induce these changes, highlighting specific inflammatory response associated with the rCNTs.

Tables 7.2, 7.3, 7.4, and 7.5 summarize the results of enrichment analysis showing top 10 significant Gene Ontology based biological processes associated with (Table 7.2) rCNT versus UT 4 h.0 h, (Table 7.3) rCNT versus UT 4 h.24 h, (Table 7.4) tCNT versus UT 4 h.0 h, and (Table 7.5) tCNT versus UT 4 h.24 h.

Enrichment analysis shows that molecular/transcriptomic outcomes of exposure vary by type of the nanoparticle (CNT) and the time post-exposure. Most significant effects have been observed in rCNT versus UT 4 h.0 h (rod-like CNT, immediately after the exposure time), as shown by several significantly enriched biological processes. This observation is also supported by Ryman et al. in the publication

Table 7.2 rCNT versus UT 4 h.0 h. Biological processes associated with rod-like carbon nanotube exposure for 4 h followed by 0 h post-exposure sacrifice of mice indicating immediate effects

Name	*p*-value	Adjusted *p*-value	Odds ratio	Combined score
Positive Regulation of Tumor Necrosis Factor Production (GO:0032760)	6.37E-07	0.002946	3.4	48.56
Positive Regulation of Tumor Necrosis Factor Superfamily Cytokine Production (GO:1903557)	2.268E-06	0.005247	3.14	40.75
Positive Regulation of Cytokine Production (GO:0001819)	0.00001268	0.01916	1.79	20.13
Regulation of Tumor Necrosis Factor Production (GO:0032680)	0.00001657	0.01916	2.37	26.13
B Cell Receptor Signaling Pathway (GO:0050853)	0.00004913	0.04545	3.62	35.94
Positive Regulation of Interleukin-6 Production (GO:0032755)	0.0001057	0.06687	2.68	24.52
Positive Regulation of Response to Biotic Stimulus (GO:0002833)	0.0001148	0.06687	3.64	33.06
T-helper 1 Cell Differentiation (GO:0045063)	0.0001156	0.06687	83675	758511
Inflammatory Response (GO:0006954)	0.0002184	0.09789	1.76	14.85
Positive Regulation of Response to External Stimulus (GO:0032103)	0.0002232	0.09789	1.98	16.65

Table 7.3 rCNT versus UT 4 h.24 h. Biological processes associated with rod-like carbon nanotube exposure for 4 h followed by 24 h post-exposure sacrifice of mice indicating delayed/sustained effects

Name	*p*-value	Adjusted *p*-value	Odds ratio	Combined score
T-helper 1 Cell Differentiation (GO:0045063)	0.00003222	0.1357	87355	903517.04
Positive Regulation of Apoptotic Cell Clearance (GO:2000427)	0.0005419	1	17.3	130.12
Cilium Movement Involved in Cell Motility (GO:0060294)	0.001294	1	11.53	76.7
Positive Regulation of Syncytium Formation by Plasma Membrane Fusion (GO:0060143)	0.001305	1	6.06	40.24
Synaptic Vesicle Fusion to Presynaptic Active Zone Membrane (GO:0031629)	0.002609	1	8.65	51.46
Response to Interferon-Alpha (GO:0035455)	0.003134	1	4.85	27.94
protein-RNA Complex Assembly (GO:0022618)	0.003727	1	1.81	10.12
Regulation of Amyloid Precursor Protein Biosynthetic Process (GO:0042984)	0.004676	1	6.92	37.13
Cellular Response to ATP (GO:0071318)	0.004931	1	5.19	27.58
Regulation of Noncanonical Wnt Signaling Pathway (GO:2000050)	0.004931	1	5.19	27.58

Table 7.4 tCNT versus UT 4 h.0 h. Biological processes associated with tangled carbon nanotube exposure for 4 h followed by 0 h post-exposure sacrifice of mice indicating immediate effects

Name	*p*-value	Adjusted *p*-value	Odds ratio	Combined score
B Cell Receptor Signaling Pathway (GO:0050853)	1.82E-09	4.39E-06	13.24	266.54
Positive Regulation of Tumor Necrosis Factor Production (GO:0032760)	1.53E-08	1.34E-05	8.23	148.02
Antigen Receptor-Mediated Signaling Pathway (GO:0050851)	1.67E-08	1.34E-05	5.87	105.06
Positive Regulation of Tumor Necrosis Factor Superfamily Cytokine Production (GO:1903557)	2.99E-08	1.73E-05	7.74	134.12
Positive Regulation of Cytokine Production (GO:0001819)	3.58E-08	1.73E-05	3.66	62.81
Regulation of Tumor Necrosis Factor Production (GO:0032680)	4.39E-08	1.77E-05	5.83	98.81
Inflammatory Response (GO:0006954)	3.57E-07	0.000123	3.9	57.85
Regulation of Angiogenesis (GO:0045765)	5.89E-07	0.000178	4.09	58.64
B Cell Activation (GO:0042113)	9.06E-07	0.000243	6.17	85.88
Dendritic Cell Chemotaxis (GO:0002407)	1.375E-06	0.000332	24.76	334.19

Table 7.5 tCNT versus UT 4 h.24 h. Biological processes associated with tangled carbon nanotube exposure for 4 h followed by 24 h post-exposure sacrifice of mice indicating delayed/sustained effects

Name	*p*-value	Adjusted *p*-value	Odds ratio	Combined score
protein-RNA Complex Assembly (GO:0022618)	0.0002567	0.8237	2.04	16.88
Spliceosomal Complex Assembly (GO:0000245)	0.0007771	0.8237	2.65	18.98
Positive Regulation of Biosynthetic Process (GO:0009891)	0.0007997	0.8237	3.1	22.12
Regulation of Phagocytosis (GO:0050764)	0.0008974	0.8237	2.7	18.94
Striated Muscle Cell Differentiation (GO:0051146)	0.0009453	0.8237	4.9	34.14
Somite Development (GO:0061053)	0.002312	1	9.98	60.55
mRNA Splicing, via Spliceosome (GO:0000398)	0.003633	1	1.63	9.16
Positive Regulation of Phagocytosis (GO:0050766)	0.00364	1	2.37	13.31
RNA Processing (GO:0006396)	0.003756	1	1.68	9.39
B Cell Receptor Signaling Pathway (GO:0050853)	0.003947	1	2.62	14.51

associated with this dataset (Rydman et al. 2014). The major effect shown by enrichment analysis is mediated by TNF signaling in case of immediate effect, whereas Th-cell signaling and positive regulation of apoptotic clearance were the two most significant biological processes associated with delayed response of rCNT exposure. We also found B Cell Receptor Signaling Pathway and Positive Regulation of Tumor Necrosis Factor Production as the significantly enriched biological processes associated with tCNT, 0 h post-exposure. However, the effects were not significant after 24 h post-exposure, showing no significantly enriched biological process after adjusting for "false discovery the rate".

In addition to the aforementioned study focusing on transcriptomic changes associated with acute response (0 h and 24 h after a 4-h exposure period), two other studies focused on gene expression changes associated with chronic exposures to MWCNT in 28 days and 56 days post-exposure, respectively. The first study from our laboratory used tangled CNTs (4 mg/kg body weight) applied via orotracheal aspiration. C57B/6 mice sacrificed on 28 days post-exposure showed more than 6K significant DEGs that were in networks of inflammatory molecules, showing IL-1β and TNF as major hub genes and "respiratory disease" as a major significant "disease function" (Frank et al. 2017). Interestingly, this result could be connected with the results of acute exposure data (Rydman et al.; GSE50176) which also indicates the involvement of TNF signaling pathway in CNT exposure outcomes (Tables 7.2 and 7.4). The network was also overlaid with the "lung cancer" disease function, suggesting association of CNT exposure with cancer promotion by CNTs under suitable conditions (Frank et al. 2017). The second study undertaken by Khaliullin et al. used a suspension of MWCNT (40 μg/mouse) for exposure via pharyngeal aspiration (suspension placed in posterior throat, tongue held until aspiration into lungs). The mice were scarified after 56 days, and lungs and blood transcriptomes were analyzed. The results showed sustained inflammation, immune cell trafficking, and phagocytotic responses. Oxidative phosphorylation, mitochondrial dysfunction, IL-6 signaling, and VEGF signaling were among the most significant perturbed pathways. Inflammatory response, cancer, and organismal injury and abnormalities were the top significant diseases associated with DEGs identified from lung tissues (Khaliullin et al. 2020). The authors also noted that CNTs induced differentially regulated genes/pathways identifiable from both whole blood and lung samples, which were associated with cancer/tumor development pathways, as well as atopy (Khaliullin et al. 2020).

In conclusion, while carbon nanomaterials offer significant benefits across various fields, their exposure risks and health effects must not be overlooked. Continued research is crucial to fully understand their impact on human health and to develop effective strategies for safe handling and use. Existing data shows that the effects of carbon nanoparticle exposure greatly vary with (1) NP type, (2) NP shape, (3) exposure duration, (4) Time elapsed after exposure, and (5) target organ. Secondary analysis of publicly available multi-origin datasets, particularly using advanced analytics including machine learning, could help identify biomarkers for diagnosis and treatment response monitoring of carbon nanoparticle exposure. Additional studies are warranted involving multi-omics approaches to capture biomarkers of exposure and personalized management.

References

Ahmed S, Hussain CM (eds) (2018) Green and sustainable advanced materials: applications, 1st edn. Wiley. https://doi.org/10.1002/9781119528463

Alshammari BH, Lashin MMA, Mahmood MA, Al-Mubaddel FS, Ilyas N, Rahman N, Sohail M, Khan A, Abdullaev SS, Khan R (2023) Organic and inorganic nanomaterials: fabrication, properties and applications. RSC Adv. 13:13735–13785. https://doi.org/10.1039/D3RA01421E

Beard JD, Erdely A, Dahm MM, de Perio MA, Birch ME, Evans DE, Fernback JE, Eye T, Kodali V, Mercer RR, Bertke SJ, Schubauer-Berigan MK (2018) Carbon nanotube and nanofiber exposure and sputum and blood biomarkers of early effect among U.S. workers. Environ Int 116:214–228. https://doi.org/10.1016/j.envint.2018.04.004

Bergamaschi E, Garzaro G, Wilson Jones G, Buglisi M, Caniglia M, Godono A, Bosio D, Fenoglio I, Guseva Canu I (2021) Occupational exposure to carbon nanotubes and carbon nanofibres: more than a cobweb. Nanomaterials (Basel) 11:745. https://doi.org/10.3390/nano11030745

Bhattacharya K, Andón FT, El-Sayed R, Fadeel B (2013) Mechanisms of carbon nanotube-induced toxicity: focus on pulmonary inflammation. Adv Drug Deliv Rev 65:2087–2097. https://doi.org/10.1016/j.addr.2013.05.012

Bhattacharya SS, Yadav B, Rosen L, Nagpal R, Yadav H, Yadav JS (2022a) Crosstalk between gut microbiota and lung inflammation in murine toxicity models of respiratory exposure or co-exposure to carbon nanotube particles and cigarette smoke extract. Toxicol Appl Pharmacol 447:116066. https://doi.org/10.1016/j.taap.2022.116066

Bhattacharya SS, Yadav B, Yadav E, Hus A, Yadav N, Kaur P, Rosen L, Jandarov R, Yadav JS (2022b) Differential modulation of lung aquaporins among other pathophysiological markers in acute (Cl2 gas) and chronic (carbon nanoparticles, cigarette smoke) respiratory toxicity mouse models. Front Physiol 13:880815. https://doi.org/10.3389/fphys.2022.880815

Boyadzhiev A, Avramescu M-L, Wu D, Williams A, Rasmussen P, Halappanavar S (2021) Impact of copper oxide particle dissolution on lung epithelial cell toxicity: response characterization using global transcriptional analysis. Nanotoxicology 15:380–399. https://doi.org/10.1080/17435390.2021.1872114

Boyadzhiev A, Wu D, Avramescu M-L, Williams A, Rasmussen P, Halappanavar S (2023) Toxicity of metal oxide nanoparticles: looking through the lens of toxicogenomics. IJMS 25:529. https://doi.org/10.3390/ijms25010529

Carbon Nanotubes Market Size, Share & Industry Analysis, By Product (Multi Walled Carbon Nanotubes (MWCNT) and Single Walled Carbon Nanotubes (SWCNT)), By Application (Plastics & Composites, Electrical & Electronics, Energy, and Others), and Regional Forecast, 2024–2032 Source (No. FBI102700) (n.d.). India

Chen M, He Y, Ye Q, Wang X, Hu Y (2019) Shape-dependent solar thermal conversion properties of plasmonic Au nanoparticles under different light filter conditions. Solar Energy 182:340–347. https://doi.org/10.1016/j.solener.2019.02.070

Chortarea S, Gupta G, Saarimäki LA, Netkueakul W, Manser P, Aengenheister L, Wichser A, Fortino V, Wick P, Greco D, Buerki-Thurnherr T (2023) Transcriptomic profiling reveals differential cellular response to copper oxide nanoparticles and polystyrene nanoplastics in perfused human placenta. Environ Int 177:108015. https://doi.org/10.1016/j.envint.2023.108015

Clough E, Barrett T (2016) The gene expression omnibus database. Methods Mol Biol 1418:93–110. https://doi.org/10.1007/978-1-4939-3578-9_5

Current intelligence bulletin 65: occupational exposure to carbon nanotubes and nanofibers (2013) U.S. Department of Health and Human Services, Public Health Service, Centers for Disease Control and Prevention, National Institute for Occupational Safety and Health. https://doi.org/10.26616/NIOSHPUB2013145

Dahm MM, Schubauer-Berigan MK, Evans DE, Birch ME, Bertke S, Beard JD, Erdely A, Fernback JE, Mercer RR, Grinshpun SA (2018) Exposure assessments for a cross-sectional epidemiologic study of US carbon nanotube and nanofiber workers. Int J Hygiene Environ Health 221:429–440. https://doi.org/10.1016/j.ijheh.2018.01.006

De Volder MFL, Tawfick SH, Baughman RH, Hart AJ (2013) Carbon nanotubes: present and future commercial applications. Science 339:535–539. https://doi.org/10.1126/science.1222453

Donaldson K, Murphy FA, Duffin R, Poland CA (2010) Asbestos, carbon nanotubes and the pleural mesothelium: a review and the hypothesis regarding the role of long fibre retention in the parietal pleura, inflammation and mesothelioma. Part Fibre Toxicol 7:5. https://doi.org/10.1186/1743-8977-7-5

Dong J, Ma Q (2018) Type 2 immune mechanisms in carbon nanotube-induced lung fibrosis. Front Immunol 9:1120. https://doi.org/10.3389/fimmu.2018.01120

Fatkhutdinova LM, Khaliullin TO, Vasil'yeva OL, Zalyalov RR, Mustafin IG, Kisin ER, Birch ME, Yanamala N, Shvedova AA (2016) Fibrosis biomarkers in workers exposed to MWCNTs. Toxicol Appl Pharmacol 299:125–131. https://doi.org/10.1016/j.taap.2016.02.016

Florek E, Witkowska M, Szukalska M, Richter M, Trzeciak T, Miechowicz I, Marszałek A, Piekoszewski W, Wyrwa Z, Giersig M (2023) Oxidative stress in long-term exposure to multi-walled carbon nanotubes in male rats. Antioxidants (Basel) 12:464. https://doi.org/10.3390/antiox12020464

Frank EA, Birch ME, Yadav JS (2015) MyD88 mediates in vivo effector functions of alveolar macrophages in acute lung inflammatory responses to carbon nanotube exposure. Toxicol Appl Pharmacol 288:322–329. https://doi.org/10.1016/j.taap.2015.08.004

Frank EA, Carreira VS, Birch ME, Yadav JS (2016) Carbon nanotube and asbestos exposures induce overlapping but distinct profiles of lung pathology in Non-Swiss Albino CF-1 mice. Toxicol Pathol 44:211–225. https://doi.org/10.1177/0192623315620587

Frank EA, Carreira VS, Shanmukhappa K, Medvedovic M, Prows DR, Yadav JS (2017) Genetic susceptibility to toxicologic lung responses among inbred mouse strains following exposure to carbon nanotubes and profiling of underlying gene networks. Toxicol Appl Pharmacol 327:59–70. https://doi.org/10.1016/j.taap.2017.04.019

Funahashi S, Okazaki Y, Ito D, Asakawa A, Nagai H, Tajima M, Toyokuni S (2015) Asbestos and multi-walled carbon nanotubes generate distinct oxidative responses in inflammatory cells. J Clin Biochem Nutr 56:111–117. https://doi.org/10.3164/jcbn.14-92

Georgakilas V, Perman JA, Tucek J, Zboril R (2015) Broad family of carbon nanoallotropes: classification, chemistry, and applications of fullerenes, carbon dots, nanotubes, graphene, nanodiamonds, and combined superstructures. Chem Rev 115:4744–4822. https://doi.org/10.1021/cr500304f

Ghosh D, Ding L, Bernstein JA, Mersha TB (2020) The utility of resolving asthma molecular signatures using tissue-specific transcriptome data. G3 (Bethesda) 10:4049–4062. https://doi.org/10.1534/g3.120.401718

Gochfeld M (2017) Sex differences in human and animal toxicology: toxicokinetics. Toxicol Pathol 45:172–189. https://doi.org/10.1177/0192623316677327

Gong X, Jadhav ND, Lonikar VV, Kulkarni AN, Zhang H, Sankapal BR, Ren J, Xu BB, Pathan HM, Ma Y, Lin Z, Witherspoon E, Wang Z, Guo Z (2024) An overview of green synthesized silver nanoparticles towards bioactive antibacterial, antimicrobial and antifungal applications. Adv Colloid Interface Sci 323:103053. https://doi.org/10.1016/j.cis.2023.103053

Grosse Y, Loomis D, Guyton KZ, Lauby-Secretan B, El Ghissassi F, Bouvard V, Benbrahim-Tallaa L, Guha N, Scoccianti C, Mattock H, Straif K, International Agency for Research on Cancer Monograph Working Group (2014) Carcinogenicity of fluoro-edenite, silicon carbide fibres and whiskers, and carbon nanotubes. Lancet Oncol 15:1427–1428. https://doi.org/10.1016/S1470-2045(14)71109-X

Guseva Canu I, Batsungnoen K, Maynard A, Hopf NB (2020) State of knowledge on the occupational exposure to carbon nanotubes. Int J Hyg Environ Health 225:113472. https://doi.org/10.1016/j.ijheh.2020.113472

Han L, Haefner V, Yu Y, Han B, Ren H, Irmler M, Beckers J, Liu Q, Feuchtinger A, Yildirim AO, Adler H, Stoeger T (2023) Nanoparticle-exposure-triggered virus reactivation induces lung emphysema in mice. ACS Nano 17:21056–21072. https://doi.org/10.1021/acsnano.3c04111

Hussain S, Sangtian S, Anderson SM, Snyder RJ, Marshburn JD, Rice AB, Bonner JC, Garantziotis S (2014) Inflammasome activation in airway epithelial cells after multi-walled carbon nanotube exposure mediates a profibrotic response in lung fibroblasts. Part Fibre Toxicol 11:28. https://doi.org/10.1186/1743-8977-11-28

Jeong YH, Kwon M, Shin S, Lee J, Kim KS (2024) Biomedical applications of CNT-based fibers. Biosensors 14:137. https://doi.org/10.3390/bios14030137

Ji Z, Zhang D, Li L, Shen X, Deng X, Dong L, Wu M, Liu Y (2009) The hepatotoxicity of multi-walled carbon nanotubes in mice. Nanotechnology 20:445101. https://doi.org/10.1088/0957-4484/20/44/445101

Kagan VE, Konduru NV, Feng W, Allen BL, Conroy J, Volkov Y, Vlasova II, Belikova NA, Yanamala N, Kapralov A, Tyurina YY, Shi J, Kisin ER, Murray AR, Franks J, Stolz D, Gou P, Klein-Seetharaman J, Fadeel B, Star A, Shvedova AA (2010) Carbon nanotubes degraded by neutrophil myeloperoxidase induce less pulmonary inflammation. Nat Nanotechnol 5:354–359. https://doi.org/10.1038/nnano.2010.44

Kasai T, Umeda Y, Ohnishi M, Mine T, Kondo H, Takeuchi T, Matsumoto M, Fukushima S (2016) Lung carcinogenicity of inhaled multi-walled carbon nanotube in rats. Part Fibre Toxicol 13:53. https://doi.org/10.1186/s12989-016-0164-2

Khaliullin TO, Yanamala N, Newman MS, Kisin ER, Fatkhutdinova LM, Shvedova AA (2020) Comparative analysis of lung and blood transcriptomes in mice exposed to multi-walled carbon nanotubes. Toxicol Appl Pharmacol 390:114898. https://doi.org/10.1016/j.taap.2020.114898

Kim J-E, Cho M-H (2023) Effects of multiwall carbon nanotubes on premature kidney aging: biochemical and histological analysis. Toxics 11:373. https://doi.org/10.3390/toxics11040373

Kinaret P, Marwah V, Fortino V, Ilves M, Wolff H, Ruokolainen L, Auvinen P, Savolainen K, Alenius H, Greco D (2017) Network analysis reveals similar transcriptomic responses to intrinsic properties of carbon nanomaterials in vitro and in vivo. ACS Nano 11:3786–3796. https://doi.org/10.1021/acsnano.6b08650

Kobayashi N, Izumi H, Morimoto Y (2017) Review of toxicity studies of carbon nanotubes. J Occup Health 59:394–407. https://doi.org/10.1539/joh.17-0089-RA

Köhler AR, Som C, Helland A, Gottschalk F (2008) Studying the potential release of carbon nanotubes throughout the application life cycle. J Clean Prod 16:927–937. https://doi.org/10.1016/j.jclepro.2007.04.007

Kolosnjaj-Tabi J, Just J, Hartman KB, Laoudi Y, Boudjemaa S, Alloyeau D, Szwarc H, Wilson LJ, Moussa F (2015) Anthropogenic carbon nanotubes found in the airways of Parisian children. EBioMedicine 2:1697–1704. https://doi.org/10.1016/j.ebiom.2015.10.012

Kooter I, Ilves M, Gröllers-Mulderij M, Duistermaat E, Tromp PC, Kuper F, Kinaret P, Savolainen K, Greco D, Karisola P, Ndika J, Alenius H (2019) Molecular signature of asthma-enhanced sensitivity to CuO nanoparticle aerosols from 3D cell model. ACS Nano 13:6932–6946. https://doi.org/10.1021/acsnano.9b01823

Krishna RH, Chandraprabha MN, Samrat K, Krishna Murthy TP, Manjunatha C, Kumar SG (2023) Carbon nanotubes and graphene-based materials for adsorptive removal of metal ions – a review on surface functionalization and related adsorption mechanism. Appl Surf Sci Adv 16:100431. https://doi.org/10.1016/j.apsadv.2023.100431

Lee JS, Choi YC, Shin JH, Lee JH, Lee Y, Park SY, Baek JE, Park JD, Ahn K, Yu IJ (2015) Health surveillance study of workers who manufacture multi-walled carbon nanotubes. Nanotoxicology 9:802–811. https://doi.org/10.3109/17435390.2014.978404

Li Z, Hulderman T, Salmen R, Chapman R, Leonard SS, Young S-H, Shvedova A, Luster MI, Simeonova PP (2007) Cardiovascular effects of pulmonary exposure to single-wall carbon nanotubes. Environ Health Perspect 115:377–382. https://doi.org/10.1289/ehp.9688

Liao H-Y, Chung Y-T, Lai C-H, Wang S-L, Chiang H-C, Li L-A, Tsou T-C, Li W-F, Lee H-L, Wu W-T, Lin M-H, Hsu J-H, Ho J-J, Chen C-J, Shih T-S, Lin C-C, Liou S-H (2014) Six-month follow-up study of health markers of nanomaterials among workers handling engineered nanomaterials. Nanotoxicology 8(Suppl 1):100–110. https://doi.org/10.3109/17435390.2013.858793

Liou S-H, Tsou T-C, Wang S-L, Li L-A, Chiang H-C, Li W-F, Lin P-P, Lai C-H, Lee H-L, Lin M-H, Hsu J-H, Chen C-R, Shih T-S, Liao H-Y, Chung Y-T (2012) Epidemiological study of health hazards among workers handling engineered nanomaterials. J Nanopart Res 14:878. https://doi.org/10.1007/s11051-012-0878-5

Liu W, Du H, Zheng T, Si C (2021) Biomedical applications of bacterial cellulose based composite hydrogels. CMC 28:8319–8332. https://doi.org/10.2174/0929867328666210412124444

Love MI, Huber W, Anders S (2014) Moderated estimation of fold change and dispersion for RNA-seq data with DESeq2. Genome Biol 15:550. https://doi.org/10.1186/s13059-014-0550-8

Matysiak M, Kapka-Skrzypczak L, Brzóska K, Gutleb AC, Kruszewski M (2016) Proteomic approach to nanotoxicity. J Proteomics 137:35–44. https://doi.org/10.1016/j.jprot.2015.10.025

Mekuye B, Abera B (2023) Nanomaterials: an overview of synthesis, classification, characterization, and applications. Nano Select 4:486–501. https://doi.org/10.1002/nano.202300038

Mercer RR, Hubbs AF, Scabilloni JF, Wang L, Battelli LA, Friend S, Castranova V, Porter DW (2011) Pulmonary fibrotic response to aspiration of multi-walled carbon nanotubes. Part Fibre Toxicol 8:21. https://doi.org/10.1186/1743-8977-8-21

Migliore L, Saracino D, Bonelli A, Colognato R, D'Errico MR, Magrini A, Bergamaschi A, Bergamaschi E (2010) Carbon nanotubes induce oxidative DNA damage in RAW 264.7 cells. Environ Mol Mutagen 51:294–303. https://doi.org/10.1002/em.20545

Mitchell LA, Lauer FT, Burchiel SW, McDonald JD (2009) Mechanisms for how inhaled multiwalled carbon nanotubes suppress systemic immune function in mice. Nat Nanotechnol 4:451–456. https://doi.org/10.1038/nnano.2009.151

Mohammed AN, Yadav N, Kaur P, Jandarov R, Yadav JS (2024) Immunomodulation of susceptibility to pneumococcal pneumonia infection in mouse lungs exposed to carbon nanoparticles via dysregulation of innate and adaptive immune responses. Toxicol Appl Pharmacol 483:116820. https://doi.org/10.1016/j.taap.2024.116820

Moreira L, Costa C, Pires J, Teixeira JP, Fraga S (2021) How can exposure to engineered nanomaterials influence our epigenetic code? A review of the mechanisms and molecular targets. Mutat Res/Rev Mutat Res 788:108385. https://doi.org/10.1016/j.mrrev.2021.108385

Muller J, Huaux F, Moreau N, Misson P, Heilier J-F, Delos M, Arras M, Fonseca A, Nagy JB, Lison D (2005) Respiratory toxicity of multi-wall carbon nanotubes. Toxicol Appl Pharmacol 207:221–231. https://doi.org/10.1016/j.taap.2005.01.008

Murjani BO, Kadu PS, Bansod M, Vaidya SS, Yadav MD (2022) Carbon nanotubes in biomedical applications: current status, promises, and challenges. Carbon Lett. 32:1207–1226. https://doi.org/10.1007/s42823-022-00364-4

Nagai H, Okazaki Y, Chew SH, Misawa N, Yamashita Y, Akatsuka S, Ishihara T, Yamashita K, Yoshikawa Y, Yasui H, Jiang L, Ohara H, Takahashi T, Ichihara G, Kostarelos K, Miyata Y, Shinohara H, Toyokuni S (2011) Diameter and rigidity of multiwalled carbon nanotubes are critical factors in mesothelial injury and carcinogenesis. Proc Natl Acad Sci U S A 108:E1330–E1338. https://doi.org/10.1073/pnas.1110013108

Nath Roy D, Goswami R, Pal A (2017) Nanomaterial and toxicity: what can proteomics tell us about the nanotoxicology? Xenobiotica 47:632–643. https://doi.org/10.1080/00498254.2016.1205762

Nowack B, Ranville JF, Diamond S, Gallego-Urrea JA, Metcalfe C, Rose J, Horne N, Koelmans AA, Klaine SJ (2012) Potential scenarios for nanomaterial release and subsequent alteration in the environment. Environ Toxicol Chem 31:50–59. https://doi.org/10.1002/etc.726

Oberdörster G, Oberdörster E, Oberdörster J (2005) Nanotoxicology: an emerging discipline evolving from studies of ultrafine particles. Environ Health Perspect 113:823–839. https://doi.org/10.1289/ehp.7339

Oh Y, Noga R, Shanov V, Ryu H, Chandra H, Yadav B, Yadav J, Chae S (2019) Electrically heatable carbon nanotube point-of-use filters for effective separation and in-situ inactivation of Legionella pneumophila. Chem Eng J 366:21–26. https://doi.org/10.1016/j.cej.2019.02.054

Onoda A, Takeda K, Umezawa M (2017) Dose-dependent induction of astrocyte activation and reactive astrogliosis in mouse brain following maternal exposure to carbon black nanoparticle. Part Fibre Toxicol 14:4. https://doi.org/10.1186/s12989-017-0184-6

Onoda A, Takeda K, Umezawa M (2018) Dysregulation of major functional genes in frontal cortex by maternal exposure to carbon black nanoparticle is not ameliorated by ascorbic acid pretreatment. Sci Total Environ 634:1126–1135. https://doi.org/10.1016/j.scitotenv.2018.04.016

Palomäki J, Karisola P, Pylkkänen L, Savolainen K, Alenius H (2010) Engineered nanomaterials cause cytotoxicity and activation on mouse antigen presenting cells. Toxicology 267:125–131. https://doi.org/10.1016/j.tox.2009.10.034

Palomäki J, Välimäki E, Sund J, Vippola M, Clausen PA, Jensen KA, Savolainen K, Matikainen S, Alenius H (2011) Long, needle-like carbon nanotubes and asbestos activate the NLRP3 inflammasome through a similar mechanism. ACS Nano 5:6861–6870. https://doi.org/10.1021/nn200595c

Patel DK, Kim H-B, Dutta SD, Ganguly K, Lim K-T (2020) Carbon nanotubes-based nanomaterials and their agricultural and biotechnological applications. Materials (Basel) 13:1679. https://doi.org/10.3390/ma13071679

Perrier F, Bertucci A, Pierron F, Feurtet-Mazel A, Simon O, Klopp C, Candaudap F, Pokrovski O, Etcheverria B, Mornet S, Baudrimont M (2020) Transfer and transcriptomic profiling in liver and brain of European Eels (Anguilla anguilla) after diet-borne exposure to gold nanoparticles. Environ Toxicol Chem 39:2450–2461. https://doi.org/10.1002/etc.4858

Peters RJB, Bouwmeester H, Gottardo S, Amenta V, Arena M, Brandhoff P, Marvin HJP, Mech A, Moniz FB, Pesudo LQ, Rauscher H, Schoonjans R, Undas AK, Vettori MV, Weigel S, Aschberger K (2016) Nanomaterials for products and application in agriculture, feed and food. Trends Food Sci Technol 54:155–164. https://doi.org/10.1016/j.tifs.2016.06.008

Petersen EJ, Huang Q, Weber WJ (2008) Bioaccumulation of radio-labeled carbon nanotubes by Eisenia foetida. Environ Sci Technol 42:3090–3095. https://doi.org/10.1021/es071366f

Pogribna M, Hammons G (2021) Epigenetic effects of nanomaterials and nanoparticles. J Nanobiotechnol 19:2. https://doi.org/10.1186/s12951-020-00740-0

Principi E, Girardello R, Bruno A, Manni I, Gini E, Pagani A, Grimaldi A, Ivaldi F, Congiu T, De Stefano D, Piaggio G, de Eguileor M, Noonan DM, Albini A (2016) Systemic distribution of single-walled carbon nanotubes in a novel model: alteration of biochemical parameters, metabolic functions, liver accumulation, and inflammation in vivo. Int J Nanomed 11:4299–4316. https://doi.org/10.2147/IJN.S109950

Rydman EM, Ilves M, Koivisto AJ, Kinaret PAS, Fortino V, Savinko TS, Lehto MT, Pulkkinen V, Vippola M, Hämeri KJ, Matikainen S, Wolff H, Savolainen KM, Greco D, Alenius H (2014) Inhalation of rod-like carbon nanotubes causes unconventional allergic airway inflammation. Part Fibre Toxicol 11:48. https://doi.org/10.1186/s12989-014-0048-2

Sager TM, Umbright CM, Mustafa GM, Roberts JR, Orandle MS, Cumpston JL, McKinney WG, Boots T, Kashon ML, Joseph P (2022) Pulmonary toxicity and gene expression changes in response to whole-body inhalation exposure to multi-walled carbon nanotubes in rats. Inhal Toxicol 34:200–218. https://doi.org/10.1080/08958378.2022.2081386

Schubauer-Berigan MK, Dahm MM, Erdely A, Beard JD, Eileen Birch M, Evans DE, Fernback JE, Mercer RR, Bertke SJ, Eye T, de Perio MA (2018) Association of pulmonary, cardiovascular, and hematologic metrics with carbon nanotube and nanofiber exposure among U.S. workers: a cross-sectional study. Part Fibre Toxicol 15:22. https://doi.org/10.1186/s12989-018-0258-0

Shamim N, Sharma VK (eds) (2013) Sustainable nanotechnology and the environment: advances and achievements, ACS symposium series. American Chemical Society, Washington, DC, pp ix–ix. https://doi.org/10.1021/bk-2013-1124.pr001

Shvedova AA, Kisin E, Murray AR, Johnson VJ, Gorelik O, Arepalli S, Hubbs AF, Mercer RR, Keohavong P, Sussman N, Jin J, Yin J, Stone S, Chen BT, Deye G, Maynard A, Castranova V, Baron PA, Kagan VE (2008) Inhalation vs. aspiration of single-walled carbon nanotubes in C57BL/6 mice: inflammation, fibrosis, oxidative stress, and mutagenesis. Am J Physiol Lung Cell Mol Physiol 295:L552–L565. https://doi.org/10.1152/ajplung.90287.2008

Shvedova AA, Yanamala N, Kisin ER, Khailullin TO, Birch ME, Fatkhutdinova LM (2016) Integrated analysis of dysregulated ncRNA and mRNA expression profiles in humans exposed to carbon nanotubes. PLoS One 11:e0150628. https://doi.org/10.1371/journal.pone.0150628

Sierra MI, Valdés A, Fernandez AF, Torrecillas R, Fraga MF (2016) The effect of exposure to nanoparticles and nanomaterials on the mammalian epigenome. IJN 11:6297–6306. https://doi.org/10.2147/IJN.S120104

Song L, Shi J, Lu J, Lu C (2015) Structure observation of graphene quantum dots by single-layered formation in layered confinement space. Chem Sci 6:4846–4850. https://doi.org/10.1039/C5SC01416F

Subramanian A, Tamayo P, Mootha VK, Mukherjee S, Ebert BL, Gillette MA, Paulovich A, Pomeroy SL, Golub TR, Lander ES, Mesirov JP (2005) Gene set enrichment analysis: a knowledge-based approach for interpreting genome-wide expression profiles. Proc Natl Acad Sci U S A 102:15545–15550. https://doi.org/10.1073/pnas.0506580102

Takagi A, Hirose A, Nishimura T, Fukumori N, Ogata A, Ohashi N, Kitajima S, Kanno J (2008) Induction of mesothelioma in p53+/- mouse by intraperitoneal application of multi-wall carbon nanotube. J Toxicol Sci 33:105–116. https://doi.org/10.2131/jts.33.105

The Gene Ontology Consortium, Aleksander SA, Balhoff J, Carbon S, Cherry JM, Drabkin HJ, Ebert D, Feuermann M, Gaudet P, Harris NL, Hill DP, Lee R, Mi H, Moxon S, Mungall CJ, Muruganugan A, Mushayahama T, Sternberg PW, Thomas PD, Van Auken K, Ramsey J, Siegele DA, Chisholm RL, Fey P, Aspromonte MC, Nugnes MV, Quaglia F, Tosatto S, Giglio M, Nadendla S, Antonazzo G, Attrill H, Dos Santos G, Marygold S, Strelets V, Tabone CJ, Thurmond J, Zhou P, Ahmed SH, Asanitthong P, Luna Buitrago D, Erdol MN, Gage MC, Ali Kadhum M, Li KYC, Long M, Michalak A, Pesala A, Pritazahra A, Saverimuttu SCC, Su R, Thurlow KE, Lovering RC, Logie C, Oliferenko S, Blake J, Christie K, Corbani L, Dolan ME, Drabkin HJ, Hill DP, Ni L, Sitnikov D, Smith C, Cuzick A, Seager J, Cooper L, Elser J, Jaiswal P, Gupta P, Jaiswal P, Naithani S, Lera-Ramirez M, Rutherford K, Wood V, De Pons JL, Dwinell MR, Hayman GT, Kaldunski ML, Kwitek AE, Laulederkind SJF, Tutaj MA, Vedi M, Wang S-J, D'Eustachio P, Aimo L, Axelsen K, Bridge A, Hyka-Nouspikel N, Morgat A, Aleksander SA, Cherry JM, Engel SR, Karra K, Miyasato SR, Nash RS, Skrzypek MS, Weng S, Wong ED, Bakker E, Berardini TZ, Reiser L, Auchincloss A, Axelsen K, Argoud-Puy G, Blatter M-C, Boutet E, Breuza L, Bridge A, Casals-Casas C, Coudert E, Estreicher A, Livia Famiglietti M, Feuermann M, Gos A, Gruaz-Gumowski N, Hulo C, Hyka-Nouspikel N, Jungo F, Le Mercier P, Lieberherr D, Masson P, Morgat A, Pedruzzi I, Pourcel L, Poux S, Rivoire C, Sundaram S, Bateman A, Bowler-Barnett E, Bye-A-Jee H, Denny P, Ignatchenko A, Ishtiaq R, Lock A, Lussi Y, Magrane M, Martin MJ, Orchard S, Raposo P, Speretta E, Tyagi N, Warner K, Zaru R, Diehl AD, Lee R, Chan J, Diamantakis S, Raciti D, Zarowiecki M, Fisher M, James-Zorn C, Ponferrada V, Zorn A, Ramachandran S, Ruzicka L, Westerfield M, Aleksander SA, Balhoff J, Carbon S, Cherry JM, Drabkin HJ, Ebert D, Feuermann M, Gaudet P, Harris NL, Hill DP, Lee R, Mi H, Moxon S, Mungall CJ, Muruganugan A, Mushayahama T, Sternberg PW, Thomas PD, Van Auken K, Ramsey J, Siegele DA, Chisholm RL, Fey P, Aspromonte MC, Nugnes MV, Quaglia F, Tosatto S, Giglio M, Nadendla S, Antonazzo G, Attrill H, Dos Santos G, Marygold S, Strelets V, Tabone CJ, Thurmond J, Zhou P, Ahmed SH, Asanitthong P, Luna Buitrago D, Erdol MN, Gage MC, Ali Kadhum M, Li KYC, Long M, Michalak A, Pesala A, Pritazahra A, Saverimuttu SCC, Su R, Thurlow KE, Lovering RC, Logie C, Oliferenko S, Blake J, Christie K, Corbani L, Dolan ME, Drabkin HJ, Hill DP, Ni L, Sitnikov D, Smith C, Cuzick A, Seager J, Cooper L, Elser J, Jaiswal P, Gupta P, Jaiswal P, Naithani S, Lera-Ramirez M, Rutherford K, Wood V, De Pons JL, Dwinell MR, Hayman GT, Kaldunski ML, Kwitek AE, Laulederkind SJF, Tutaj MA, Vedi M, Wang S-J, D'Eustachio P, Aimo L, Axelsen K, Bridge A, Hyka-Nouspikel N, Morgat A, Aleksander SA, Cherry JM, Engel SR, Karra K, Miyasato SR, Nash RS, Skrzypek MS, Weng S, Wong ED, Bakker E, Berardini TZ, Reiser L, Auchincloss A, Axelsen K, Argoud-Puy G, Blatter M-C, Boutet E, Breuza L, Bridge A, Casals-Casas C, Coudert E, Estreicher A, Livia Famiglietti M, Feuermann M, Gos A, Gruaz-Gumowski N, Hulo C, Hyka-Nouspikel N, Jungo F, Le Mercier P, Lieberherr D, Masson P, Morgat A, Pedruzzi I,

Pourcel L, Poux S, Rivoire C, Sundaram S, Bateman A, Bowler-Barnett E, Bye-A-Jee H, Denny P, Ignatchenko A, Ishtiaq R, Lock A, Lussi Y, Magrane M, Martin MJ, Orchard S, Raposo P, Speretta E, Tyagi N, Warner K, Zaru R, Diehl AD, Lee R, Chan J, Diamantakis S, Raciti D, Zarowiecki M, Fisher M, James-Zorn C, Ponferrada V, Zorn A, Ramachandran S, Ruzicka L, Westerfield M (2023) The gene ontology knowledgebase in 2023. Genetics 224:iyad031. https://doi.org/10.1093/genetics/iyad031

Vlaanderen J, Pronk A, Rothman N, Hildesheim A, Silverman D, Hosgood HD, Spaan S, Kuijpers E, Godderis L, Hoet P, Lan Q, Vermeulen R (2017) A cross-sectional study of changes in markers of immunological effects and lung health due to exposure to multi-walled carbon nanotubes. Nanotoxicology 11:395–404. https://doi.org/10.1080/17435390.2017.1308031

Wang X, Xia T, Duch MC, Ji Z, Zhang H, Li R, Sun B, Lin S, Meng H, Liao Y-P, Wang M, Song T-B, Yang Y, Hersam MC, Nel AE (2012) Pluronic F108 coating decreases the lung fibrosis potential of multiwall carbon nanotubes by reducing lysosomal injury. Nano Lett 12:3050–3061. https://doi.org/10.1021/nl300895y

Warheit DB, Laurence BR, Reed KL, Roach DH, Reynolds GAM, Webb TR (2004) Comparative pulmonary toxicity assessment of single-wall carbon nanotubes in rats. Toxicol Sci 77:117–125. https://doi.org/10.1093/toxsci/kfg228

Wong BSE, Hu Q, Baeg GH (2017) Epigenetic modulations in nanoparticle-mediated toxicity. Food Chem Toxicol 109:746–752. https://doi.org/10.1016/j.fct.2017.07.006

Yadav B, Yadav JS (2024) Carbon nanotube immunotoxicity in alveolar epithelial type II cells is mediated by physical contact-independent cell–cell interaction with macrophages as demonstrated in an optimized air–liquid interface (ALI) coculture model. Nanomaterials 14:1273. https://doi.org/10.3390/nano14151273

Yadav B, Bhattacharya SS, Rosen L, Nagpal R, Yadav H, Yadav JS (2024) Oro-respiratory dysbiosis and its modulatory effect on lung mucosal toxicity during exposure or co-exposure to carbon nanotubes and cigarette smoke. Nanomaterials (Basel) 14:314. https://doi.org/10.3390/nano14030314

Zheng W, McKinney W, Kashon ML, Pan D, Castranova V, Kan H (2018) The effects of inhaled multi-walled carbon nanotubes on blood pressure and cardiac function. Nanoscale Res Lett 13:189. https://doi.org/10.1186/s11671-018-2603-5

Chapter 8
Characterizing Nanoparticle Aerosols for Better Air Quality Management

Manish Joshi, B. K. Sapra, and D. K. Aswal

Abstract Characterizing nanoparticle aerosols is crucial for understanding their behavior during generation, evolution, and deposition. Among the various methods available for generating these aerosols, electrical heating of metallic wires stands out as the simplest and most widely used technique, particularly for metallic particles. The aerosols generated through this method are used for calibrating nanosizers and conducting specialized experiments to validate theoretical models. Additionally, these nanoparticles, generated at high number concentrations (>10^6 cm^{-3}), have been shown to demonstrate fascinating features such as fractal shapes, a sharp peaking profile, and sporadic oscillations in the concentration profile. These properties find applications in characterizing aerosol sources by determining their number emission rates and in validating aerosol models related to nucleation and coagulation processes. Managing air quality effectively requires efficient mechanisms to mitigate airborne particles. Alongside conventional filtration-based techniques, charging-based methods are also employed. Unipolar ionizers, for example, are commonly used in small, enclosed environments. However, the efficiency of aerosol removal is size-dependent, with nanoparticles presenting a particular challenge due to their lower charging efficiencies. This limitation hinders the overall efficacy of charging-based mitigation strategies. To enhance efficiency, it is crucial to identify the most sensitive parameters influencing charging-based particle removal. Two key factors that define the charge characteristics of an ionizer are "charge concentration" and "aerosol current". Experimental studies using single and multi-array ionizers have provided valuable insights into these parameters. Such investigations allow for better optimization of particle charging, ultimately improving removal efficacy, particularly in the nanoscale size range. Recent advancements in nanoparticle scavenging include the use of electrohydrodynamic atomizers, which enhance cleaning efficiency by charging nanodroplets. This innovative approach may provide an alternative to conventional electrostatic precipitators in industrial applications. Another promising development is the incorporation of CNTs into filtration

M. Joshi · B. K. Sapra · D. K. Aswal (✉)
Health, Safety and Environment Group, Bhabha Atomic Research Centre, Mumbai, India

N. Ghosh et al. (eds.), *Nanotechnology in Air Quality Management*, Sustainable Development and Biodiversity 39, https://doi.org/10.1007/978-981-96-8339-0_8

systems. Because of their high surface-to-volume ratio, CNTs offer excellent air-cleaning capabilities, provided that pressure drops are effectively managed. The production of CNTs in reactors is influenced by the properties of precursors in the aerosol phase. Controlling the characteristics of nucleated particles is key to improving reactor throughput and optimizing CNT production. In this chapter, we have outlined the processes of generating and characterizing nanoparticles, emphasizing their role in refining applications for air quality management. These advancements underscore the importance of tailoring nanoparticle behavior for enhanced environmental and industrial outcomes.

Keywords CNT · EHDA · Hot wire generator · Ionizers · Nanoparticle aerosols

8.1 Introduction

Nanoparticle aerosols, defined as particles with at least one dimension smaller than 100 nm, play a vital role across diverse fields, including environmental science, medicine, industry, and materials research. Over the past few decades, there has been growing interest in exploring their remarkable catalytic, optical, mechanical, and biochemical properties. This progress can be attributed to the advancement of specialized experimental and theoretical techniques that have deepened our understanding of their formation and behavior. In environmental studies, the measurement of number size distributions in the ambient atmosphere reveals that nanoparticle aerosols can constitute a significant fraction of airborne particulates (Joshi et al. 2016). This highlights their relevance not only in understanding atmospheric processes but also in addressing broader challenges related to air quality and health impacts.

As the surfaces of these particles carry relatively higher toxicity, their lung deposition is a matter of concern. This raises significant health concerns, particularly in indoor environments where anthropogenic nanoparticle aerosols are more prevalent due to higher occupant densities. Picking and characterizing new particle formation (NPF) events in the atmosphere still holds a key toward better understanding regional and global climate patterns (Kulmala et al. 2022). Since the use of gold colloids for medical purposes in the seventeenth century, the scope of nanoparticle aerosol research in thrust areas other than the environment has also broadened extensively. Favorable results obtained from the study of special materials such as high entropy alloy (HEA) nanoparticles (Yang et al. 2020), carbon nanotubes (CNT), drug nano formulations, and so on have regalvanized the interest of industry in recent times. Studies on the evolution of nanoparticle aerosols under controlled laboratory conditions have not only provided valuable insights for validating aerosol microphysical models but have also revealed key characteristics essential for the future characterization of aerosol sources. Such studies have been assisted by the use of aerosol generators having precise control over sizing and concentration. Air purification technologies, evolving from electrostatic and dynamic principles, aim to achieve efficient particle charging in the nano size range. Mitigators based on these principles rely on optimizing the electrical drift to enhance particle removal efficacy. The characterization of devices, particularly through charge-based parameters during both the optimization and operational phases, is critical for transitioning from controlled laboratory settings to real-world field applications. A significant focus in materials research is to increase the yield of CNT production (Kaushal et al. 2024). Controlling and modifying the generation characteristics of nanotubes in the nuclei-mode size range is key to achieving improvements in production efficiency and application. Overall, the growth of nanoparticle aerosol research has been accelerating in recent decades, with ongoing advancements continuously refining our understanding of their properties. This progress is propelling their application in various fields.

This chapter provides a comprehensive overview of research conducted at the *Bhabha Atomic Research Center (BARC)* in the field of nanoparticle aerosol studies. It details the "parametrization of the Glowing/Hot Wire Generator (HWG)" and its application in investigating the formation and evolution of nanoparticle aerosols under controlled, closed-chamber conditions. The chapter also highlights the latest advancements in aerosol mitigation technologies, with a particular emphasis on "unipolar ionizers" and "electrohydrodynamic atomizers (EHDA)." In its concluding section, the chapter discusses the role of aerosol nucleation in influencing the generation characteristics of "carbon nanotubes" and explores their potential application in aerosol filtration.

8.2 Generation of Nanoparticle Aerosols

Methods for generating nanoparticle aerosols in the liquid phase include techniques such as the atomization of solutions (Yonezawa and Kunitake 1999) and the crystallization of drops (Ranz et al. 1952). These approaches, commonly referred to as wet chemistry methods, have limitations in applications requiring high purity (Boies et al. 2011a, b; Brust et al. 1994). The use of solvents in wet chemistry methods can result in uncontrolled surface chemistry and contamination within the nanoparticles, undermining their quality and making them unsuitable for high-purity applications. In contrast, direct generation of nanoparticles in the aerosol phase offers several advantages, including the absence of liquid waste, high purity of the product, and the feasibility for continuous, high-throughput operations. This approach eliminates the challenges associated with solvent use and enables more efficient production of nanoparticles, particularly for applications requiring stringent purity standards.

Nanoparticles generated through aerosol routes offer the advantage of well-defined chemical composition, density, and morphology, which are essential for controlling their dynamics for specific applications. Several methods exist for directly generating nanoparticles in the aerosol phase, including spark discharge (Hontanon et al. 2013), plasma reactors (Kim 2005), and laser ablation (Supronowicz

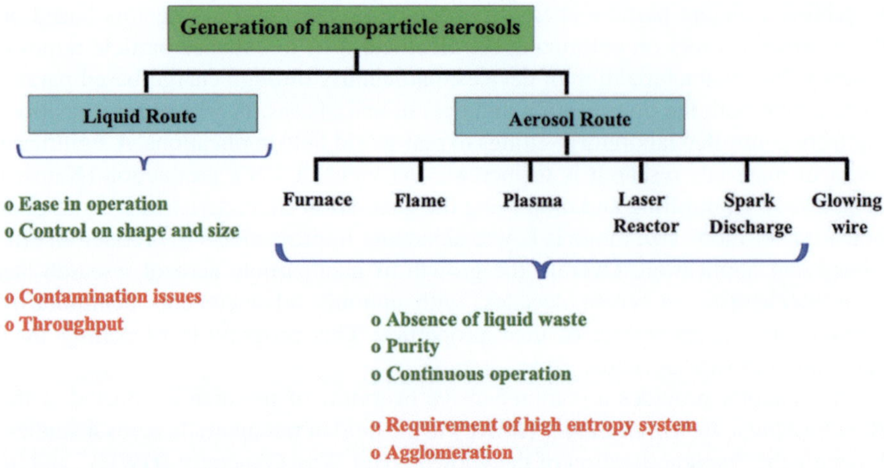

Fig. 8.1 Generation methods for nanoparticle aerosols: advantages and disadvantages

et al. 2002). While these methods are effective, they also present challenges in both laboratory and industrial settings. These challenges include the need for high-temperature furnaces, maintaining appropriate dilution to prevent agglomeration, and difficulties in achieving spherical nanoparticle shapes (Khan et al. 2014). Addressing these issues is crucial for improving the efficiency and quality of aerosol-based nanoparticle generation for various applications. Figure 8.1 compares the advantages and disadvantages of different methods used to generate nanoparticle aerosols.

Electrical generation of nanoparticle aerosols is achieved by directly heating a metallic substrate (such as a wire) and purging the resulting vapors. Under optimal conditions, these vapors nucleate, leading to the formation of nanoparticle aerosols. The method offers several advantages, including precise control over particle size and contamination (Peineke et al. 2009), high emission rates (Khan et al. 2014), and tunable properties of the generated particles (Chen and Zhang 2017), making it a preferred choice for nanoparticle generation. Extensive research into the characteristics of electrically generated nanoparticle aerosols has enabled their effective application in various fields, including advanced research (Fomenko et al. 2021), code validation (Ghosh et al. 2020), and industry (Sabbagh et al. 2020).

8.3 Hot Wire Generator

The Glowing/Hot Wire Generator, based on the evaporation–condensation process, was introduced by Schmidt-Ott et al. in 1980. This method has since evolved, with various researchers developing tailored versions to meet specific needs and applications. Over the years, these advancements have contributed to the refinement of

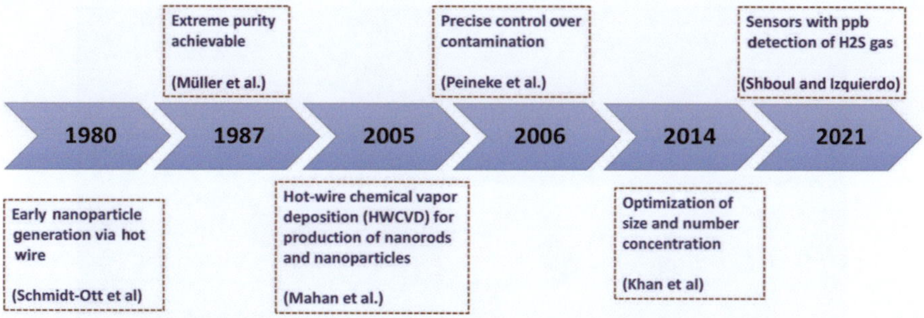

Fig. 8.2 Important milestones pertaining to HWG-based research

nanoparticle generation techniques. A timeline of key milestones in this development is presented in Fig. 8.2, illustrating the progress made in optimizing the glowing wire method for diverse research and industrial applications.

In most of the methods mentioned above, the material is evaporated from a resistively heated wire, and the resulting vapor stream is rapidly quenched by a carrier gas flow. These techniques stand out compared to other aerosol generation methods due to their ability to generate particles within the nanoparticle size range. In addition to offering precise control over particle size and purity, these methods are characterized by significantly higher emission rates. This makes them ideal for use as a preferred nanoparticle generator in laboratory settings, enabling the study of aerosol dynamics over a practical timescale. The high emission rates allow for the efficient generation of nanoparticles, making these methods particularly valuable for experiments that require large numbers of particles in a short period (Khan et al. 2014). The characteristics of the wire, including its chemical composition, electrical resistance, electronic stability, and the carrier gas flow, significantly influence the generation and stability of nanoparticle concentration under dynamically varying temperature conditions. To optimize the performance of HWGs, studies are conducted to understand and interpret these effects, aiming to fine-tune the operating parameters and minimize fluctuations in particle concentration. Despite the challenges associated with the generation and control phases of HWGs, they remain the most viable and effective option for the generation of nanoparticle aerosols. Their ability to generate high-quality aerosols with precise control continues to make them the preferred choice in both laboratory and industrial applications.

The HWG developed by BARC consists of a metallic coil that is electrically connected to a constant current power supply (0–5 A). The power supply delivers the necessary energy to the wire for heating, which ultimately causes the material of the wire to evaporate. The metallic coil is surrounded by a reaction chamber, which is equipped with ports for sampling and/or control. Figure 8.3 illustrates the HWG setup, showing the power supply and reaction chamber in the glowing condition, where the wire is heated and the material begins to vaporize. This configuration enables efficient nanoparticle generation through the evaporation–condensation process.

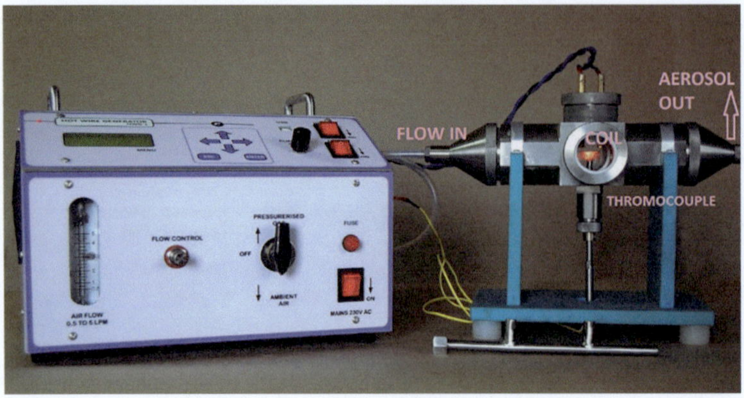

Fig. 8.3 Hot wire generator: power supply + reaction chamber

In the HWG described above, the control module plays a crucial role by providing electrical connectivity to the wire and facilitating the flow of the carrier gas, which is essential for both quenching the vapor and transporting the nucleated nanoparticles to the other end of the chamber. A thermocouple is positioned at a vertical port of the chamber to monitor the temperature of the wire or the surrounding environment, ensuring precise control over the heating process. The outlet stream of nanoparticles can be directed to a measuring instrument or an associated setup for further analysis. Alternatively, the nanoparticle stream can be transported to an external experimental chamber or reactor, depending on the specific application. The choice of wire material and carrier gas is flexible and depends on the user's requirements. Either clean air from the in-built pump or externally connected inert gas can be used for generating the oxidized/unoxidized nucleates. Controlled nucleation governs the formation of the particles, while coagulation influences the evolution of the size spectrum. Together, these processes result in a stable and tunable stream of nanoparticle aerosols at the outlet port, offering a versatile tool for generating particles with specific properties tailored to the needs of the experiment or application.

The vapor source term, or emission rate, for a typical electrical aerosol generation process can be estimated using the Hertz–Knudsen equation (Eq. 8.1Kb). This equation takes into account the saturation vapor pressure of the wire's constituent material, which is a key input in determining the rate at which the material vaporizes and subsequently condenses to form nanoparticles. The saturation vapor pressure represents the pressure at which a material's vapor is in equilibrium with its liquid or solid phase, and it is a crucial parameter for understanding the vaporization and nucleation processes involved in aerosol generation.

$$r = \lambda \sqrt{\frac{1}{2\pi M N_a k_a T}} \left(p_s(T) - p_0 \right) \tag{8.1}$$

where λ = accommodation coefficient, M = molar mass, T = temperature, $p_s(T)$ = saturation vapor pressure, p_0 = vapor pressure, N_a = Avogadro number, and k_b = Boltzmann constant. Accommodation coefficient in the above equation accounts for vapor scattering back to the wire surface and should be <1. However, under the generation phase, λ can be approximated to a value of 1.

For a wire made of an alloy or a mixture of elements, the evaporation rate is calculated for each individual component of the wire. The effective evaporation rate of the wire is then estimated by incorporating the proportional weighting of the wire's constituent materials. This allows for an accurate estimation of the overall emission rate of the wire, accounting for the different rates of vaporization for each element in the alloy. $p_s(T)$ needed for the above equation can be obtained for the elements of the wire using look-up tables. For example, $p_s(T)$ for Cr, Mn, Fe, and Ni (constituents of nichrome wire) at their melting temperatures is 739, 142, 3.39, and 0.104 Pa, respectively. This consequently translates to an evaporation rate of 9.6, 2.15, 4.66×10^{-2}, and 1.42×10^{-3} mole $m^{-2}\ s^{-1}$, respectively.

The source or vapor emission rate depends on several factors: the evaporation rate of the material, the surface area of the evaporating material, and the generator volume. These factors are combined to estimate the emission rate of nanoparticles in the aerosol generation process.

$$\mathrm{SE} = \frac{(r \times S)}{V} \tag{8.2}$$

where S is the total surface area of the evaporating material and V is the generator volume.

For commercial-grade nichrome wire, the source emission rate can be estimated by considering only the constituent with the highest evaporation rate—chromium (Cr)—since it dominates the vaporization process in the wire. Given the volume of the generator as 450 cm^3 and assuming only chromium contributes to the emission, the number emission rate comes out to be $\approx 2.0 \times 10^{25}$ molecules $m^{-3}\ s^{-1}$.

The generator has been standardized by optimizing its operating parameters to achieve a stable concentration of nanoaerosols. Two critical parameters, viz. carrier gas flow rate and applied voltage, have been identified as key factors influencing the output. The carrier gas flow rate plays a crucial role in determining the residence time of vapors within the hot zone, which directly impacts the nucleation rate of nanoparticles. At flow rates above 4 L min^{-1}, a significant reduction in number concentration is observed. Under high flow rate conditions, the wire temperature decreases due to the enhanced cooling effect caused by the increased carrier gas flow. This reduction in wire temperature leads to a decline in the evaporation rate of the material, resulting in a lower concentration of vapor in the chamber. Additionally, the reduction in temperature gradient within the generator hinders the supersaturation of the vapor, a critical condition for nucleation. Without sufficient supersaturation, the formation of nanoparticles is suppressed, leading to a significant drop in concentration of the generated aerosols. The joule heating of the wire, which governs its temperature and evaporation rate, is directly influenced by the wattage supplied to the system. As the applied voltage is varied, it modulates the energy input, thereby controlling the wire's heating dynamics. As the voltage increases, the wire temperature rises due to enhanced joule heating. This leads to a higher evaporation rate of the material, resulting in an increased concentration of nanoparticles in the output stream. However, beyond a certain threshold voltage, the number concentration reaches a saturation point stabilizing at a near equilibrium value. Above this voltage, further increases in voltage do not produce a proportional rise in particle concentration due to material's constrained evaporation rate.

After conducting measurements over a range of operating voltages (5–9 V) and carrier gas flow rates (0.5–5 L min^{-1}), the following optimized parameters have been determined for the generator to achieve stable and efficient nanoparticle aerosol production:

Applied voltage: 6–9 V
Equivalent power: 30–50 W
Carrier gas flow rate: 1–1.5 L min^{-1}

8.3.1 Characteristics of Nanoparticle Aerosols Generated from HWG

Peineke et al. (2006) provides a classification of materials that can be used as substrates for evaporation–condensation-based aerosol generation, based on their evaporation rates. This classification compares the evaporation rates of materials at their melting temperatures to determine their suitability for such processes. However, in practical applications, the actual evaporation rates can vary due to differences in temperature conditions, which are influenced by the wire and reaction chamber configurations. Apart from the evaporation rate, electrical resistance and structural integrity of wire at red hot conditions become key considerations for the selection of wire material. Resistance changes with temperature, and only materials with appropriate resistive properties are suitable for prolonged use. In addition, materials and wire geometries prone to breaking or deforming at high temperatures are unsuitable.

The characterization of nanoparticle aerosols produced by the HWG has been carried out by analyzing its short-term and long-term performance. For this purpose, experiments were conducted using a nichrome wire under optimized operating conditions, i.e., 1 L min^{-1} carrier flow and applied voltage of 7 V. Figure 8.4 shows the concentration and number size distribution of generated particles. The figure also demonstrates the stability of the number concentration and the geometric mean size of the particles over an extended duration.

In Fig. 8.4a, the time series of the number concentration of the generated nanoparticles is presented. The measurements were conducted using a GRIMM condensation particle counter (CPC), which was connected to the outlet port of the

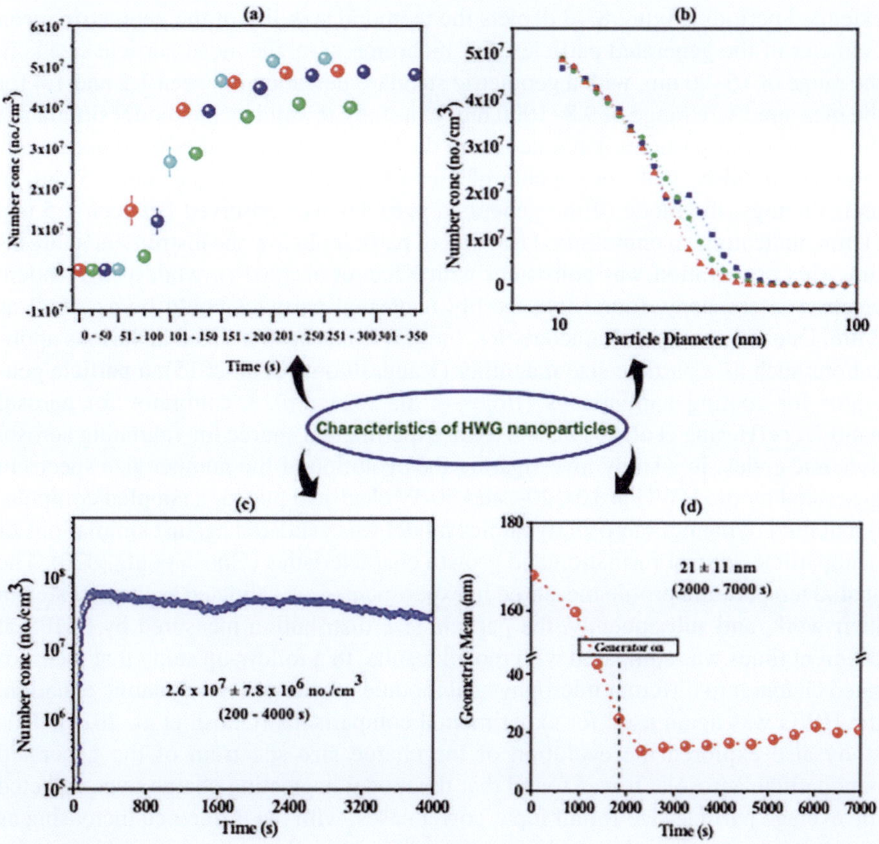

Fig. 8.4 Characteristics of nanoparticle aerosols generated with nichrome wire in HWG: **(a)** Temporal number concentration for 5-7 V, **(b)** Number size distribution for 5-7 V, **(c)** Stability of number concentration and **(d)** Stability of mean size

HWG. After an initial warm-up period, the concentration of nanoparticles can be seen to be rising sharply, eventually stabilizing at levels between 3 and 5×10^7 cm^{-3}. As the concentration levels are higher than those of conventional aerosol generators, the HWG can be used for various applications that require high nanoparticle concentrations. The size distribution of nanoparticle aerosols was measured using a GRIMM scanning mobility particle sizer (SMPS), which has a size range of >9.8 nm, connected to the outlet under similar operating conditions. The typical size distribution pattern, shown in Fig. 8.4b, indicates that a significant fraction of the particles had sizes below 30 nm, confirming the suitability of the HWG as an effective nanoparticle aerosol generator.

The long-term stability of the HWG output can be assessed by tracking both the number concentration and mean geometric diameter of the particles over extended measurement periods. In Fig. 8.4c, the temporal behavior of the number concentration at the outlet shows that the concentration remains above $\approx 2 \times 10^7$ cm^{-3} for extended periods. Figure 8.4d depicts the temporal stability of the geometric mean diameter of the generated particles. For nichrome wire, the mean particle size is in the range of 15–20 nm, with a geometric standard deviation between 1.2 and 1.4 for the measured size range of 9.8–1000 nm. In a study to validate the model simulating the evolution of particles generated from the HWG, Ghosh et al. (2020) used a differential mobility analyzer capable of measuring particles larger than 4.5 nm. In their findings, the mode of the generated particles was observed between 4.5 and 10 nm, indicating an unmeasured fraction of particles below the instrument's threshold. This observation was consistent with Khan et al. (2014), who, using tandem electron microscopy, found that freshly nucleated particles could be as small as 3 nm. Due to its unique characteristics, the HWG has been utilized in various applications such as a particle size magnifier (Kangasluoma et al. 2015), a particle generator for coating applications (Boies et al. 2011a, b), a calibrator for aerosol nanosizers (Hering et al. 2017), and as an experimental source for validating aerosol dynamic codes. In a study investigating the evolution of the number size spectrum generated by the HWG at 10-, 20-, and 30-W electrical power, a coupled computational fluid dynamics–aerosol dynamics model was validated against simulations of nanoparticle aerosol formation and growth characteristics (Ghosh et al. 2020). The spatial temperature profile measured in experiments was validated as the first step in their work, and subsequently, the particle size distribution measured by SMPS at different times was compared with model results. In a follow-up study that incorporated charge into different microphysical modules of the general dynamic equation, the HWG was again used for experimental comparisons (Ghosh et al. 2021). This study also explored the evolution of the charge size spectrum of the generated nanoparticle aerosols. It was found that the model neglecting charge overpredicted the average particle size for all input power cases, with the difference increasing as input power rose. The observed formation of a modified bipolar or pseudo-unipolar charge condition, which resulted in lower coagulation rates, was attributed to the discrepancies between experimental results and theoretical predictions.

8.3.2 Hot Wire Generated Nanoparticle Aerosols: Special Effects

The evolution of nanoparticle aerosol characteristics in closed chamber conditions leads to several interesting observations. Under continuous source conditions, freshly formed particles start to coagulate with either the background aerosol sizes or the evolving particle sizes within the chamber. Hetero-coagulation between particles of highly unequal sizes results in a faster depletion of the number concentration. When the source is on, the temporal behavior of the total number concentration is governed by the interplay between the source rate, coagulation rate, and background deposition rate. Such dynamic behavior often leads to the formation of a peak in the time series of number concentration, a phenomenon predicted by models

(Pratsinis 1988). Similar effect is referred to as the "coagulation peaking effect," where the peak becomes a function of aerosol and process parameters such as background deposition and chamber volume. The "characteristic coagulation peak" is defined by two key parameters: peak concentration and peaking time. The behavior of injected nanoparticle aerosols in a closed chamber, resulting in this coagulation-induced peak point, can be utilized to estimate the emission rate of aerosol sources. However, only a few studies have explored this approach due to the experimental challenges and the rapid timescale of the dynamics involved.

The change of concentration of aerosols ($n(u, t)$) in the chamber can be mathematically represented as

$$\frac{\partial n(u,t)}{\partial t} = \left(\frac{1}{V}\right) S^*(u,t) + \frac{1}{2}\int_0^u K(u',u-u')n(u',t)n(u-u',t)\,du' - n(u,t)\int_0^\infty K(u',u)n(u',t)\,du' - \{\lambda_v + \lambda_d(u)\}n(u,t) \tag{8.3}$$

where u is particle volume, $K(u, u')$ is coagulation kernel, $\lambda_v$is ventilation rate, $\lambda(u)$ is removal rate, $S^*(u, t)$ is particle number emission rate (number s^{-1}), and V is the chamber volume. The concentration of particles and time, in the context of coagulation peaking, can be mathematically scaled in terms of the emission rate and the primary coagulation coefficient. Such models enable the estimation of the emission rate from the observed number concentration dynamics and coagulation processes, providing a quantitative approach to understanding aerosol behavior in closed chamber settings (Anand et al. 2012). This scaling can be achieved using numeric and semi-analytical solutions such as:

$$N^*(t^*) = N(t)/N_c \tag{8.4}$$

$$t^* = t/t_c \tag{8.5}$$

where $N_c = \sqrt{S^*/VK(u_0,u_0)}$ is characteristic number concentration, $t_c = 1/\sqrt{S^*K(u_0,u_0)/V}$ is characteristic time, and u_0 is the volume of the emitted primary particle. Under this framework, $N^*(t^*)$ always peaks at $N^* \approx 1$ for $t^* \approx 2$ regardless of the source strength. Using the above equations, number emission rate estimated from an experimentally measured peak number concentration (N_p) at time (t_p) can be written as

$$S^* = VN_c/t_c = 2VN_p/t_p \tag{8.6}$$

For continuously injected nanoparticles from the HWG in a 40 L stirred chamber, the temporal response of the number concentration measured by SMPS is shown in Fig. 8.5. In this case, the peak concentration was observed to be 2.16×10^6 cm^{-3}, and the peak time was 496 s. These values illustrate the typical coagulation peaking

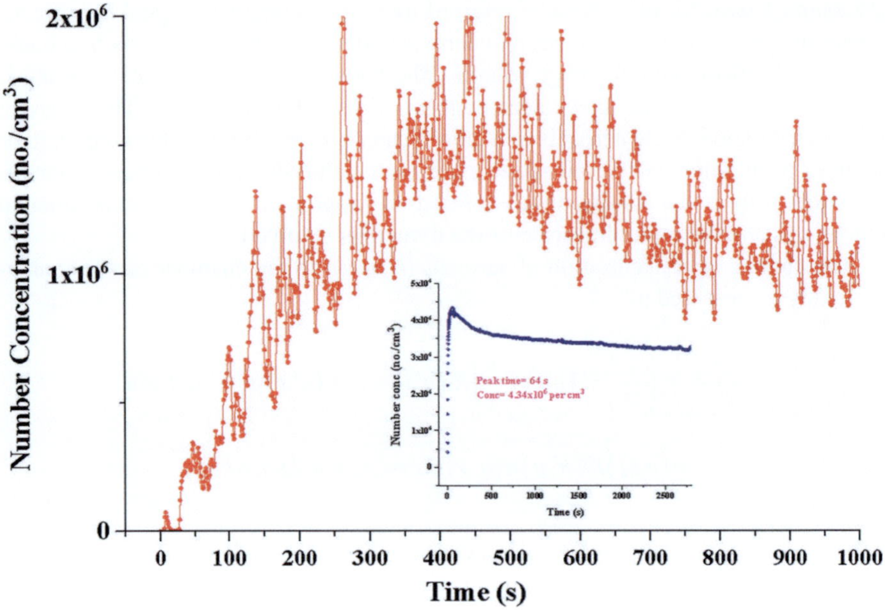

Fig. 8.5 Coagulation peaking effect for HWG-generated nanoparticle aerosols

effect in the chamber, reflecting the dynamic balance between particle generation, coagulation, and deposition processes. As a comparison, similar measures were also conducted using smoke aerosols generated from incense sticks in the same chamber. The inset of Fig. 8.5 shows the time series profile of the evolution of the number concentration of smoke particles. This comparison helps in understanding the differences in dynamics between nanoparticles generated by the HWG and larger smoke particles, highlighting variations in concentration patterns and temporal behavior under the same experimental conditions. The difference in the peaking behavior can be observed in Fig. 8.5, where the concentration of smoke particles increases sharply and reaches a higher value ($>4 \times 10^6$ cm^{-3}) compared to the HWG-generated nanoparticle aerosols. This indicates that smoke particles exhibit a more pronounced and faster concentration rise compared to the nanoparticles.

Corresponding to the peak concentration and the peaking time observed in temporal profile for nanoaerosols and smoke particles, emission rate calculated from Eq. (8.6) comes out to be 3.5×10^8 s^{-1} and 8–9 $\times 10^9$ s^{-1} for HWG and incense stick, respectively.

The effect of hetero-coagulation for nanoparticle aerosols generated by HWG and continuously injected into a homogenized chamber is evident in the evolution of the number size distribution spectrum. As shown in Fig. 8.6, measured by SMPS (5–350 nm) in a chamber volume of 500 L, the distribution evolves over time, with a secondary peak forming at later times. This secondary peak shifts to larger sizes as time progresses, accompanied by a simultaneous decrease in the number concentration of particles in the primary peak. This behavior highlights the ongoing

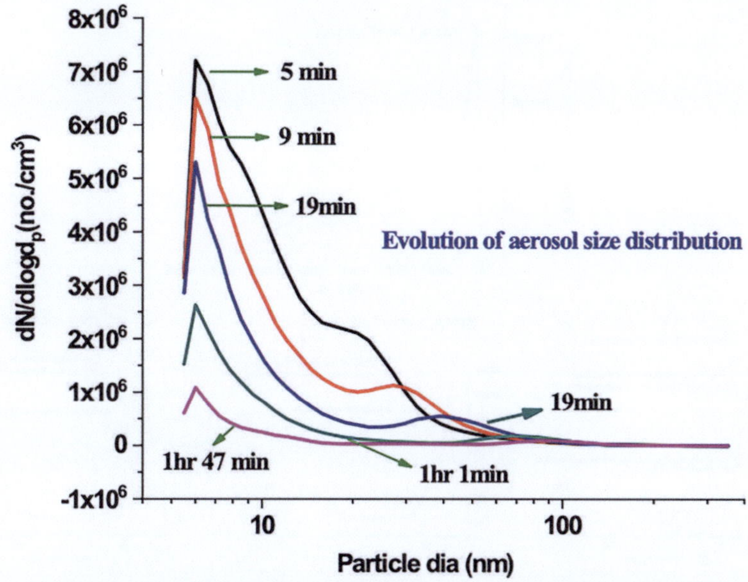

Fig. 8.6 Typical evolution of size spectrum for HWG-generated nanoparticle aerosols

coagulation processes within the chamber, where smaller particles aggregate, leading to an increase in the size of the larger particles and a reduction in the total number of smaller particles.

An intriguing feature predicted by analytical models is the appearance of oscillations in the number concentration profile. These oscillations arise from the complex interplay of multiple processes, including nucleation, coagulation, vapor condensation, wall deposition, and transport/removal via ventilation. When these processes are mathematically treated and coupled within a computational domain, sustained oscillations in the number concentration can occur. Pratsinis' model (1988), which incorporates a free molecular coagulation kernel, predicted such oscillations, illustrating the dynamic and sometimes unstable behavior of nanoparticle aerosols in confined environments. These oscillations could be indicative of nonequilibrium conditions, where the rates of generation and removal processes fluctuate in a way that causes periodic peaks and troughs in particle concentration.

To experimentally capture the oscillatory behavior and other long-term dynamics of nanoparticle aerosols, the glowing wire was placed inside a well-stirred 500 L chamber. Simultaneous measurements were made for total number concentration using a CPC and number size distribution using a SMPS. Given that the peak intensity for secondary peaks is expected to be low, it was crucial to ensure that operating characteristics (such as applied voltage) and the chamber environment remained consistent throughout the measurement period. Figure 8.7 shows the long-term temporal profile of the number concentration for an experiment conducted with nichrome wire.

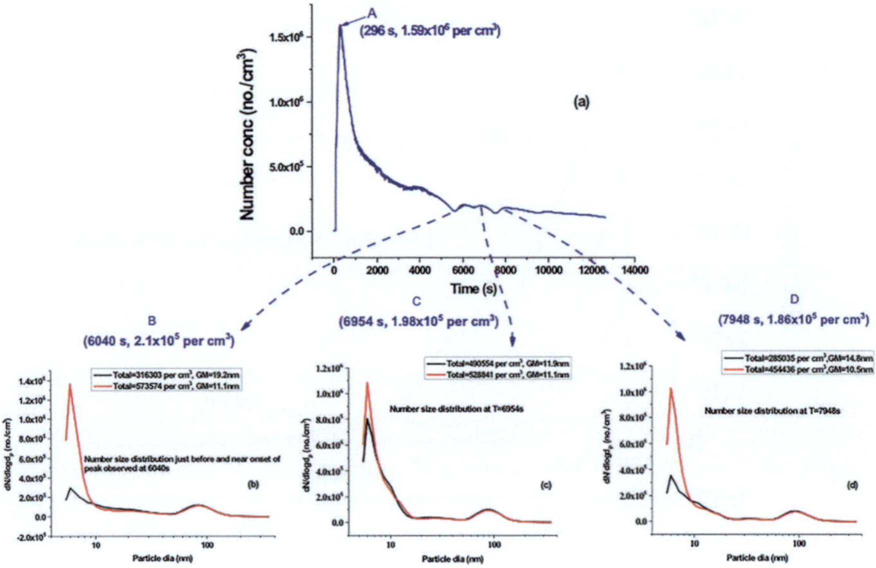

Fig. 8.7 Secondary peaks: **(a)** Nucletion peaks captured by SMPS , **(b)** Size distribution change close to the peak B, **(c)** Size distribution change close to the peak B and **(d)** Size distribution change close to the peak D

As shown in Fig. 8.7, the number concentration exhibited a few minor spikes/ peaks (peak B, peak C) at later times, in addition to the characteristic short-term peak (peak A). During these secondary spikes, the number concentration in the nuclei mode size ranges increased significantly, prompting their statistical treatment as "peaks" in subsequent analysis.

Figure 8.7 shows representative results of the analysis where the size distribution is compared for the times just before and near the onset of the peaks observed at 6040, 6954, and 7948 s. As clearly seen, the changes in the number concentration distribution occurred primarily in the nuclei mode size ranges (<10 nm), which significantly affected the total number concentration. These shifts in the size distribution suggest that the observed peaks were primarily driven by processes affecting the smallest particles, such as nucleation or rapid coagulation, leading to the observed fluctuations in concentration.

The modification in the size distribution, particularly in the lower size ranges, indicated that nucleation-driven generation was playing a key role in these observed phenomena. The small peaks in the decaying concentration likely correspond to the build-up of vapor pressure, which then triggers new particle formation. However, as the process was continuous, the newly formed particles soon began to undergo hetero-coagulation, leading to a decrease in the overall number concentration. These sporadic peaks, occurring at regular intervals, are a result of continuous vapor particle formation followed by coagulation, highlighting the dynamic behavior of particle generation and evolution in the chamber over time. The changes in the number concentration profile under continuous source conditions, as observed in the experiments, exemplify the phenomenon known as "aerosol oscillations." These oscillations occur due to the dynamic interplay between various processes, such as nucleation, coagulation, vapor condensation, and wall deposition.

8.4 Tuning Properties of Nanoparticles for Aerosol Mitigation

Aerosol mitigation in indoor environments can be achieved through various methods, primarily ventilation and filtration. Ventilation helps to dilute indoor air pollutants by introducing fresh outdoor air, while filtration captures particles from the air. Mechanical air mitigators, such as HEPA filters, require regular maintenance and face challenges related to their size, power consumption, and performance degradation over time. On the other hand, electronic mitigators, like electrostatic precipitators, utilize electrical fields to attract and remove particles from the air. However, their clean air delivery rates are typically limited, particularly in household applications (Mølgaard et al. 2014). Despite their advantages in energy efficiency, these electronic systems may not be as effective in environments with high aerosol concentrations or where high airflow is needed.

Ionizers, which operate based on the principle of aerosol charging and induced electrical drift, have been widely studied for their effectiveness in various indoor environments. These devices, particularly negative ion generators, are designed to charge aerosol particles, causing them to drift toward surfaces or collection plates, thereby reducing airborne particulate concentrations. Studies have explored the efficiencies, limitations, and characterization of ionizers, providing valuable insights into their performance (Aishwarya et al. 2023). For low ventilation environments, effectiveness of negative ion generators has been proven for different indoor environment settings (Grinshpun et al. 2005; Joshi et al. 2010). However, despite their benefits, the widespread use of ionizers is limited by several factors.

While much of the research on ionizers has focused on size-integrated characterization of their efficacy, fewer studies have addressed the particle size dependence of ionizer performance. One of the key challenges in using ionizers for aerosol mitigation is their reduced charging efficiency for nanoparticles, which are typically in the sub-50 nm range. Since nanoparticles make up a significant fraction of airborne particulate matter in many environments, this inefficiency in charging poses a difficulty in cleaning environments that contain substantial nanoparticle concentrations. Research focusing on the spatial profile of ion concentration and aerosol current could help improve the design and performance of ionizers (Khandare et al. 2019). In their study performed in a typical office environment (unoccupied, unventilated room of 80 m^3 volume), commercial negative ion generators in different orientations were placed at few designated locations. Ion concentration (measured by a handheld ion meter) and aerosol current (measured by Charme aerosol electrometer) were measured for these experimental settings. Aerosol electrometer used in this work measured current carried by the particles having diameter in the range of 2 nm- 10 μm. These measurements were performed at a height of ~1 m from the ground level, which was same as the height of the platform on which ionizer assembly was placed.

Figure 8.8a illustrates the experimental setup used to investigate the spatial distribution of ion concentration and aerosol current in the room (for ionizer's needle direction as shown in the figure). Measurements were taken at various nodes, as indicated in the figure. Figure 8.8b and c shows the spatial profiles of ion concentration and aerosol current, respectively, with the data interpolated between the nodes. Similar measurements were performed for a circular array of ionizers (5 in number) kept in center of the similar setup. The average ion concentration measured under the room's background conditions was approximately 10^3 ions cm^{-3}. However, when the ionizer was under operation, the ion concentration increased significantly, reaching concentrations on the order of 10^6 ions cm^{-3} at a distance of 0.5 m in front of the unit.

As observed, although the zone of maximum charge was located in front of the ionizer's needles, a near-isotropic distribution of ions was achieved. This pattern

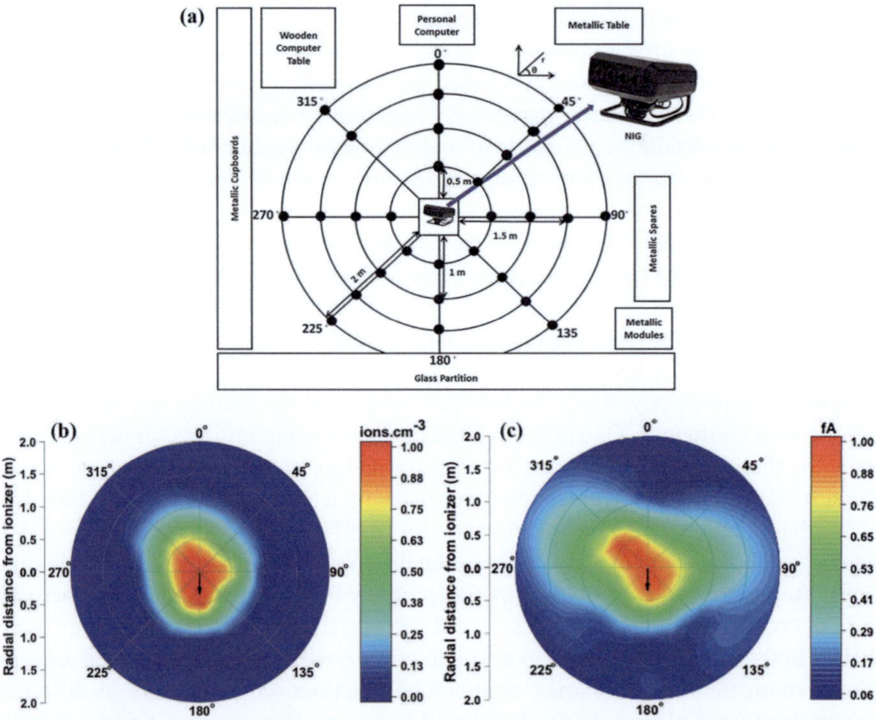

Fig. 8.8 (**a**) Experimental setup, (**b**) spatial distribution of normalized ion concentration, and (**c**) spatial distribution of aerosol current

remained consistent even when the direction of the ionizer's needles was adjusted. It is important to note that a sufficient number of ions were also present behind the ionizer in all experimental configurations. Regarding aerosol current, measurements showed levels of 3–4 fA under background conditions, approximately 380 fA at 0.5 m in front of the unit, about 190 fA at 0.5 m behind the unit, and 60–80 fA at a distance of 2 m. For a multi-ionizer setup, although the ion levels in the central zone were similar, a wider range (more spread) of ions was observed. This behavior, which occurred in the absence of air circulation patterns, aligns with predictions from certain theoretical studies and suggests that adjusting charge characteristics through both intrinsic and extrinsic factors is crucial for enhancing ionizer performance in air purification applications. For an average aerosol concentration of $2.1 \times 10^{10} \pm 3.1 \times 10^{9}$ m^{-3} and an average geometric mean diameter of 102 nm, the estimated average charge per particle was found to be 11.2, 7.5, 6.5, and 2.5 at radial distances of 0.5, 1, 1.5, and 2 m from the ionizer, respectively.

Another method for aerosol reduction is through the use of charged droplets generated by electrohydrodynamic atomization (EHDA), also known as electrospraying. EHDA is a technique that produces ultrafine, monodispersed droplets from a liquid by applying electrical forces (Tang and Gomez 1994). The liquid is passed through a nozzle at a flow rate ranging from microliters per hour (μL/h) to milliliters per hour (mL/h), with an electric field applied between the nozzle and a counter electrode. This electric field induces a surface charge on the liquid as it exits the nozzle. By adjusting the liquid flow rate and the electric field, the surface tension of the liquid can be overcome, causing it to break into droplets. By varying these parameters, droplets with a narrow size distribution can be produced, with mean diameters ranging from nanometers to a few micrometers. EHDA has a wide range of applications across various fields due to its ability to produce fine, monodispersed droplets with precise control over size. Some of the key applications of EHDA include environmental remediation, nanoparticle production, coating, and drug delivery.

Singh et al. (2013) carried out a systematic study at BARC with a laboratory-developed prototype EHDA. In order to standardize the design, response studies for current–voltage and current-flow rate were conducted. Electrospray-generated droplets, which typically carry around 10,000 elementary charges (Juan et al. 1997), are particularly useful for enhancing the removal efficiency of electrostatic precipitators. Kim et al. (2010) demonstrated that particle removal efficiency increased by 21–36% for the same size range when electrospray was used as a pre-charger for the electrostatic precipitator. Jaworek et al. (2010) explored the use of a multi-nozzle electrospray system for removing smoke aerosols, finding that removal efficiency for particles smaller than 1 μm was as high as 80–90%. Singh et al. (2021) tested an EHDA-based air purification system on various aerosol types, including atomized NaCl particles, smoke particles, and metallic aerosols generated by a hot wire. The system was experimentally tuned in terms of its optimization parameters for effective mitigation. The study found that the removal efficiency for smoke aerosols was strongly size-dependent. For particles around 30 nm, the removal efficiency was as low as 15%, but for larger particles (above 200 nm), it increased significantly, reaching up to 90%. In the case of NaCl aerosols, the removal efficiency remained relatively stable across particle sizes, ranging from 70% to 90% for particles larger than 100 nm. The system also demonstrated good performance in removing metallic nanoparticles generated by the hot wire, with high efficiency for particles larger than 30 nm.

8.5 Aerosol Measurements toward CNT Research

CNT due to its unique properties such as high mechanical strength, large surface area, high electrical conductivity, and so on can be utilized for various applications ranging from energy storage, electronics devices to aerosol filtration (Rathinavel et al. 2021). Its formation involves synthesis of CNTs in gas phase and subsequent condensation of these nanotubes to aerosol phase, which are then directed into a chamber for further processing in desired form. CNTs can be formed through various processes such as chemical vapor deposition (CVD), laser ablation, and arch discharge. CVD, one of the most common methods involves a carbon containing precursor gas and a catalyst. The precursor gas is decomposed under high temperature and the carbon atoms are deposited onto the catalyst causing the growth of CNTs. The growth depends on composition of precursor gas, catalyst, temperature, and pressure. However, CVD process has some limitation such as co-synthesis of amorphous carbon, poor utilization of catalyst, and so on, leading to low efficiency/low yield. Studies show the catalyst utilization is as poor as only 1% and removal of amorphous carbon and residual catalyst pose another challenge. Recycling catalysts and removing impurities are crucial but often economically unfeasible. Ongoing research aims to minimize these issues, which is essential for improving yield and efficiency in industrial-scale production.

Since CNT formation involves the aerosol phase, aerosol measurements supplement the efforts to understand the process dynamics. Hoecker et al. (2016) studied the role of nucleation of catalyst and indicated that the process of re-nucleation of catalyst outside high temperature zone is the major limiting factor in increasing the yield of CNT. Kaushal et al. (2024) expanded on the role of nucleation and studied parameters such as role of purge flow rate, different temperature zone, and so on by employing frontogenesis. Nearly four-fold increase in yield was obtained due to catalyst re-nucleation through the optimization of cold purge gas. The increased purge gas flow led to the push of catalyst nucleation in the high temperature zone where the rate of reaction and CNT growth rate are elevated. However, a higher purge gas flow also leads to a reduction in both residence time and reaction temperature, which may influence the overall process efficiency. Therefore, variation of number concentration and particle diameter at various purge flow rates were measured using SMPS. The number concentration and diameter of particles increased with the purge flow rate up to 30 L min^{-1}. However, beyond this point, a reduction in yield was observed, which correlates with a decrease in aerosol number concentration. This decline was attributed to a reduction in surface area, as increased

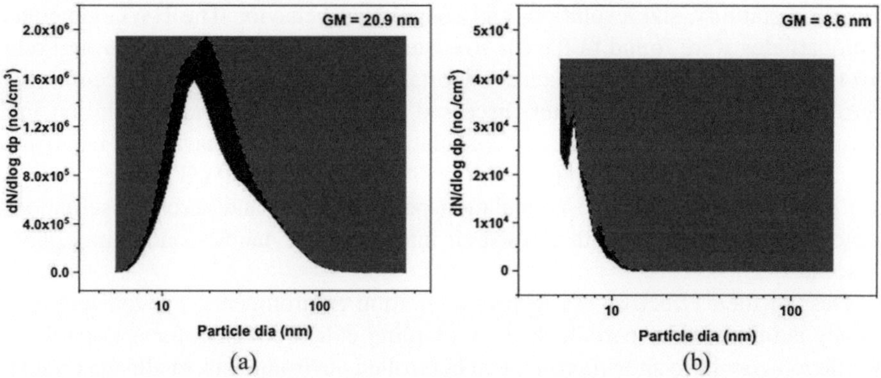

Fig. 8.9 Aerosol size distribution measured by SMPS: (**a**) before (**b**) after CNT formation

particle concentration leads to enhanced agglomeration of catalyst particles. Figure 8.9 shows the average size distribution of aerosol particles (along with the standard deviation) of particles before and after the CNT formation. After the formation of CNT, average aerosol concentration decreased from 9.7×10^5 cm^{-3} to 6.2×10^3 cm^{-3}, whereas geometric mean size decreased from 20.9 to 8.57 nm. This indicates the consumption of precursor and the catalyst in the formation CNT.

In another domain, CNTs possessing large surface area, high mechanical strength, electrical conductivity, and high porosity make them ideal candidates for improving the performance aerosol filtration system. High surface area to volume ratio and nano porous structure results in efficient trapping of aerosol particles, especially small sized. It also exhibits electrical conductivity, which can be exploited to capture charged particles via electrostatic attraction. At the same time, high strength makes CNT-based filters resistant to wear and tear and therefore can last longer. Filters made with CNTs in BARC have been tested in terms of their particle filtration efficiencies and were found to possess efficiencies >95%. Currently, efforts are undergoing to additionally optimize pressure drop, thus enhancing the quality factor of the CNT-based filters.

8.6 Summary and Conclusions

The chapter discussed the generation and characterization of nanoparticle aerosols produced by Hot Wire Generator. It is an effective method for producing nanoparticle aerosols by heating a wire (nichrome in this case) to high temperatures, causing evaporation and subsequent condensation of the vapor into nanoparticles. The operating parameters of HWG, such as the applied voltage and carrier gas flow rate, were optimized to achieve stable and tunable nanoparticle generation. The produced particles exhibited a narrow size distribution, with significant fractions falling in the nanoparticle range (<30 nm). The generated particles were also analyzed for their temporal stability, size evolution, and coagulation behavior. The HWG-generated nanoparticles were found to have a consistent geometric mean diameter and relatively stable concentration, which is important for applications requiring controlled aerosol generation. The chapter discussed the effects of particle evolution in a closed chamber, demonstrating the coagulation and growth behaviors of nanoparticle aerosols. The generation and behavior of aerosols from HWG showed interesting phenomena such as the coagulation peaking effect and aerosol oscillations, which can be used for further research into aerosol dynamics and source emission rates.

Despite their effectiveness in low-ventilation environments, the ionizer's efficiency is impacted by particle size, as charging is less effective for nanoparticles. While ionizers have shown promise in controlled environments, challenges remain in applying them effectively in larger or dynamic settings where airflow and particle size distributions vary. Research into improving ionizer designs, particularly for nanoparticle removal, is essential to broaden their use in real-world air-cleaning applications. Measurement and optimization of charge-based parameters for ionizer holds the key in these directions. Studies into aerosol currents and ion concentration profiles are important for better understanding and enhancing ionizer performance, especially for nanoparticle-rich environments. By enhancing the charging of aerosol particles, Electro Hydrodynamic Atomizer can significantly improve the performance of electrostatic precipitators and other air purification technologies. However, further optimization and research are necessary to improve the system's effectiveness for nanoparticle removal and its performance in dynamic environments.

Carbon nanotubes offer promising advancements in air-cleaning applications due to their exceptional material properties. Their integration into filtration systems, electrostatic precipitators, and other air purification technologies could significantly enhance the removal of particulate matter, including nanoparticles. In another direction, accurate measurements in nuclei-mode size ranges promise to increase the yield of CNT production.

In summary, nanoparticle aerosols, with their unique properties and dynamic behavior, present both challenges and opportunities in air quality management. Advancements in generation, characterization, and mitigation technologies, such as hot wire generators, ionizers, and electrohydrodynamic systems, offer innovative solutions, paving the way for cleaner and healthier environments.

References

Aishwarya K, Malyan V, Sahu M (2023) Air pollution control technologies for indoor particulate matter pollution: a review. Aerosol Sci Eng 2:261–282

Anand S, Mayya YS, Yu M, Seipenbusch M, Kasper G (2012) A numerical study of coagulation of nanoparticle aerosols injected continuously into a large, well stirred chamber. J Aerosol Sci 52:18–32

Boies AM, Lei P, Calder S, Girshick SL (2011a) Gas-phase production of gold-decorated silica nanoparticles. Nanotechnology 22(31):315603

Boies AM, Lei P, Calder S, Shin WG, Girshick SL (2011b) Hot wire synthesis of gold nanoparticles. Aerosol Sci Technol 45(5):654–663

Brust M, Walker M, Bethell D, Schiffrin DJ, Whyman R (1994) Synthesis of thiol-derivatised gold nanoparticles in a two phase liquid–liquid system. J Chem Soc Chem Commun 7(7):801–802

Chen H, Zhang Z (2017) Condensation-type quasi monodisperse aerosol generator with WOx nanoparticles as nuclei produced by glowing wire method. Powder Technol 313:344–352

Fomenko E, Altman I, Boskovic L, Agranovski IE (2021) Nanoparticle generation in glowing wire generator: insight into nucleation peculiarities. Materials 14:7775

Ghosh K, Tripathi SN, Joshi M, Mayya YS, Khan A, Sapra BK (2020) Particle formation from vapors emitted from glowing wires: theory and experiments. Aerosol Sci Technol 54(3):243–261

Ghosh K, Tripathi SN, Joshi M, Mayya YS, Khan A, Sapra BK (2021) Effect of charge on aerosol microphysics of particles emitted from a hot wire generator: theory and experiments. Aerosol Sci Technol 55(9):1084–1098

Grinshpun SA, Mainelis G, Trunov M, Adhikari A, Reponen T, Willeke K (2005) Evaluation of ionic air purifiers for reducing aerosol exposure in confined indoor spaces. Indoor Air 15:235–245

Hering SV, Lewis GS, Spielman SR, Eiguren-Fernandez A, Kreisberg NM, Kuang C, Attoui M (2017) Detection near 1-nm with a laminar-flow, water based condensation particle counter. Aerosol Sci Technol 51(3):354–362

Hoecker C, Smail F, Bajada M, Pick M, Boies A (2016) Catalyst nanoparticle growth dynamics and their influence on product morphology in a CVD process for continuous carbon nanotube synthesis. Carbon 96:116–124

Hontanon E, Palomares JM, Stein M, Guo X, Engeln R, Nirschl H, Kruis FE (2013) The transition from spark to arc discharge and its implications with respect to nanoparticle production. J Nanopart Res 15:1957

Jaworek A, Balachandran W, Lackowski M, Kulon J, Krupa A (2010) Multi-nozzle electrospray system for gas cleaning processes. J Electrost 64(3–4):194–202

Joshi M, Sapra BK, Khan A, Kothalkar PS, Mayya YS (2010) Thoron (Rn-220) decay products removal in poorly ventilated environments using unipolar ionizers: dosimetric implications. Sci Total Environ 408(23):5701–5706

Joshi M, Khan A, Anand S, Sapra BK (2016) Size evolution of ultrafine particles: differential signatures of normal and episodic events. Environ Pollut 208:354–360

Juan DL, Fernandez De La Mora J, Loscertales IG (1997) Charge and size distribution of electrospray drops. J Colloid Interface Sci 186:280–293

Kangasluoma J, Attoui M, Junninen H, Lehtipalo K, Samodurov A, Korhonen F, Sarnela N, Schmidt-Ott A, Worsnop Kulmala M, Petaja T (2015) Sizing of neutral sub 3 nm tungsten oxide clusters using airmodus particle size magnifier. J Aerosol Sci 87:53–62

Kaushal A, Alexander R, Joshi M, Mariam, Singh J, Dasgupta K (2024) Frontogenesis-inspired efficient synthesis of dense SWCNT fiber through in-situ boosting of catalyst re-nucleation. Chem Eng J 484:149254

Khan A, Modak P, Joshi M, Khandare P, Koli A, Gupta A, Anand S, Sapra BK (2014) Generation of high concentration nanoparticles using glowing wire technique. J Nanopart Res 16(12):2776

Khandare P, Joshi M, Nakhwa A, Khan A, Mariam, Sapra BK (2019) Unexplored aspects of unipolar ionizer characteristics in context of indoor air cleaning. Environ Sci Pollut Res 26:18191–18199

Kim K (2005) Plasma synthesis and characterization of nanocrystalline aluminum nitride particles by aluminum plasma jet discharge. J Cryst Growth 283(3–4):540–546

Kim JH, Lee HS, Kim HH, Ogata A (2010) Electrospray with electrostatic precipitator enhances fine particles collection efficiency. J Electrost 68:305

Kulmala M, Junninen H, Dada L, Salma I, Weidinger T, Thén W, Vörösmarty M, Komsaare K, Stolzenburg D, Cai R, Yan C, Li X, Deng C, Jiang J, Petäjä T, Nieminen T, Kerminen V-M (2022) Quiet new particle formation in the atmosphere. Front Environ Sci 10:912385

Mahan AH, Parilla PA, Jones KM, Dillon AC (2005) Hot-wire chemical vapor deposition of crystalline tungsten oxide nanoparticles at high density. Chem Phys Lett 413(1–3):88–94
Mølgaard B, Koivisto AJ, Hussein T, Hämeri K (2014) A new clean air delivery rate test applied to five portable indoor air cleaners. Aerosol Sci Technol 48(4):409–417
Müller U, Schmidt-Ott A, Burtscher H (1987) First measurement of gas-adsorption to free ultrafine particles: O_2 on Ag. Phys Rev Lett 58:1684–1686
Peineke C, Attoui MB, Schmidt-Ott A (2006) Using a glowing wire generator for production of charged, uniformly sized nanoparticles at high concentrations. J Aerosol Sci 37:1651–1661
Peineke C, Attoui MB, Robles R, Reber A, Khanna S, Schmidt-Ott A (2009) Production of equal sized atomic clusters by a hot wire. J Aerosol Sci 40(5):423–430
Pratsinis SE (1988) Simultaneous nucleation, condensation and coagulation in aerosol reactors. J Colloid Interface Sci 124:416–427
Ranz W, Marshall W et al (1952) Evaporation from drops. Chem Eng Prog 48(3):141–146
Rathinavel S, Priyadharshini P, Panda D (2021) A review on carbon nanotube: an overview of synthesis, properties, functionalization, characterization, and the application. Mater Sci Eng B 268:115095
Sabbagh F, Kiarostami K, Khatir NM, Rezania S, Muhamad II (2020) Green synthesis of $Mg_{0.99}$ $Zn_{0.01}O$ nanoparticles for the fabrication of κ-Carrageenan/NaCMC hydrogel in order to deliver catechin. Polymers 12:861
Schmidt-Ott A, Schurtenberger P, Siegmann H (1980) Enormous yield of photoelectrons from small particles. Phys Rev Lett 45(15):1284–1287
Shboul AM, Izquierdo R (2021) Printed chemiresistive In_2O_3 nanoparticle-based sensors with ppb detection of H_2S gas for food packaging. ACS Appl Nano Mater 4:9508–9517
Singh S, Khan A, Sapra BK, Mayya YS (2013) Parameterization of an electro-hydrodynamic atomization based aerosol generator. Part Sci Technol 31(5):494–500
Singh S, Khan A, Nakhwa A, Sapra BK, Mayya YS (2021) Scavenging of submicron aerosol particles by cloud of charged droplets generated from electro-hydrodynamic atomizer (EHDA). Aerosol Sci Eng 5:223–232
Supronowicz P, Ajayan P, Ullmann K, Arulanandam B, Metzger D, Bizios R (2002) Novel current-conducting composite substrates for exposing osteoblasts to alternating current stimulation. J Biomed Mater Res 59(3):499–506
Tang K, Gomez A (1994) Generation by electrospray of monodisperse water droplets for targeted drug delivery by inhalation. J Aerosol Sci 25:1237–1249
Yang Y, Song B, Ke X, Xu F, Bozhilov KN, Hu L, Shahbazian-Yassar R, Zachariah MR (2020) Aerosol synthesis of high entropy alloy nanoparticles. Langmuir 36:1985–1992
Yonezawa T, Kunitake T (1999) Practical preparation of anionic mercapto ligand-stabilized gold nanoparticles and their immobilization. Colloids Surf A Physicochem Eng Asp 149(1–3):193–199

Chapter 9
Biosensors for Detecting Air Pollutants and Hazardous Gases

Utkarsh Kumar, Wen-Min Huang, B. C. Yadav, and Chiu-Hsien Wu

Abstract This chapter provides a comprehensive overview of recent developments and applications in the field of biosensors tailored for toxic gas sensing. Biosensors, leveraging the specificity and sensitivity of biomolecular recognition elements, play a pivotal role in the continuous evolution of gas sensing technology. The diverse range of biomaterials employed in biosensors, including enzymes, antibodies, DNA, and whole cells, each tailored to detect specific toxic gases with heightened precision. Emphasis is placed on the advancements in sensor design, fabrication techniques, and signal transduction mechanisms, which collectively contribute to the enhanced selectivity and sensitivity of biosensors. Additionally, it also addresses the challenges encountered in toxic gas sensing applications, such as cross-reactivity and environmental factors, and discusses innovative strategies employed to overcome these hurdles. The applications of biosensors in various industries, ranging from industrial safety to environmental monitoring, are highlighted, underscoring the critical role these devices play in mitigating risks associated with toxic gas exposure. This chapter is a valuable resource for researchers, engineers, and practitioners involved in advancing biosensor technology by providing insights into the current state of the field and avenues for future research.

Keywords Biomaterials · Detection · Toxic gas · Gas exposure

U. Kumar · W.-M. Huang
Department of Physics, National Chung Hsing University, Taichung, Taiwan

B. C. Yadav
Nanomaterials and Sensor Research Laboratory, Department of Physics, Babasaheb Bhimrao Ambedkar University, Lucknow, India

C.-H. Wu (✉)
Department of Physics, National Chung Hsing University, Taichung, Taiwan

Institute of Nanoscience, National Chung Hsing University, Taichung, Taiwan
e-mail: chwu@phys.nchu.edu.tw

N. Ghosh et al. (eds.), *Nanotechnology in Air Quality Management*, Sustainable Development and Biodiversity 39, https://doi.org/10.1007/978-981-96-8339-0_9

9.1 Introduction

Air pollutants and hazardous gases pose significant threats to both human health and the natural environment (Anderson 2001; Birks et al. 2020). These issues have wide-ranging impacts, encompassing respiratory problems, ecosystem degradation, and broader environmental consequences. Fine particles have the ability to infiltrate the respiratory system, potentially leading to the onset or worsening of conditions such as asthma, bronchitis, and other respiratory diseases. Nitrogen dioxide (NO_2) gas irritates the respiratory system and can lead to respiratory infections, especially in vulnerable populations (Zhang et al. 2019; Kumar et al. 2022, 2023a, b). Ground-level ozone exposure is linked to respiratory problems, and long-term exposure can result in decreased lung function. Air pollutants, including fine particulate matter and certain gases, are associated with cardiovascular diseases, heart attacks, and strokes. Extended exposure to certain air pollutants, such as benzene and formaldehyde, has been associated with an elevated likelihood of developing cancer. Emerging research suggests that air pollution may contribute to neurological disorders, including cognitive decline and conditions such as Alzheimer's disease.

Children are particularly vulnerable to the effects of air pollution, which can lead to developmental issues, impaired lung function, and an increased risk of respiratory infections. Acid rain, caused by pollutants such as sulfur dioxide (SO_2) and nitrogen oxides (NO_x) harms aquatic ecosystems, damages forests, and erodes soil quality. Air pollution can contribute to habitat degradation, threatening plant and animal species and disrupting ecosystems. Specific air pollutants, such as carbon dioxide (CO_2) (Kumar and Yadav 2019) and methane (CH_4) (Kumar et al. 2021) are contributors to climate change, resulting in elevated temperatures, changes in weather patterns, and the escalation of sea levels. Ozone and other pollutants can damage crops, affecting agricultural yields and food security. Some air pollutants, when deposited, can contaminate water bodies and soil, leading to adverse effects on aquatic life and terrestrial ecosystems. Particulate matter and pollutants can reduce visibility, impacting navigation and recreational activities. Addressing these challenges requires comprehensive strategies, including the adoption of cleaner technologies, stringent emission standards, and global cooperation to mitigate the adverse effects of air pollutants and hazardous gases on both human health and the environment. The enhancement in the level of toxic gases also affect the human brains as shown in Fig. 9.1 (Onyancha et al. 2021).

The integration of biomaterials into gas sensing applications has emerged as a promising avenue, leveraging the unique properties of biological components for enhanced sensitivity, selectivity, and sustainability in detecting various gases. Biomaterials, such as enzymes, proteins, and DNA, exhibit inherent specificity and recognition capabilities that can be harnessed to create advanced sensing platforms. Enzymes can selectively catalyze reactions with specific gases, enabling the development of enzymatic gas sensors. For example, glucose oxidase-based sensors have been explored for detecting gases such as hydrogen peroxide. Biomolecules such as antibodies and aptamers serve as natural receptors, offering high specificity for

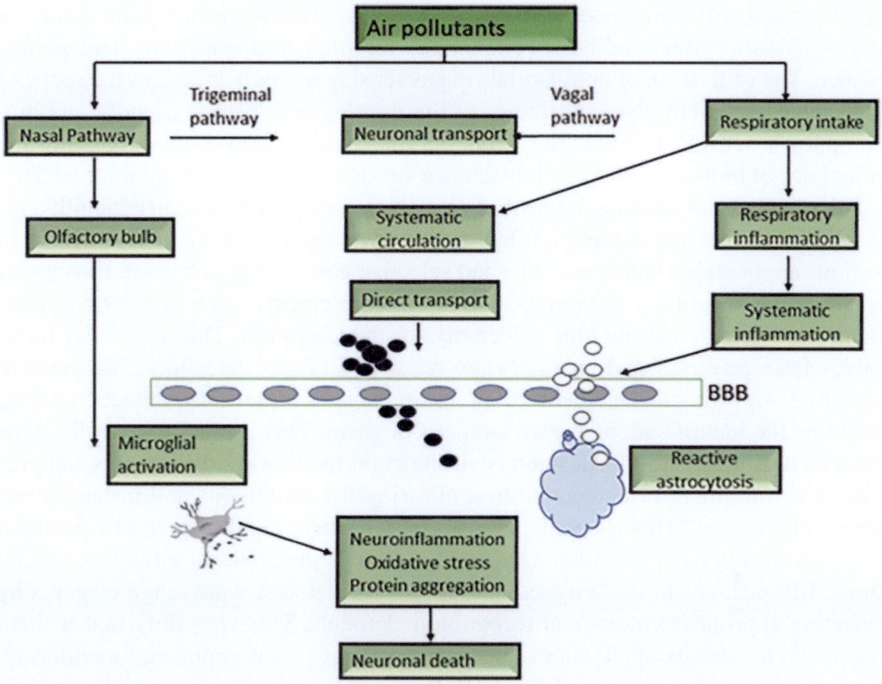

Fig. 9.1 Effect of air pollutant on human brain. (Onyancha et al. 2021. © Elsevier)

target gases. Functionalizing sensor surfaces with these receptors enhances selectivity in gas detection. Living cells or microorganisms can be employed as sensitive elements in gas sensors. Their physiological responses, such as changes in metabolism or membrane potential, can be transduced into measurable signals indicative of gas exposure. DNA sequences can be engineered to bind specifically with target gases, leading to conformational changes that can be detected electrochemically or optically. DNA-based sensors offer versatility in designing selective probes (Hwang et al. 2023). Engineered protein nanomaterials, such as bacteriorhodopsin and viral capsids, have demonstrated potential for gas sensing applications. These nanomaterials exhibit unique electronic and optical properties that can be exploited for signal transduction. The use of biodegradable materials aligns with sustainable sensing practices. Biopolymers and naturally derived matrices offer environmentally friendly alternatives for sensor fabrication. Biomaterials often possess biocompatible properties, reducing the toxicity associated with sensing devices. This is especially important for applications where sensors come into direct contact with biological systems. Certain biomaterials respond selectively to specific gases, undergoing changes in their structural or electronic properties. This responsiveness forms the basis for sensitive and selective gas detection. Leveraging the innate sensing mechanisms found in biological systems, biomaterial-based gas sensors can emulate and enhance the exquisite specificity observed in nature. Biomaterials can be integrated with nanomaterials to create hybrid sensing platforms, capitalizing on the synergistic effects of both components for improved sensitivity and performance. The utilization of biomaterials in gas sensing not only broadens the scope of detectable gases but also contributes to the development of eco-friendly and biocompatible sensing technologies. This interdisciplinary approach, combining the principles of biology with materials science, holds tremendous potential for advancing the field of gas sensing in terms of sensitivity, selectivity, and sustainability.

Biosensors for gas detection offer several advantages, making them essential in various applications where accurate and selective gas sensing is crucial. Biosensors leverage the specificity of biological recognition elements, such as enzymes, antibodies, or DNA, ensuring high selectivity for the target gas. This selectivity minimizes false positives and enhances the reliability of gas detection. The inherent sensitivity of biological elements allows biosensors to achieve low detection limits, enabling the identification of trace amounts of gases. This is crucial in applications such as healthcare diagnostics and environmental monitoring. Biosensors can provide real-time monitoring capabilities, allowing for continuous and instantaneous detection of gases. This is beneficial in scenarios where rapid response to changing gas concentrations is essential, such as in industrial processes or emergency situations. Biosensors can be designed and tailored to detect a wide range of gases by selecting appropriate biological recognition elements. This versatility makes them adaptable to various applications and environments. In environmental monitoring, biosensors offer the ability to detect pollutants and hazardous gases with high precision. They play a crucial role in assessing air quality and ensuring compliance with environmental regulations. Biosensors are extensively used in medical diagnostics for the detection of specific biomarkers associated with diseases. In the context of gas sensing, they can be employed for monitoring gases related to respiratory conditions or metabolic processes. Biological recognition elements often exhibit minimal interference from other gases or environmental factors, contributing to the accuracy of gas detection. This is particularly advantageous in complex gas mixtures. Biosensors can be integrated into remote sensing devices, facilitating the monitoring of gases in challenging or inaccessible locations. This is valuable in applications such as space exploration or monitoring of remote industrial sites. The use of biological components in biosensors aligns with sustainable and green technology practices. The biodegradable nature of these sensors reduces environmental impact. Biosensors are particularly effective in detecting volatile organic compounds (VOCs) and other volatile gases, playing a crucial role in industrial and indoor air quality monitoring (Mitsubayashi et al. 2022). In summary, the need for biosensors in gas detection arises from their unique capabilities, including high selectivity, biocompatibility, low detection limits, real-time monitoring, and versatility.

In this comprehensive exploration, the emphasis is on leveraging the unique properties of biomaterials for the selective and sensitive detection of toxic gases. Biosensors, employing biological recognition elements such as enzymes, antibodies, DNA, and whole cells, offer a targeted approach to identifying hazardous gases with high precision. This chapter delves into the design principles, mechanisms, and applications of these biosensors, elucidating their potential in mitigating risks associated with toxic gas exposure. It addresses the challenges and advancements in the field, showcasing how biosensors not only enhance detection capabilities but also contribute to sustainable and eco-friendly sensing practices. This chapter provides a valuable resource for researchers, practitioners, and industries, seeking innovative solutions for the reliable detection and monitoring of toxic gases in diverse environments (Kumar et al. 2017, 2023c, 2024; Mishra et al. 2021; Cheng et al. 2023).

9.2 Different Types of Biomaterials Used for the Gas Sensing

Various biomaterials are employed in gas sensing applications, leveraging the specific properties and functionalities of biological elements for enhanced sensitivity and selectivity. Here are different types of biomaterials commonly used for gas sensing.

9.2.1 Enzymes

Enzymes function as highly specialized biocatalysts, orchestrating specific chemical transformations. In the realm of sensor technology, enzyme-based sensors are strategically engineered to harness the catalytic prowess of enzymes, facilitating the selective and discerning detection of particular gases. This molecular orchestration is particularly exemplified in scenarios where the targeted identification of gases, such as hydrogen peroxide or ammonia, is paramount. The enzymatic catalysis inherent in these sensors not only ensures a heightened level of specificity but also amplifies the sensitivity of the detection mechanism, thereby establishing a robust platform for precise gas sensing applications as shown in Fig. 9.2 (Huang et al. 2022).

9.2.2 Antibodies

Antibodies, being proteins characterized by their exceptional affinity and specificity, serve as formidable molecular entities in immunosensor configurations. These sensors strategically deploy antibodies as recognition elements, capitalizing on their innate ability to selectively bind to specific molecules or antigens. This targeted molecular recognition imparts a heightened degree of sensitivity to immunosensors, facilitating the discerning detection of target gases or antigens (Ma et al. 2023). The utilization of antibodies as recognition elements in immunosensors thus not only ensures precision in molecular identification but also enhances the overall sensitivity of the sensing platform, rendering it adept for sophisticated applications demanding a high level of specificity in gas detection.

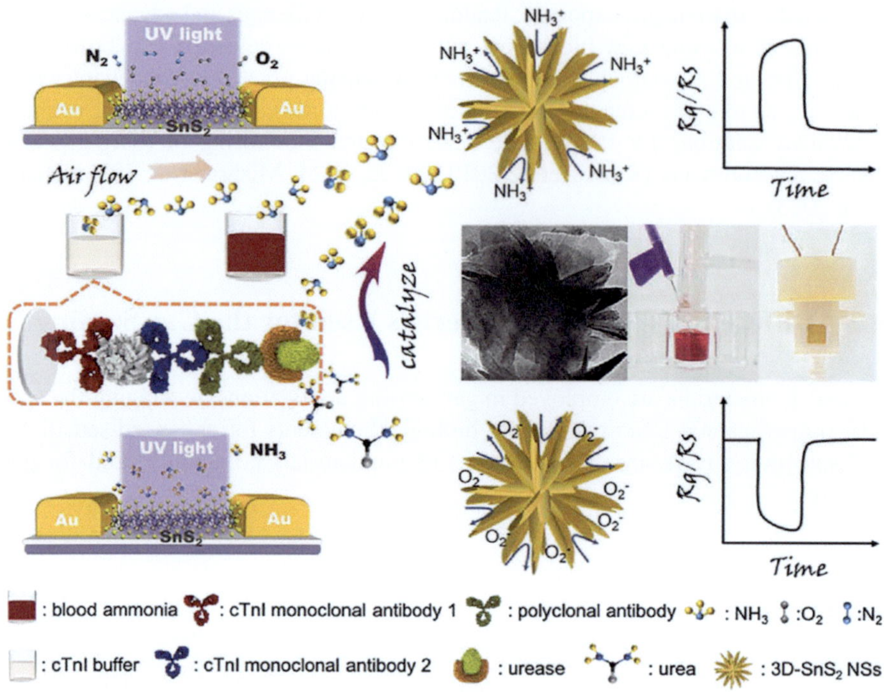

Fig. 9.2 Schematic diagram for UV-assisted NH_3 detection by using biocatalyst. (Huang et al. 2022. © ACS Publications)

9.2.3 DNA (Deoxyribonucleic Acid)

DNA-based sensors leverage the distinctive binding characteristics inherent in DNA sequences for the targeted detection of specific gases. Through molecular engineering, DNA strands are systematically designed to form specific interactions with target gas molecules, instigating conformational alterations in the DNA structure (Reddeppa et al. 2019). The induced conformational changes serve as discernible signals that can be detected through various analytical methods. This intricate molecular design allows DNA-based sensors to achieve a high degree of specificity and sensitivity in gas detection applications, establishing a versatile and precise platform for molecular recognition and signaling in response to the presence of target gases.

9.2.4 Aptamers

Aptamers, comprising truncated single-stranded DNA or RNA sequences, undergo a meticulous selection process to attain a heightened specificity for target molecules, encompassing gases (Gaggiotti et al. 2020). Within the realm of sensor technology, aptamer-based sensing platforms manifest as versatile and selective systems, providing an intricate and discerning methodology for gas detection. The aptamer's intrinsic ability to form specific and high-affinity interactions with target gases endows these sensors with a nuanced and targeted approach, underscoring their efficacy in achieving selective and precise detection capabilities within diverse gas sensing applications.

9.2.5 Whole Cells and Microorganisms

Utilizing living cells or microorganisms as discerning constituents in gas sensors involves leveraging their intricate physiological dynamics for sensitive detection. Perturbations in cellular metabolism or alterations in membrane potential serve as responsive indicators to gas exposure. These nuanced physiological responses are then transduced into quantifiable signals, offering a sophisticated mechanism for the detection and quantification of gases (Poma et al. 2023). The integration of living entities into gas sensing platforms thus capitalizes on their inherent biological reactivity, providing a complex and responsive framework for translating environmental gas interactions into measurable and analytically valuable signals.

9.2.6 Proteins and Peptides

Engineered protein nanomaterials and peptides, distinguished by their specific gas-binding attributes, represent versatile candidates for gas sensing applications (Wasilewski et al. 2022). These materials exhibit distinctive electronic and optical properties that can be strategically harnessed for signal transduction mechanisms. The tailored gas-binding properties of these nanomaterials confer a high degree of selectivity, while their inherent electronic and optical features provide a robust foundation for transducing gas interactions into discernible signals. This sophisticated integration of engineered protein nanomaterials and peptides into gas sensing platforms showcases their potential to contribute to highly sensitive and selective detection methodologies with advanced signal processing capabilities.

9.2.7 Nanobodies

Nanobodies, derived from camelids and characterized by their single-domain antibody structure, demonstrate exceptional stability and specificity, rendering them particularly noteworthy for advanced applications in gas sensing. The diminutive size of nanobodies, coupled with their robust binding affinity, positions them as optimal candidates for integration into gas sensor platforms. Leveraging their unique molecular characteristics, nanobodies provide an intricate and efficient means of binding to target molecules. The utilization of nanobodies in gas sensors underscores their potential to contribute significantly to sensor technology, enhancing both sensitivity and specificity in the detection of specific gases.

9.2.8 Microbial Fuel Cells (MFCs)

Microbial fuel cells (MFCs) harness the metabolic activity of microorganisms to initiate an electrochemical process, resulting in the generation of an electric current. In the context of gas sensing, MFCs are investigated as intricate sensing devices, wherein alterations in microbial activity, triggered by exposure to specific gases, lead to quantifiable variations in the generated electrical signals (Yao et al. 2023). The dynamic interplay between microbial metabolic processes and the electrochemical response within MFCs constitutes a sophisticated mechanism for transducing gas-induced biological changes into precise and measurable electrical outputs, showcasing the potential of MFCs as advanced gas sensing platforms.

9.2.9 Metalloproteins

Metalloproteins, exemplified by bacteriorhodopsin, present an intriguing prospect for integration into gas sensing devices owing to their distinctive electronic attributes. The incorporation of these proteins is motivated by their unique electronic properties, which undergo modulation upon interaction with specific gases (Sousa and Tian 2023). This modulation induces discernible structural changes within the metalloprotein, facilitating an intricate mechanism for signal transduction. The utilization of such metalloproteins in gas sensors signifies a sophisticated approach to transducing gas-induced structural alterations into quantifiable electronic signals, emphasizing their potential as pivotal components in advanced gas sensing technologies.

9.2.10 Phytochemicals and Plant Extracts

Phytochemicals and plant extracts harbor a repertoire of bioactive compounds capable of interacting selectively with specific gases. The exploration of plant-based sensors is underpinned by the unique ability of these bioactive compounds to modulate their properties in response to gas interactions. This has prompted investigations into the potential of plant-based sensors for applications in environmental monitoring and agriculture (Ouandaogo et al. 2023). The intricate molecular interactions between plant-derived bioactive compounds and target gases underscore the viability of these sensors as precise and environmentally conscious tools for detecting and monitoring specific gases in diverse settings.

9.2.11 MicroRNA (miRNA)

MicroRNA (miRNA)-based sensors leverage the sequence-specificity inherent in miRNA molecules for the targeted detection of particular gases. The binding event between miRNA and its designated target gas serves as a molecular recognition event that can be transduced into quantifiable signals, constituting a robust approach for gas sensing (Chen et al. 2019). The utilization of miRNA sequences in sensor design capitalizes on their unique selectivity, allowing for a nuanced and sensitive response to specific gas molecules. This advanced molecular recognition strategy underscores the potential of miRNA-based sensors as intricate and selective tools for gas sensing applications.

The choice of biomaterial depends on the specific requirements of the gas sensing application, including the type of gas to be detected, desired sensitivity, and environmental conditions. Integrating biomaterials into gas sensors enables the development of highly selective and efficient devices for various industries, including healthcare, environmental monitoring, and industrial safety.

9.3 Detection of Different Gases Using Biosensors

Since last few years, several biosensors have been used for the detection of various toxic gases. In this section, different kinds of gases, which were efficiently detected using biosensors, have been discussed.

9.3.1 Ethanol Detection

Ethanol is commonly found in alcoholic beverages and certain industrial products. While moderate and responsible consumption of ethanol is generally considered safe for many adults, excessive or chronic intake can lead to various health problems. Ethanol is a central nervous system depressant. Overconsumption can result in intoxication, leading to impaired coordination, slowed reaction times, and compromised decision-making abilities. Chronic alcohol abuse, including excessive ethanol consumption, can lead to liver diseases such as fatty liver, alcoholic hepatitis, and cirrhosis. The liver metabolizes ethanol, and excessive intake can overwhelm its capacity, causing damage over time. Prolonged and heavy ethanol consumption has been linked to cardiovascular problems, including high blood pressure, irregular heart rhythms, and an increased risk of heart disease. Ethanol can irritate the digestive tract, leading to issues such as gastritis, ulcers, and an increased risk of gastrointestinal bleeding. Ethanol can have neurotoxic effects, potentially leading to cognitive impairment, memory loss, and an increased risk of neurological disorders. Chronic alcohol abuse, including heavy ethanol consumption, is associated with an increased risk of mental health disorders such as depression and anxiety.

It is important for individuals to be aware of the potential health risks associated with ethanol consumption and to consume alcoholic beverages responsibly. If someone is facing challenges related to alcohol use, seeking professional help and support is crucial for their well-being. Additionally, individuals with certain medical conditions or those taking specific medications should consult with healthcare professionals to understand how ethanol consumption may interact with their health status. Alcohol oxidase (AOD) has been used to detect ethanol but the LOQ (limit of quantification) is very low. Furthermore, when ADH (alcohol dehydrogenase) is used as sensing material, then the LOQ increases to 2×10^3 but still the detection time is large. When the sensing material was changed to AOD and HRP (horse radish peroxidase), the sensor with low LOQ and good dynamic sensing range was achieved. The results of intensity based and other ethanol detector has been depicted in Fig. 9.3a–c (Arakawa et al. 2013; Mitsubayashi et al. 2005; Park 1995). The comparison of results obtained by different groups of scientists for the detection of ethanol using biosensors has been shown in Table 9.1.

9.3.2 Formaldehyde Detection

Formaldehyde is a colorless gas with a pungent odor that is widely used in various industries and products, including building materials, household products, and certain medical treatments. While low levels of formaldehyde are naturally present in the environment, exposure to elevated levels can pose health risks. Formaldehyde is a respiratory irritant and can cause irritation of the eyes, nose, and throat. Short-term

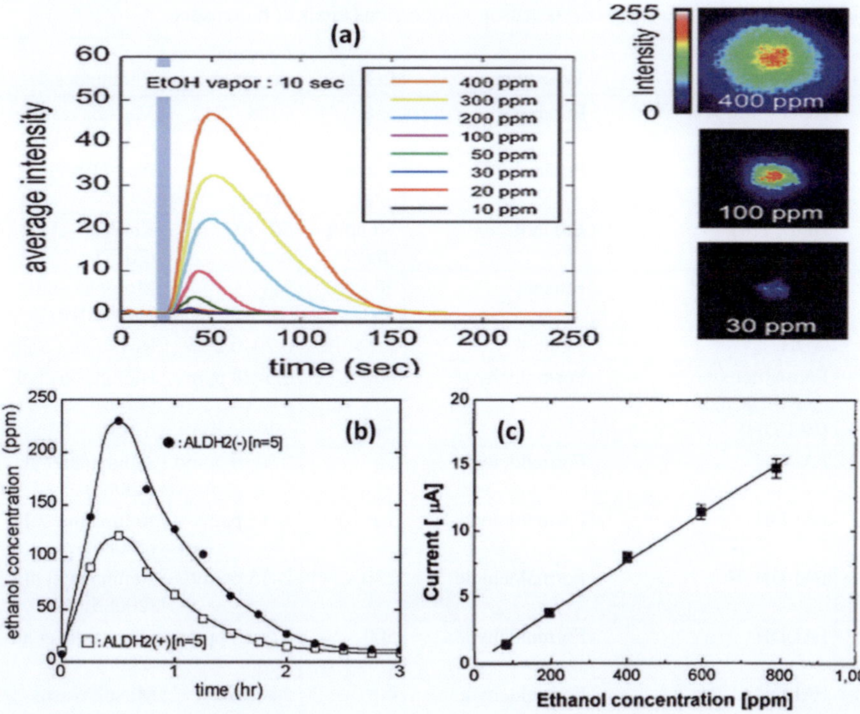

Fig. 9.3 (**a**) The temporal fluctuations in gaseous ethanol concentrations ranging from 10 to 400 ppm were observed. The emission duration of ethanol vapor was set at 10 s. (Arakawa et al. 2013. © Elsevier). (**b**) Contrasts in the variations of ethanol and acetaldehyde concentrations over time in exhaled breath were made between individuals with ALDH2 [−] (filled circle) and ALDH2 [+] (open square) statuses following the consumption of alcohol. (Mitsubayashi et al. 2005. © Elsevier). (**c**) Biosensor-activated ethanol curve for 50 mL of ethanol injection. (Park 1995. © Elsevier)

exposure to elevated levels may result in symptoms such as coughing, wheezing, and difficulty breathing. Some individuals may develop allergic reactions to formaldehyde, leading to skin rashes, itching, and other dermatological symptoms upon contact with products containing the substance. Individuals with asthma may experience exacerbation of symptoms when exposed to formaldehyde (Zhu et al. 2022). This can include increased wheezing, chest tightness, and difficulty breathing. Prolonged or chronic exposure to formaldehyde has been associated with respiratory problems, including an increased risk of developing respiratory diseases such as bronchitis and asthma. Formaldehyde has been classified as a Group 1 human carcinogen by the International Agency for Research on Cancer (IARC). Long-term exposure to high levels of formaldehyde has been linked to an increased risk of nasopharyngeal cancer and respiratory tract cancers. Some studies suggest that formaldehyde exposure may be associated with neurological effects, including headaches, dizziness, and cognitive impairment. However, more research is needed

Table 9.1 Comparative study of gas sensor using various kinds of biosensors

S. no.	Enzyme	Type of gas	LOQ	Dynamic range	References
1	AOD	Ethanol	358	0358–1242 ppmv	Mitsubayashi et al. (1994)
2	ADH	Ethanol	2×10^3	20–800 ppmv	Park (1995)
3	AOD, HRP	Ethanol	50 ppm (m/v)	50–500 ppm (m/v)	Kuretake et al. (2017)
4	AOD and HRP	Ethanol	20×10^3	NA	Motooka and Uno (2018)
5	ADH	Ethanol	7.1×10^3	0.70–0.2%v	Roy et al. (2020)
6	Formaldehyde dehydrogenase (FALDH)	Formaldehyde	300	0.3–18 ppmv	Hämmerle et al. (1996)
7	FALDH	Formaldehyde	76	0.5–10 ppmv	Achmann et al. (2008b)
8	FALDH	Formaldehyde	2×10^3	2–15 ppmv	Achmann et al. (2008c)
9	FALDH	Formaldehyde	130	2–15 ppmv	Achmann et al. (2008d)
10	FALDH	Formaldehyde	500	0.5–15 ppmv	Hämmerle et al. (2008)
11	FALDHALDH	Formaldehyde	40	40–2000 ppbv	Mitsubayashi et al. (2008)
12	Dye-linked FALDH	Formaldehyde	NA	NA	Achmann et al. (2008a)
13	FALDH	Formaldehyde	88	86 ppbv–12.8 ppmv	Hämmerle et al. (2010)
14	Glutathione-dependent formaldehyde	Formaldehyde	NA	NA	Sigawi et al. (2011)
15	Polyphenol oxidase	Phenol	30	0.03–13 ppmv	Dennison et al. (1995)
16	Tyrosinase	Phenol	4	4–400 ppbv	Hämmerle et al. (2010)
17	Flavin-containing monooxygenase type 3	Trimethylamine	10^4	10–98 ppmv	Mitsubayashi and Hashimoto (2000)
18	Flavin-containing monooxygenase type 3	Methyl mercaptan	10^3	1–4000 ppmv	Mitsubayashi and Hashimoto (2002)
19	Peroxidase	Hydrogen peroxide	500	0.5–30 ppbv	Otsuka et al. (2006)
20	Butyrylcholinesterase, choline oxidase	Nicotine	10	10–1000 ppmv	Mitsubayashi et al. (2006)
21	Tyrosinase	Carbamate and organophosphate	6	6–10 ppbv	Tanimoto de Albuquerque and Ferreira (2007)

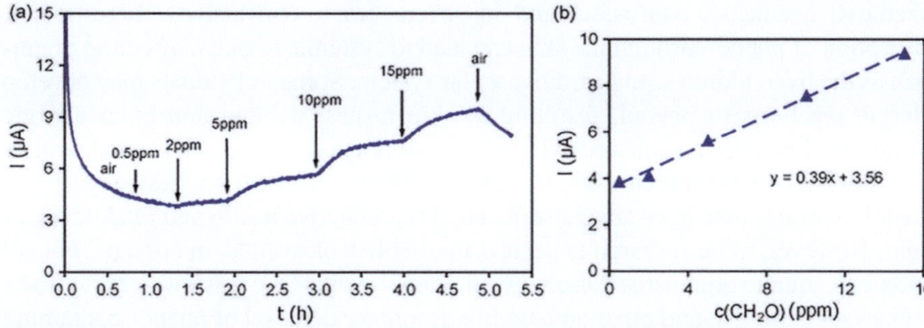

Fig. 9.4 Improvement in sensitivity achieved by quantitatively increasing the enzyme load. (**a**) 300 μg of protein included in the P60-fraction in a 200 μL buffer, (**b**) 600 μg of protein in the P60-fraction in a 200 μL buffer. Sensor response to formaldehyde and methanol vapor above aqueous sample solutions. (Achmann et al. 2008a, b, c, d. © Elsevier)

to establish a clear link. Workers in certain industries, such as the furniture and construction industries, may be at an increased risk of formaldehyde exposure. Occupational exposure can lead to a range of health issues, including respiratory and skin problems.

To minimize formaldehyde exposure, it is important to be aware of products that may release formaldehyde, such as some building materials, furniture, and certain household items. Adequate ventilation, using formaldehyde-free products, and following safety guidelines in occupational settings can help reduce the risk of formaldehyde-related health problems. If individuals experience persistent symptoms or have concerns about formaldehyde exposure, seeking medical advice is recommended. Different kinds of biomaterials can detect formaldehyde at even lower concentrations. They employed formaldehyde dehydrogenase (FALDH) as the active sensing material and through strategic structural modification and optimization of the synthesis process, successfully enabled formaldehyde detection at room temperature. The result of FALDH-based formaldehyde detector has been shown in Fig. 9.4 (Achmann et al. 2008a, b. A comparative study of different biosensor-based formaldehyde detectors has been reported in Table 9.1.

9.3.3 Phenol Detection

Phenol, also known as carbolic acid, is a chemical compound with various industrial applications. Exposure to phenol can have several effects on both humans and the environment. Phenol is corrosive and can cause irritation or burns upon contact with the skin. Eye exposure can lead to severe eye irritation or damage. Inhalation of phenol vapors or aerosols can irritate the respiratory tract, leading to symptoms such as coughing, shortness of breath, and throat irritation. High levels of phenol exposure may affect the central nervous system, leading to symptoms such as dizziness, headache, confusion, and in severe cases, convulsions. Ingestion or absorption of phenol through the skin can lead to systemic toxicity, affecting organs such as the liver, kidneys, and cardiovascular system. Some individuals may develop allergic reactions to phenol, resulting in skin rashes, itching, and other allergic symptoms. Long-term exposure to high levels of phenol has been associated with an increased risk of certain cancers, including leukemia. Animal studies suggest that phenol exposure may have adverse effects on reproductive health and fetal development. However, more research is needed to establish clear links in humans. Phenol is toxic to aquatic organisms. Discharge of phenol-containing wastewater into water bodies can harm fish and other aquatic life. Improper disposal of phenol-containing substances can lead to soil contamination, affecting the health of plants and soil-dwelling organisms. Phenol can be released into the air during industrial processes. Prolonged exposure to elevated levels of phenol in the air may contribute to air pollution. Phenol has the potential to bioaccumulate in the tissues of living organisms, leading to higher concentrations as it moves up the food chain.

Workers handling phenol or working in environments where phenol is present should follow proper safety protocols, including the use of personal protective equipment. Regulatory measures are in place to limit the discharge of phenol into air and water to protect human health and the environment. Proper disposal and treatment of phenol-containing waste are crucial to prevent environmental contamination. It is essential to handle phenol with care, implement safety measures in industrial settings, and adhere to environmental regulations to mitigate the potential adverse effects on both humans and the environment. The measurement of phenol concentration by using various biosensor has been depicted in Fig. 9.5a, b (Hämmerle et al. 2010; Dennison et al. 1995).

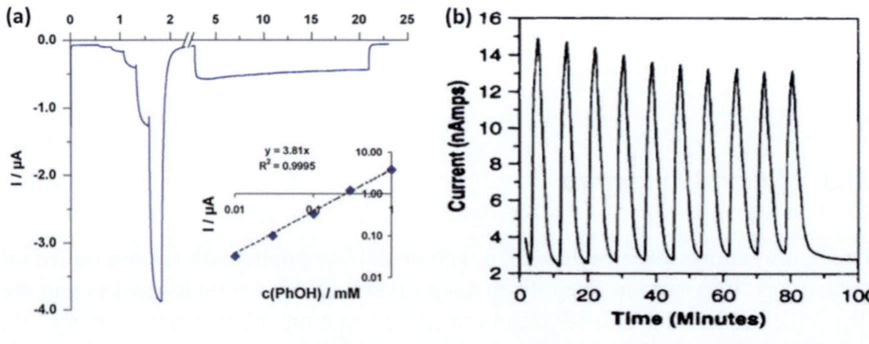

Fig. 9.5 (**a**) The phenol sensor exhibits a standard response to phenol vapor at concentrations of 0.01, 0.03, 0.1, 0.3, and 1 mM, with a prolonged exposure at 0.1 mM. The insets display the corresponding calibration curve for reference. (Hämmerle et al. 2010. © Elsevier). (**b**) Amperometric response of phenol biosensors to repeated exposure to 1.4 ppm phenol vapor, with each exposure lasting 100 s. (Dennison et al. 1995. © ACS Publications)

9.3.4 Gas Sensing Mechanism

The gas sensing mechanism of biosensors are quite similar to the sensing mechanism of metal oxide-based sensor. On the basis of adsorption of analytes, the gas sensing mechanism has been categorized in different ways.

9.3.4.1 DNA/Aptamer Hybridization, Recognition, and Cleavage

The DNA or aptamer hybridization, recognition, and cleavage gas sensing mechanism involves the use of DNA or aptamer sequences as the sensing element to detect the presence of specific gases as shown in Fig. 9.6 (Hwang et al. 2023). First, DNA sequences or aptamers are designed with specific base sequences that allow them to selectively bind to target gas molecules. The sequence is designed to have high affinity and specificity for the particular gas of interest. Furthermore, The DNA or aptamer sequences are immobilized onto a sensor surface, which can be made of materials such as gold, silicon, or graphene. This surface acts as a platform for gas detection. When the sensor is exposed to the target gas, the DNA or aptamer sequences selectively bind or hybridize with the gas molecules. This binding event induces a conformational change in the DNA or aptamer structure. The conformational change triggered by the binding event is detected by the sensor. This change is often transduced into a measurable signal, such as a change in electrical conductivity, fluorescence, or mass. The sensor's transducer converts the conformational

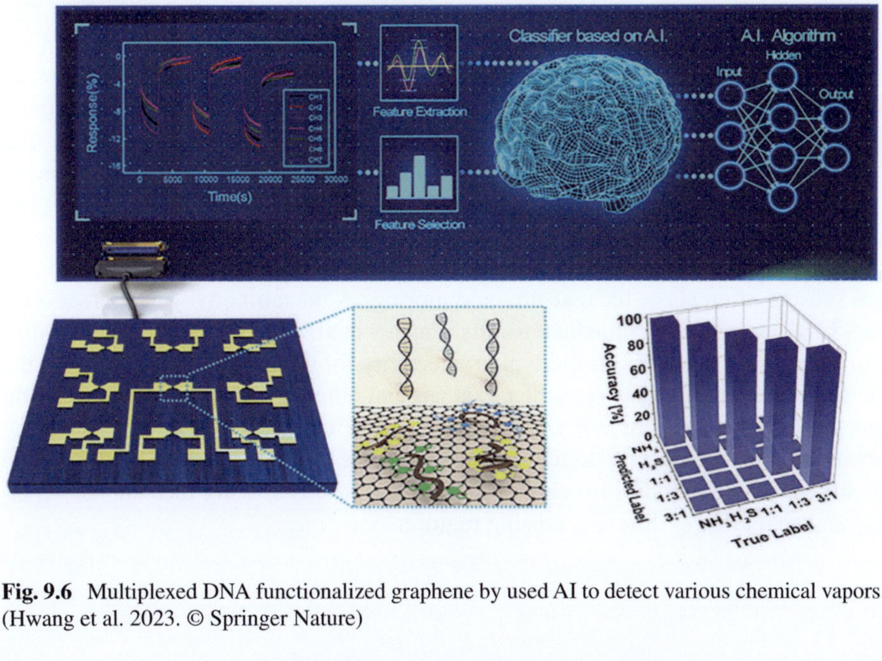

Fig. 9.6 Multiplexed DNA functionalized graphene by used AI to detect various chemical vapors. (Hwang et al. 2023. © Springer Nature)

change into an output signal. For example, if the sensor is based on electrical conductivity, the binding event may lead to a change in the electrical properties of the sensor surface.

In the case of DNA sensors, there might be a cleavage mechanism involved. The binding of the gas to the DNA sequence may induce a cleavage event, leading to the release of a fragment. This cleavage event can also be used as a signal for gas detection. One of the advantages of DNA or aptamer-based gas sensors is their potential for real-time monitoring. The rapidity of the hybridization event allows for immediate detection and response. The DNA or aptamer sequences are designed to be highly specific to the target gas, providing a high level of selectivity. This specificity allows for the discrimination of the target gas from other substances. This hybridization, recognition, and cleavage gas sensing mechanism offer a sensitive and specific approach for the detection of various gases. The ability to tailor DNA or aptamer sequences for different gases makes this sensing mechanism versatile and applicable in diverse fields, including environmental monitoring and industrial safety.

9.3.4.2 Immunobinding

The immunobinding gas sensing mechanism utilizes antibodies or immunoglobulins as selective sensing elements to detect the presence of specific gas molecules. These antibodies are engineered or chosen for their high affinity and specificity towards the target gas, enabling precise recognition and binding, which triggers a measurable sensor response. This process often involves immunizing animals with the target gas conjugated to a carrier protein to generate specific antibodies. The selected antibodies are immobilized onto a sensor surface. Common sensor surfaces include gold, silicon, or other materials that facilitate antibody attachment. The sensor is exposed to the environment containing the target gas. If the target gas is present, it will interact with the immobilized antibodies on the sensor surface. The specific antibodies selectively bind to the target gas molecules. This binding event is highly specific due to the unique affinity between the antibody and its target. The binding of antibodies to the target gas induces a conformational change in the antibody structure. This change can be detected as a signal by the sensor. Immunobinding gas sensors often allow for real-time monitoring, as the antibody–gas interaction is rapid and can provide immediate results (Pandey et al. 2011). The use of antibodies provides a high level of specificity and selectivity for the target gas. This allows the sensor to discriminate between the target gas and other substances. Immunobinding gas sensing mechanisms are widely employed in various applications, including environmental monitoring, industrial safety, and healthcare. The ability to generate highly specific antibodies for different gases makes this sensing mechanism versatile and adaptable to different sensing requirements.

9.3.4.3 Enzymatic Reactions of Proteases

The enzymatic reactions involving proteases in gas sensing mechanisms typically utilize protease activity as a sensing element to detect the presence of specific gases. Proteases, also known as peptidases or proteinases, are enzymes that cleave peptide bonds in proteins. A specific protease is selected or engineered based on its ability to interact with a particular substrate or target gas. The selected protease is immobilized onto a sensor surface. Common sensor surfaces include materials such as gold, silicon, or other suitable substrates that facilitate enzyme attachment. A substrate containing a peptide or protein sequence susceptible to protease cleavage is introduced to the sensor. This substrate is often labeled or modified in a way that the enzymatic cleavage event can be easily detected. The sensor is exposed to the environment containing the target gas (Zhang et al. 2023). If the target gas is present, it may interact with the immobilized protease or the substrate. The target gas may induce changes in the activity of the immobilized protease. This could include alterations in the enzyme's conformation, binding affinity, or catalytic activity, depending on the specific mechanism of interaction. If the target gas interacts with the protease or substrate, it may influence the enzymatic cleavage of the substrate. The protease cleaves the peptide or protein sequence in the substrate in the presence of the target gas. The cleavage of the substrate generates specific cleavage products or fragments. The detection of these products serves as a signal for the presence of the target gas. The sensor's transducer converts the enzymatic cleavage event into a measurable signal. This can include changes in electrical conductivity, fluorescence, or mass, depending on the sensor design. Enzymatic gas sensors, including those involving proteases, can offer real-time monitoring capabilities due to the rapid enzymatic reactions. Enzymatic reactions involving proteases in gas sensing mechanisms find applications in various fields, including environmental monitoring, food safety, and biomedical diagnostics. The adaptability of proteases to different substrates allows for the development of sensors tailored to specific gas targets.

9.3.4.4 Functional Group Transformation

Functional group transformation gas sensing mechanisms involve the use of specific functional groups that undergo chemical transformations in the presence of target gases. These transformations are then detected and quantified to provide information about the gas concentration. Functional groups susceptible to chemical reactions with the target gas are chosen. These functional groups may be part of a sensing material or surface. The chosen functional groups are incorporated or functionalized onto a sensor material. The sensing material may be a thin film, nanoparticles, or a surface coating. The sensor is exposed to the environment containing the target gas. If the target gas is present, it interacts with the functionalized sensing material. The target gas induces a chemical reaction with the functional group. This reaction results in a change in the chemical structure of the functional group. The chemical transformation of the functional group leads to a structural change in the sensing material. This change may include alterations in bond lengths, angles, or electronic configurations. The structural changes induced by the chemical reaction are detected. This detection can be accomplished through various means, such as changes in electrical conductivity, optical properties, mass, or other physicochemical parameters. The specificity of the gas sensor is determined by the selectivity of the functional group for the target gas. The choice of functional groups and their reactivity govern the sensor's selectivity. Functional group transformation gas sensing mechanisms are versatile and can be tailored for various gases based on the reactivity of chosen functional groups. These sensors find applications in environmental monitoring, industrial safety, and gas detection in various fields.

9.4 Conclusion and Future Prospective

In conclusion, biosensors have emerged as powerful tools for the detection of air pollutants and hazardous gases, offering unique advantages such as high specificity, sensitivity, and real-time monitoring capabilities. The integration of biological elements, such as enzymes, antibodies, aptamers, and living cells, into sensing platforms has paved the way for innovative and reliable solutions in environmental monitoring and industrial safety. The biosensor-based approach has shown great promise in addressing the challenges associated with traditional gas detection methods, providing faster response times, lower detection limits, and enhanced selectivity.

The comprehensive review of biosensors for detecting air pollutants and hazardous gases highlights the diversity of biological sensing elements and transduction mechanisms employed in sensor designs. Enzymatic reactions, immunobinding, DNA/aptamer hybridization, and other innovative approaches showcase the adaptability of biosensors to a wide range of gas targets. These sensors have demonstrated their effectiveness in identifying and quantifying pollutants such as volatile organic compounds (VOCs), NOx, SO_2, and various other toxic gases.

9.5 Future Prospects

- The future prospects of biosensors for detecting air pollutants and hazardous gases present exciting avenues for research and development. Several key areas warrant exploration and advancement.
- Developments in biosensors capable of detecting multiple gases simultaneously will be crucial for comprehensive air quality monitoring. Integration of diverse biological sensing elements within a single platform can enhance versatility.
- The miniaturization of biosensor devices and the creation of portable, handheld systems will facilitate on-site monitoring, enabling quick and efficient detection in various environments, including industrial settings, urban areas, and remote locations.

- Continued efforts to enhance the sensitivity and selectivity of biosensors will contribute to their applicability in detecting trace concentrations of pollutants. This involves refining biological elements and optimizing transduction methods.
- Ensuring the long-term stability and reliability of biosensors is crucial for their practical deployment. Research should focus on improving the durability of biological elements and the overall robustness of the sensing platforms.
- Integration of biosensors with the Internet of Things (IoT) and data analytics platforms can enable real-time data sharing, remote monitoring, and predictive modeling. This integration will enhance the overall efficiency of air quality management.
- Exploring biohybrid systems that combine biological sensing elements with advanced materials or electronic components can open new possibilities for gas sensing with improved performance characteristics.
- Addressing the detection challenges posed by emerging pollutants, such as nanoparticles and new chemical compounds, will be crucial. Biosensors need to adapt to evolving environmental threats.
- Efforts to make biosensor technology more affordable and accessible will democratize air quality monitoring, enabling a broader range of stakeholders, including individuals and communities, to actively participate in pollution control efforts.

In conclusion, the ongoing advancements and future prospects of biosensors for detecting air pollutants and hazardous gases hold tremendous potential for shaping the next generation of environmental monitoring solutions, contributing to cleaner air and healthier living environments. Continued interdisciplinary collaboration, technological innovation, and a commitment to sustainability will drive the evolution of biosensor applications in air quality management.

References

Achmann S, Hermann M, Hilbrig F, Jérôme V, Hämmerle M, Freitag R, Moos R (2008a) Direct detection of formaldehyde in air by a novel NAD+- and glutathione-independent formaldehyde dehydrogenase-based biosensor. Talanta 75:786–791

Achmann S, Hämmerle M, Kita J, Moos R (2008b) Miniaturized low temperature co-fired ceramics (LTCC) biosensor for amperometric gas sensing. Sensors Actuators B Chem 135:89–95

Achmann S, Hämmerle M, Moos R (2008c) Amperometric enzyme-based biosensor for direct detection of formaldehyde in the gas phase: dependence on electrolyte composition. Electroanalysis 20:410–417

Achmann S, Hämmerle M, Moos R (2008d) Amperometric enzyme-based gas sensor for formaldehyde: impact of possible interferences. Sensors 8:1351–1365

Anderson W (2001) Asthma admissions and thunderstorms: a study of pollen, fungal spores, rainfall, and ozone. QJM 94:429–433

Arakawa T, Wang X, Kajiro T, Miyajima K, Takeuchi S, Kudo H, Yano K, Mitsubayashi K (2013) A direct gaseous ethanol imaging system for analysis of alcohol metabolism from exhaled breath. Sensors Actuators B Chem 186:27–33

Birks JW, Turnipseed AA, Andersen PC, Williford CJ, Strunk S, Carpenter B, Ennis CA (2020) Portable calibrator for NO based on the photolysis of N2O and a combined NO2/NO/O3 source for field calibrations of air pollution monitors. Atmos Meas Tech 13:1001–1018

Chen M, Li H, Zhou D, Peng W, Lian G, Gao W, Gong X, Chang J (2019) Reusable bioluminescent sensor for ultrasensitive microRNA detection based on a target-introducing "fuel-loading" mechanism. ACS Appl Mater Interfaces 11:38586–38594

Cheng L-H, Kumar U, Deng Z-Y, Wu C-H (2023) Improved charge transfer for NO2 gas sensors by using 0D SnS quantum dot/2D WSe2 heterostructures. ACS Appl Nano Mater 6:9506–9514

Dennison MJ, Hall JM, Turner APF (1995) Gas-phase microbiosensor for monitoring phenol vapor at ppb levels. Anal Chem 67:3922–3927

Gaggiotti S, DellaPelle F, Mascini M, Cichelli A, Compagnone D (2020) Peptides, DNA and MIPs in gas sensing. From the realization of the sensors to sample analysis. Sensors 20:4433

Hämmerle M, Hall EA, Cade N, Hodgins D (1996) Electrochemical enzyme sensor for formaldehyde operating in the gas phase. Biosens Bioelectron 11:239–246

Hämmerle M, Achmann S, Moos R (2008) Gas diffusion electrodes for use in an amperometric enzyme biosensor. Electroanalysis 20:2279–2286

Hämmerle M, Hilgert K, Achmann S, Moos R (2010) Direct monitoring of organic vapours with amperometric enzyme gas sensors. Biosens Bioelectron 25:1521–1525

Huang L, Cai G, Zeng R, Yu Z, Tang D (2022) Contactless photoelectrochemical biosensor based on the ultraviolet–assisted gas sensing interface of three-dimensional SnS2 nanosheets: from mechanism reveal to practical application. Anal Chem 94:9487–9495

Hwang YJ, Yu H, Lee G, Shackery I, Seong J, Jung Y, Sung S-H, Choi J, Jun SC (2023) Multiplexed DNA-functionalized graphene sensor with artificial intelligence-based discrimination performance for analyzing chemical vapor compositions. Microsyst Nanoeng 9:28

Kumar U, Yadav BC (2019) Synthesis of carbon nanotubes by direct liquid injection chemical vapor deposition method and its relevance for developing an ultra-sensitive room temperature based CO2 sensor. J Taiwan Inst Chem Eng 96:652–663

Kumar R, Goel N, Kumar M (2017) UV-activated MoS2 based fast and reversible NO2 sensor at room temperature. ACS Sens 2:1744–1752

Kumar P, Chiu Y-H, Deng Z-I, Kumar U, Chen K-L, Huang W-M, Wu C-H (2021) Surface modification of ZnO nanopillars to enhance the sensitivity towards methane: the studies of experimental and first-principle simulation. Appl Surf Sci 568:150817

Kumar U, Hsieh H-W, Liu Y-C, Deng Z-Y, Chen K-L, Huang W-M, Wu C-H (2022) Revealing a highly sensitive sub-ppb-level NO2 gas-sensing capability of novel architecture 2D/0D MoS2/SnS heterostructures with DFT interpretation. ACS Appl Mater Interfaces 14:32279–32288

Kumar U, Wu C-H, Singh K, Yadav BC, Huang W-M (2023a) Low-dimensional advanced functional materials as hazardous gas sensing. In: Advanced functional materials for optical and hazardous sensing, pp 31–45. https://doi.org/10.1007/978-981-99-6014-9_2

Kumar U, Liu Y-C, Hsieh H-W, Deng Z-Y, Huang W-M, Wu C-H (2023b) Revealing ppb order NO2 adsorption capability of 2D/0D p-p heterojunctions based on Sb/SnS nanocomposite. J Alloys Compd 969:172451

Kumar P, Peng H-Y, Yang Y-H, Kumar U, Deng Z-Y, Chen K-L, Lee M-W, Wu C-H (2023c) Highly sensitive CO gas sensor based on ternary metal sulfides PbSbS quantum dots: experimental and DFT study. J Alloys Compd 967:171688

Kumar U, Yang C-X, Deng Z-Y, Lin C-E, Yadav BC, Wu C-H (2024) Atomic-level insights of enhanced interfacial charge transfer and active sites on self-assembly p-p heterostructure based on BiVO4/SnO2 for highly selective NO2 sensing. Sensors Actuators B Chem 403:135144

Kuretake T, Kawahara S, Motooka M, Uno S (2017) An electrochemical gas biosensor based on enzymes immobilized on chromatography paper for ethanol vapor detection. Sensors 17:281

Ma M, Yang X, Ying X, Shi C, Jia Z, Jia B (2023) Applications of gas sensing in food quality detection: a review. Food Secur 12:3966

Mishra RK, Choi GJ, Mishra YK, Kaushik A, Sohn Y, Lee SH, Gwag JS (2021) A highly stable, selective, and high-performance VOC sensor using a SnS2 nano-lotus structure. J Mater Chem C 9:7713–7725

Mitsubayashi K, Hashimoto Y (2000) Development of a gas-phase biosensor for trimethylamine using a flavin-containing monooxygenase 3. Electrochemistry 68:901–903

Mitsubayashi K, Hashimoto Y (2002) Bioelectronic nose for methyl mercaptan vapor using xenobiotic metabolizing enzyme: flavin-containing monooxygenase. Sensors Actuators B Chem 83:35–40

Mitsubayashi K, Yokoyama K, Takeuchi T, Karube I (1994) Gas-phase biosensor for ethanol. Anal Chem 66:3297–3302

Mitsubayashi K, Matsunaga H, Nishio G, Toda S, Nakanishi Y (2005) Bioelectronic sniffers for ethanol and acetaldehyde in breath air after drinking. Biosens Bioelectron 20:1573–1579

Mitsubayashi K, Nakayama K, Taniguchi M, Saito H, Otsuka K, Kudo H (2006) Bioelectronic sniffer for nicotine using enzyme inhibition. Anal Chim Acta 573–574:69–74

Mitsubayashi K, Nishio G, Sawai M, Kazawa E, Yoshida H, Saito T, Kudo H, Otsuka K, Takao M, Saito H (2008) A biochemical sniffer-chip for convenient analysis of gaseous formaldehyde from timber materials. Microchim Acta 160:427–433

Mitsubayashi K, Toma K, Iitani K, Arakawa T (2022) Gas-phase biosensors: a review. Sensors Actuators B Chem 367:132053

Motooka M, Uno S (2018) Improvement in limit of detection of enzymatic biogas sensor utilizing chromatography paper for breath analysis. Sensors 18:440

Onyancha RB, Ukhurebor KE, Aigbe UO, Osibote OA, Kusuma HS, Darmokoesoemo H, Balogun VA (2021) A systematic review on the detection and monitoring of toxic gases using carbon nanotube-based biosensors. Sens Bio-Sens Res 34:100463

Otsuka K, Goto T, Amagai H, Ishii N, Endo H, Mitsubayashi K (2006) Biochemical sniffer for odourless hydrogen peroxide vapour. Int J Environ Anal Chem 86:1049–1056

Ouandaogo HS, Diallo S, Odari E, Kinyua J (2023) Phytochemical screening and GC-MS analysis of methanolic and aqueous extracts of Ocimum kilimandscharicum leaves. ACS Omega 8:47560–47572

Pandey P, Srivastava JK, Mishra VN, Dwivedi R (2011) Pd-gate MOS sensor for detection of methanol and propanol. J Nat Gas Chem 20:123–127

Park J (1995) Biosensing with surface plasmon resonance—how it all started. Biosens Bioelectron 10:587–594

Poma N, Bonini A, Vivaldi F, Biagini D, DiLuca M, Bottai D, DiFrancesco F, Tavanti A (2023) Biosensing systems for the detection and quantification of methane gas. Appl Microbiol Biotechnol 107:5627–5634

Reddeppa M, Mitta SB, Park B-G, Kim S-G, Park SH, Kim M-D (2019) DNA-CTMA functionalized GaN surfaces for NO2 gas sensor at room temperature under UV illumination. Org Electron 65:334–340

Roy N, Mitra S, Das NM, Mandal N, Bandyopadhyay D, Nemade HB, Mandal TK (2020) Paper based enzymatic chemiresistor for POC detection of ethanol in human breath. IEEE Sensors J 20:2278–2286

Sigawi S, Smutok O, Demkiv O, Zakalska O, Gayda G, Nitzan Y, Nisnevitch M, Gonchar M (2011) Immobilized formaldehyde-metabolizing enzymes from Hansenula polymorpha for removal and control of airborne formaldehyde. J Biotechnol 153:138–144

Sousa EHS, Tian S (2023) Editorial: metalloproteins as sensors of gaseous small molecules—from bench to bed and beyond. Front Mol Biosci 10:1140392

Tanimoto de Albuquerque YD, Ferreira LF (2007) Amperometric biosensing of carbamate and organophosphate pesticides utilizing screen-printed tyrosinase-modified electrodes. Anal Chim Acta 596:210–221

Wasilewski T, Neubauer D, Kamysz W, Gębicki J (2022) Recent progress in the development of peptide-based gas biosensors for environmental monitoring. Case Stud Chem Environ Eng 5:100197

Yao H, Xiao J, Tang X (2023) Microbial fuel cell-based organic matter sensors: principles, structures and applications. Bioengineering 10:886

Zhang C, Luo Y, Xu J, Debliquy M (2019) Room temperature conductive type metal oxide semiconductor gas sensors for NO2 detection. Sensors Actuators A Phys 289:118–133

Zhang W, Chen X, Xing Y, Chen J, Guo L, Huang Q, Li H, Liu H (2023) Design and construction of enzyme-based electrochemical gas sensors. Molecules 29:5

Zhu Z, Chung C-L, Kumar U, Chung M-H, Wu C-H, Wu R-J (2022) Facial construction of Pd/BiPO4 toward efficient photocatalytic oxidation of formaldehyde under UV light irradiation. J Photochem Photobiol A Chem 427:113840

Chapter 10
Nanoscale Concepts for Air Pollution Sensors

Marcio Peron Franco de Godoy and Yina Julieth Onofre

Abstract In last two centuries, air quality got worse due to several factors related to the modern human development. The presence of pollutants in atmosphere makes people passive agents subject to many types of diseases. To monitor prejudicial gases, the development of fast, sensitive, and selective sensors is a rewarding field that matches with the acquired knowledge of semiconducting surfaces, in particular, of semiconducting metal oxides. This chapter discusses the principles and structures of advanced materials employed in the development of gas sensors, including electronic and the nanoscale approaches, and the strategies to improve them.

Keywords Air quality · Nanoscale approaches · Semiconductors · Volatile organic compounds

10.1 Introduction

Air quality refers to the cleanliness of the air, specifically, to an atmosphere that is either free from or contains minimal concentrations of harmful substances (pollutants) that can adversely affect human health, living organisms, and the environment. Air pollutants are primarily emitted through combustion processes, including the burning of fossil fuels and biomass for energy production. Additionally, industrial processes and transportation methods—land, air, and sea—along with the burning of waste and both controlled and uncontrolled forest and savannah fires, contribute to air pollution (World Health Organization 2021). Two types of pollutants are found in the atmosphere: primary pollutants that result from direct emissions of their sources and secondary pollutants that are formed by physicochemical reactions with other pollutants already present in the atmosphere. A high concentration

M. P. F. de Godoy (✉) · Y. J. Onofre (✉)
Universidade Federal de São Carlos – UFSCar, São Carlos, Brazil
e-mail: mgodoy@ufscar.br; yina.onofre@ufscar.br

N. Ghosh et al. (eds.), *Nanotechnology in Air Quality Management*, Sustainable Development and Biodiversity 39, https://doi.org/10.1007/978-981-96-8339-0_10

(even low concentrations) of one or more of these pollutants in the atmosphere indicates poor air quality and becomes a health risk. According to the World Health Organization (WHO), premature deaths and a wide range of diseases affecting several organs are caused by short- and long-term exposure to pollutants (World Health Organization 2021). Among the main pollutants, volatile organic compounds have received extensive attention due to the vast number of compounds present in the air and their high chemical reactivity, which enables them to act as precursors to other pollutants such as ground-level ozone and particulate matter (PM) (Koppmann 2007). This chapter is devoted to show how nanotechnology can be applied to detect harmful molecules present in atmosphere in an almost real time, including some strategies employed to improve the response of chemical-based gas sensors.

10.2 Volatile Organic Compounds (VOCs)

Volatile organic compounds are present in both indoor and outdoor environments. According to the United States Environmental Protection Agency (US-EPA), VOCs can be divided into two groups based on their emphasis on these environments, their negative effects on human health, and the formation of photochemical smog, respectively. In general, VOCs are defined as carbon-based chemical compounds that include alkanes, alkynes, aromatic hydrocarbons, halogenated hydrocarbons, and aldehydes. This definition excludes carbon monoxide, carbon dioxide, carbonic acid, metallic carbides or carbonates, and ammonium carbonate. Thus, outdoor VOCs are organic compounds with significant photochemical reactivity, primarily impacting the environment, though they can also pose risks to human health (US Environmental Protection Agency 2024a). The most common VOCs in outdoor air include benzene, toluene, ethylbenzene, and xylene, known as BTEX, as well as 1,2,4-trimethylbenzene and isoprene, which can contribute to the formation of ground-level ozone (Huang et al. 2011; Kumar et al. 2018; Garg and Gupta 2019; Pinthong et al. 2022). Indoor VOCs are organic compounds that have a relatively high vapor pressure and can be vaporized readily into the air under normal indoor atmospheric conditions of temperature and pressure (depending on the type of building and its geographic location) (US Environmental Protection Agency 2024a). Indoor VOCs are classified based on their volatility, which is directly correlated to the boiling point temperature (World Health Organization. Regional Office for Europe 1989):

- *Very volatile organic compounds (VVOCs)*: Gaseous compounds with a lower range of boiling point (<0 °C to 50–100 °C), such as propane, butane, and methyl chloride.
- *Volatile organic compounds (VOCs)*: Group of compounds with a boiling point range of 50–100 °C to 240–260 °C that include formaldehyde, vinyl chloride, ethanol, carbon tetrachloride, toluene, acetone, isopropyl alcohol, and hexanal.

- *Semi-volatile organic compounds (SVOCs)*: Compounds with a higher range of boiling point (240–260 °C to 380–400 °C) such as pesticides (chlordane dichlorodiphenyltrichloroethane (DDT)), plasticizers (phthalates), and fire retardants (polychlorinated biphenyls (PCBs) and polybrominated biphenyls (PBBs)).

The emissions of VOCs into the atmosphere originate from both natural processes (biogenic sources) and human activities (anthropogenic sources). Biogenic sources of VOCs include vegetation, oceans, and soils. In contrast, anthropogenic sources comprise the combustion of biofuels, the use and production of fossil fuels, industrial processes, agricultural activities, biomass burning, and waste treatment (Koppmann 2007; Kumar et al. 2018; Pinthong et al. 2022). Furthermore, in indoor environments, household products such as aerosol sprays, cleansers, disinfectants, air fresheners, pesticides, stored fuels, automotive products, wood preservatives, paints, paint strippers, and solvents, as well as building materials and furnishings, graphics, and craft materials, and office equipment, contribute to VOC emissions (Zhou et al. 2023; US Environmental Protection Agency 2024b). The impact of VOCs on human health and the environment, generally, depends on factors such as the level of toxicity, chemical reactivity, atmospheric lifetime, concentration, and exposure time. Exposure to VOCs can cause a wide range of symptoms, from mild—such as eye, nose, and throat irritation, headaches, skin irritation, nausea, fatigue, and dizziness, to coughing and rapid breathing—to severe effects, including pneumonia, bronchitis, liver and kidney damage, central nervous system issues, cardiovascular and respiratory diseases, reproductive problems, and cancer (World Health Organization 2021; Zhou et al. 2023). Table 10.1 shows the limit and exposition time for some pollutants considering human health risks. Environmentally, some VOCs promote the formation of organic aerosols and acid deposition (Zhou

Table 10.1 The recommended limits considering human health risks for representative air pollutants

Pollutant	Averaging exposition time	Level (ppm)	References
Nitrogen dioxide (NO_2)	1 h	0.1	US Environmental Protection Agency (2024c)
Ozone (O_3)	8 h	0.07	US Environmental Protection Agency (2024c)
Carbon monoxide (CO)	1 h	35	US Environmental Protection Agency (2024c)
Sulfur dioxide (SO_2)	1 h	0.075	US Environmental Protection Agency (2024c)
Ammonia (NH_3)	30 min	300	AIHA (1971)
Formaldehyde (CH_2O)	30 min	0.08	World Health Organization (2010)
Ethanol (C_2H_5OH)	10 h	1000	NIOSH (1992)
Acetone (C_3H_6O)	10 h	250	NIOSH (1992)
Benzene (C_6H_6)	30 min	7500	NIOSH (1992)
Toluene (C_7H_8)	8 h	200	NIOSH (1992)

et al. 2023). Also, VOCs can aggravate global warming by acting as greenhouse gases that absorb infrared radiation from the Earth (Murrells T, Derwent RG 2007). Moreover, VOCs participate in photochemical reactions with nitrogen oxides (NOx) in the atmosphere, leading to the formation of ozone. This ground-level ozone contributes to global warming, damages vegetation, and reduces agricultural yields (Murrells T, Derwent RG 2007; Koppmann 2007; Zhou et al. 2023).

The detection of VOCs can be realized through a device, named sensor, with the ability to recognize the presence of these gases in the atmosphere and provide a signal that can be quantified and interpreted as a response. An intuitive idea of a sensor is the emulation of human body senses (taste, smell, vision, hearing, and touch) that allow us to perceive and identify specific characteristics of our surroundings.

The gas detection process can be divided into two parts: receptor and transducer functions, as illustrated in Fig. 10.1. Receptor function is associated with the recognition of gases in the air (analytes) by the sensor, which occurs due to intermolecular interactions of the analytes with the active sensing element. Materials such as oxide metals, polymers, one-dimensional (1D) and two-dimensional (2D) structures are considered as sensing elements (Khatib and Haick 2022). The recognition of analytes on the sensing element can be potentiated by modifying the sensing element through the reduction of particle size (nanometers) (Yamazoe 1991), the combination of materials with different properties (hybridization) (Andre et al. 2018), the introduction of other atoms (doping) (Postica et al. 2019), and covering the surface with atoms or molecules with a specific property (functionalization) (Postica et al. 2018). Furthermore, strong intermolecular interactions of analyte-sensing

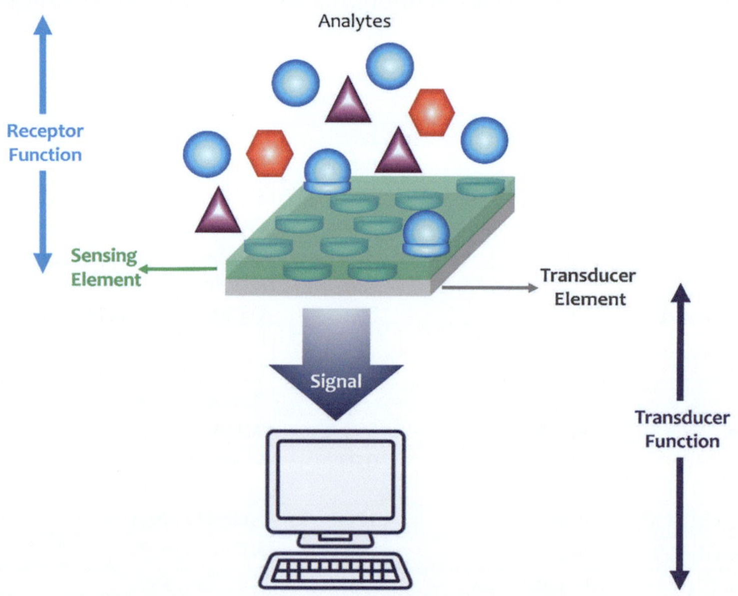

Fig. 10.1 Scheme of gas detection process: receptor and transducer functions

elements can enhance the receptor function, facilitating the selective discrimination of the target analyte (Li et al. 2019). The transducer function is responsible for converting the energy from chemical reactions into an analytical signal that can be measured (Bochenkov and Sergeev 2010). Electrical, optical, and gravimetric transduction mechanisms facilitate the VOCs detection. Electrical transducers measure variations in current/resistance and are widely used due to their direct signal conversion and compatibility with existing devices. Whereas the optical transducers can measure a broad range of variables, including absorbance, fluorescence, refractive index, scattering, and luminescence (Bochenkov and Sergeev 2010; Khatib and Haick 2022). Finally, gravimetric transducers quantify the mass variation of an analyte through changes in the resonant frequency of a quartz crystal (piezoelectric effect) and surface acoustic waves (SAW) (Cali et al. 2020).

One advantage of electrical response detection is the quite simple setup needed to measure the drop or gain in electrical voltage or current as well the possibility of integration to the conventional microelectronics. In this sense, the evaluation of electrical resistance changes is performed by a measurement between two electrodes or interdigitated electrodes (Fig. 10.2a). The measurement of electrical signals can be easily integrated into the available electronics, allowing an effective way to amplify weak signal responses given by low levels of analyte content. The basis of a chemiresistive sensor involves the characterization of surface phenomena that includes the role of adsorption/desorption of analyte and the formation of an electronic layer at a solid surface.

The basic parameters, also illustrated in Fig. 10.2b for the chemiresistive gas sensor, can be resumed as:

- *Sensitivity or response* (*S*): The changes in electrical resistance due to the interaction of target analyte at a defined concentration. Some authors define it in terms of relative change and others in terms of resistance ratio;

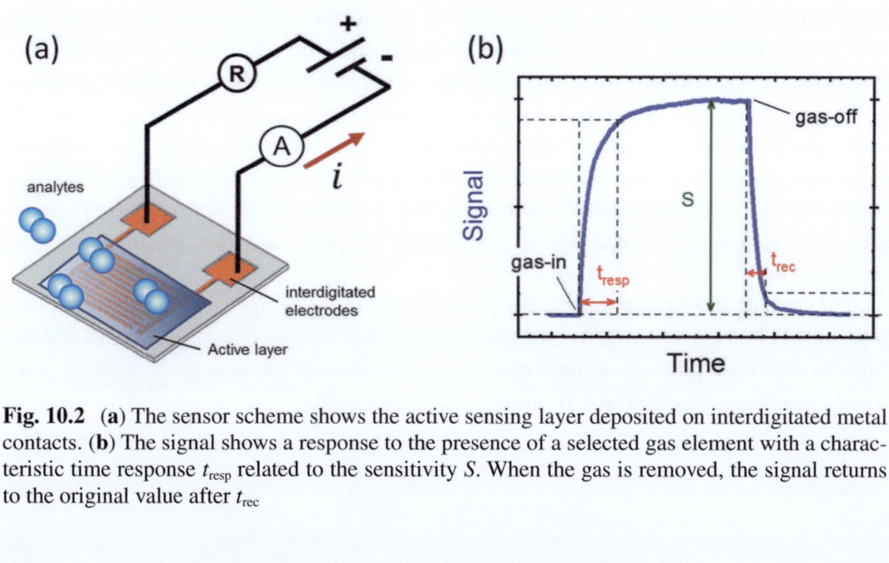

Fig. 10.2 (**a**) The sensor scheme shows the active sensing layer deposited on interdigitated metal contacts. (**b**) The signal shows a response to the presence of a selected gas element with a characteristic time response t_{resp} related to the sensitivity *S*. When the gas is removed, the signal returns to the original value after t_{rec}

- *Response time* (t_{resp}): At a specific concentration, the period needed to achieve the 90% maximum sensitivity or response S from the absence of analyte;
- *Recovery time* (t_{rec}): With the removal of the analyte, it is the period needed to recover the initial state (before exposition). In quantitative terms, it is equivalent to recovery to the 10% of maximum sensitivity or response S;
- *Selectivity*: The property to be sensitive for a restrict number of analytes.

10.3 Adsorption/Desorption Mechanisms

The adsorption of a specific analyte at a surface in a real environment depends on several factors that also compete with them. While the material volume, or bulk, consists of well-established and almost stoichiometric chemical bonds between former atoms, the surface is characterized by non-completed chemical bonds or dangling bonds. This makes the surface a rich place to provide all kinds of chemical interactions, such as Van der Walls, hydrogen, covalent, or ionic bonds.

The Van der Waals bonds refer to the weakest type of bonds between covalent molecules, described by any dipole–dipole interactions, and rapidly vanish when the distance between the interacting molecules increases. Additionally, although also being weak, hydrogen bonds are favored due to the presence of hydroxyls in the air environment. Both kinds are considered a physisorption mechanism, with energies between 10 and 100 meV, and without charge transfer. At higher temperatures, strong chemical bonds with associated energies higher than 1 eV cause significant charge transference. The covalent bonds are determined by the ability of atoms to share their electrons concerning their electronegativity, whereas ionic bonds are the strongest bond between oppositely charged ions through electrostatic attraction.

The air compounds can interact with the surface, i.e., the last layer of arranged atoms in a solid system, and the chemical bond depends strongly on the affinity between the analyte and the specific site on this surface. This is the most important specificity for sensing devices because it guarantees the selectivity of a target analyte. Usually, adsorption mechanisms can be favorable even in room temperature conditions; however, sensing is a dynamic process that decreases for small analyte concentrations. As we explain below, the aim is to measure this dynamical chemical bond in terms of an electrical signal that also decreases in a reasonable scale of time when the analyte concentration decreases. This is achieved by breaking the bond by an additional energy source, usually the device temperature, allowing the desorbing mechanism.

This makes the first compromise concerning device temperature T: the temperature should be high enough to guarantee an efficient desorption process, but not too high to avoid the adsorption mechanism responsible for the change of surface electrical resistance as a function of analyte content in air.

10.4 Active Layer Response: The Role of Depletion/Accumulation Layer

Concerning electrical properties, semiconductor materials present great versatility due to their electronic structure characteristics. In an ideal crystalline arrangement, semiconductors behave as an insulator at room temperature with empty electronic conduction bands and fully electronic valence band separated by an energy gap E_g in order of electron-volts (eV). Depending on defects or impurities presence, additional localized electronic levels are created in the bandgap region and, when the separation between these levels and the electronic bands are in order of tens of meV, the conduction or valence bands are occupied by electrons or holes respectively at room temperature (the thermal energy at 300 K is approximately 26 meV). In the case of donor levels, promoting electrons to the conduction band, the semiconductor is defined as n-type, while in the case of acceptor level, the semiconductor is defined as p-type (Sze and Ng 2006; Yu and Cardona 2010). This behavior can be managed with two strategies: during materials preparation by the doping process, in the case of impurities, or by defect engineering (Hanada 2009; Yu and Cardona 2010; Zhang et al. 2019; Silva ALC et al. 2022), controlling the vacancies or interstitial sites in the crystalline lattice.

Along the bulk material, these electronic bands are flat but at the interface between another material, atmosphere or metal, they are bending at the surface removing or accumulating the electric carriers in some extension W, denoted as the depletion layer. The concept of the depletion layer is widely employed in p–n junctions to fabricate optoelectronic devices such as rectifying diodes and light emitting diodes (LEDs), favoring the electron–hole recombination at the materials interface, as well as in solar cells, separating photocreated electron and hole pairs to be collected in the electrodes. Another interesting application, more correlated to our context, is the use of electronic bending in field effect transistors (FET), such as the most used MOSFET devices, where the application of an external electrical potential V_G to a metal–insulator–semiconductor (MIS) structure allows, for instance, the opening of an electronic channel in a p-type semiconductor.

The extension of the depletion layer W can achieve easily tens or hundreds of nanometers and depends on the ionized doped/defects levels N^* and the dielectric constant k. For a comparison, in a metal–semiconductor interface, the depletion layer in an n-type semiconductor can be estimated by the relation:

$$W = \sqrt{\frac{2k\varepsilon_0}{qN^*}\left(\psi_{bi} - V - \frac{k_B T}{q}\right)} \tag{10.1}$$

where ε_0 is the vacuum electrical permittivity, ψ_{bi} is the built-in potential (responsible for the discontinuity in the band diagram), q is the elementary charge, k_B is the Boltzmann constant, and T is the absolute temperature. The built-in potential ψ_{bi} is generated by the spaced-charged density of ionized doped/defects levels that changes abruptly at the interface and V is an external applied voltage bias. This expression shows how the extent of this layer is managed by changing the magnitude of applied potential V. And this is the mechanism that governs the gas detection in semiconductor gas sensors. In that case, instead of an applied voltage, the agent is the concentration of analyte specimens adsorbed on the surface and their oxidant or reductive nature.

Let us first consider the effect on n-type semiconductors for simplicity. The nature of electrical layer governs if the band bending is upward or downward, as displayed in Fig. 10.3, where the concentration of electrical charge on the surface is Q_S. For n-type semiconductors, the majority of free carriers are electrons (negative charge). In some n-type materials, the adsorption of oxygen species in the atmosphere presents a reactive behavior as compared to vacuum conditions or other oxygen-free environments (Scolfaro et al. 2018) (Fig. 10.3a). If an oxidant molecule bends to the surface (oxygen, ozone, or nitrous oxide, for example), it will capture the electrons from the material, increasing the localized negative charges at the surface. It causes, consequently, a depletion mechanism of free carriers in the semiconductor electronic band. These charged adsorbed molecules act as an external potential V. As a result, the free carrier concentration on the surface decreases, increasing the electrical resistance, as illustrated in Fig. 10.3b. However, if the analyte presents a reducing character, such as carbon monoxide that donates electrons, the superficial layer will accumulate electrons, decreasing the electrical resistance (Fig. 10.3c).

In analogy, when the active layer is a p-type semiconductor, as the majority of carriers are holes, the band bending occurs in the opposite way, resulting in the decrease of electrical resistance for oxidant analytes (Fig. 10.4a). Nevertheless, when the analyte presents a reducing character, the electrical resistance rises (Fig. 10.4b). However, in any case, these described effects due to the chemical reactions occur nearby the surface, letting the parallel conduction of bulk absent of response. The competition between the conductance of these two channels limits the sensor response S. Therefore, for convenience, S is usually employed as positive number and these definitions are adapted concerning the p- or n-type nature of sensing surface as well as the oxidizing or reducing nature of analyte, comparing the resistance (or conductivity) in the presence and in absence of target analyte. Thus, we call here S_{rel} as the relative change of resistance and S_{rat} as the ratio between the resistances. For example, in the case of an n-type surface sensing and oxidizing, $S_{rel} = \frac{R_s - R_0}{R_0}$ and $S_{rat} = \frac{R_s}{R_0}$, where R_s is the measurement in the presence of theanalyte and R_0 in its absence.

Among many materials systems, semiconductor oxides have shown reliable characteristics for gas sensors to detect pollutant gases such as NO_2, CO_2, HCl, NH_3, $N(CH_3)_3$ and VOCs as toluene (C_7H_8), ethanol (C_2H_5OH), formaldehyde (CH_2O), acetone (C_3H_6O), methanol (CH_3OH), isoprene (C_5H_8), liquid petroleum gas (LPG) (John and Ruban Kumar 2021) considering their oxidizing and reducing features. In addition, oxides present some advantages associated with synthesis processes as

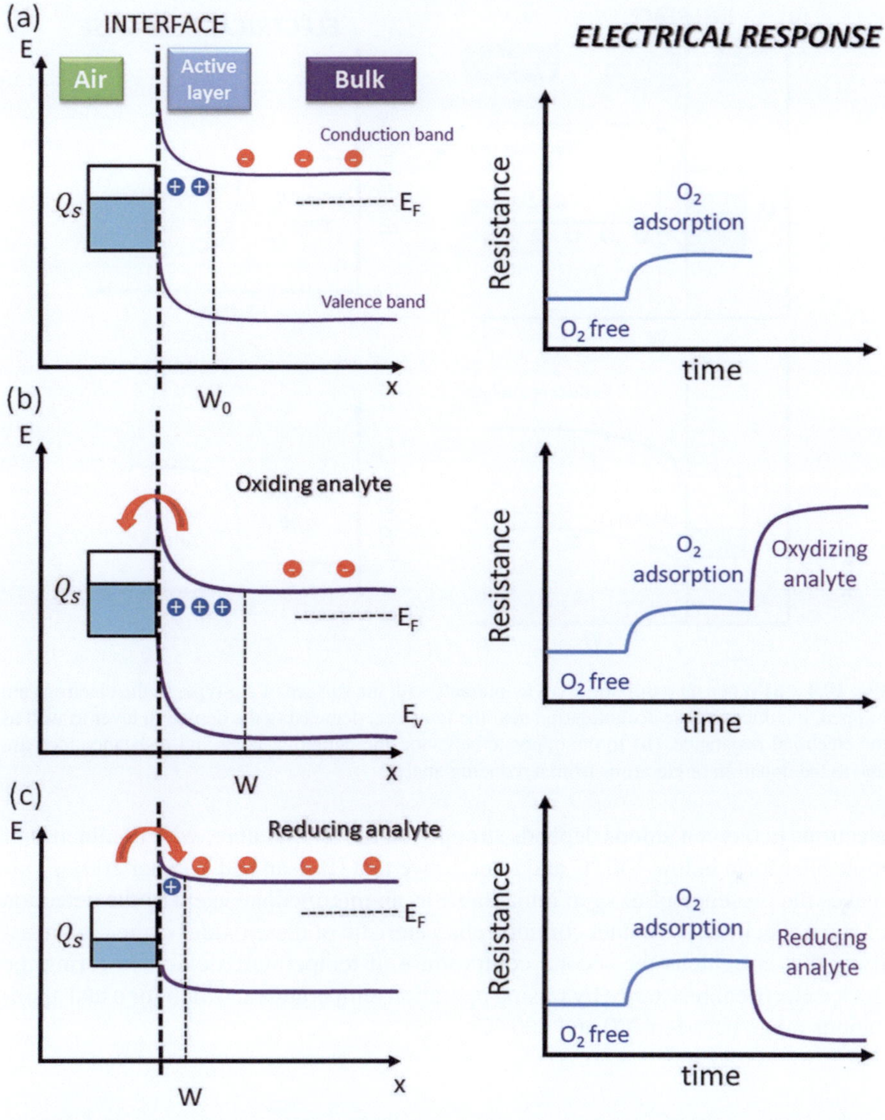

Fig. 10.3 (**a**) The effect of oxygen adsorption in a conventional atmosphere induces a depletion layer and increases the superficial resistance. (**b**) If an oxidizing analyte interacts with the surface, electrons are trapped, and the depletion layer increases. (**c**) In the presence of a reducing analyte, which donates electrons, the depletion layer decreases as well as the electrical resistance

simplicity and cost-effectiveness. Some widely employed oxides are ZnO (Wang 2020; Franco et al. 2022), SnO_2 (Li et al. 2022), WO_3 (Li et al. 2024), In_2O_3 (Shah et al. 2024), and TiO_2 (Haidry et al. 2024). It is interesting that, in the case of semiconductor oxides, the oxygen present in the atmosphere also plays an important role, which occurs in parallel with other interactions. For instance, the number of

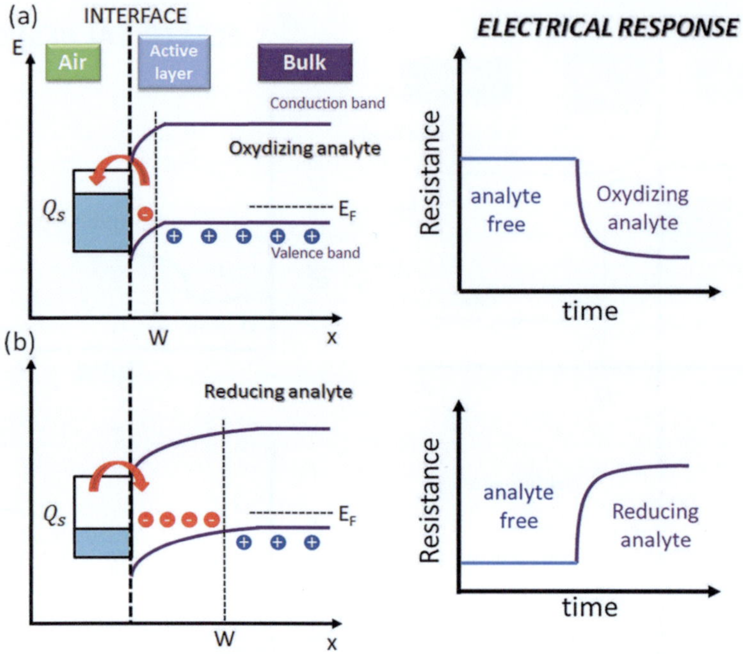

Fig. 10.4 (**a**) When an oxidizing analyte interacts with the surface of a p-type, as the electrons are trapped, it induces a hole accumulation near the interface, decreasing the depletion layer as well as the electrical resistance. (**b**) In the opposite behavior, the depletion layer and resistance increase due to the donation of electrons from a reducing analyte

electrons in oxygen anions depends strongly on the temperature, with dominance of molecular form below 150 °C and ionic above that (Barsan and Weimar 2001). This makes the presence of oxygen remarkable as an intermediate agent in the detection of some specimens. Another common characteristic of these oxides is the high resistivity that enlightens the second compromise of temperature device: lowering the device electrical resistance by raising operation temperatures, which for usual applications range between 200 and 400 °C.

10.5 Improving Detection at Nanoscale

The increased of available surface area for analyte adsorption makes nanoestructured materials strong candidates for high-sensitivity sensors, due to their high surface-area ratio. This characteristic is observed in porous layers, formed by interconnected polycrystalline grains with sizes larger than the depletion layer width as well as in nanoflowers (Wang 2020) and nanowires (Cui et al. 2001; de Araújo et al. 2022) nanoestructures, which can be employed in *ensemble* or individually. For many purposes, extending the discussion here, the conduction model developed by

Barsan and Weimar (Barsan and Weimar 2001) provided a suitable comprehension of sensing mechanisms for users, developers, and basic researcher scientists, treating the principle of electronic band bending for grains. Synthesis engineering can be used to tailor the types and dimensions of nanostructures. The versatility of chemical processes can be explored as in the case of In_2O_3 grain sizes, which show an improvement of twofold to fourfold for butane (C_4H_{10}) and ethanol (C_2H_5OH) sensing as particle sizes decreases from hundred to tens of nanometer (John and Ruban Kumar 2021). Additional strategies to enhance sensor response include the formation of 3D nanostructures on the surface of microstructures, the so-called hierarchical structures, such as on ZnO nanodisks and nanowires synthesized via hydrothermal method for acetone detection, which also exhibit response and recovery times as short as 2 s (Alenezi et al. 2013).

An additional functionalization mechanism is the doping with strategic elements, such as noble or transition metals onto its surface or into its crystal structure (host), which leads to changes in the surface chemistry and electronic properties of the layer. Noble metals, by their catalytic properties, act as sites of selective reaction on the surface, allowing efficient adsorption and molecule activation of specific gases that improve the interaction between the target gas and the sensing material. Furthermore, the metal/oxide interface creates a Schottky barrier that controls the charge transfer between the gas and the layer, favoring the adsorption of specific gases over interfering ones (Barsan et al. 2019). Conversely, the addition of transition metals into the host (doping) also introduces energy levels in the bandgap, influencing the free carrier concentration, as well as the availability and nature of active sites on its surface, which can promote or inhibit the adsorption of certain gas molecules (Barsan 2019).

The selectivity of the sensing layer toward a target gas is also driven by the adequate choice of noble or transition metal and its concentration. Noble metals such as Ag, Au, and Pt on the surface of ZnO improve the selectivity of pristine ZnO toward NO_2, acetone, and triethylamine (TEA), respectively (Yin et al. 2017; Postica et al. 2019; Huang et al. 2020; Liu et al. 2021). Common semiconducting metal oxides, ZnO, SnO_2, and WO_3, when doped with optimal concentrations of transition metals or other elements, usually low concentration up to 7 wt% of impurity, become highly selective toward acetaldehyde (C_2H_4O) for ZnO with 5% of Co content (Thomas et al. 2024), ethanol for SnO_2 with 7% of Tb content (Wang et al. 2021), xylene (C_8H_{10}) for WO_3 with 3% of Cr content (Parne et al. 2022), and acetone for ZnO with 3% of Al content (Benamara et al. 2023).

Other interesting examples of this functionalization include the case of Pt-doping in ZnO hierarchical microspheres achieving responses up to $S_{rat} = 242$ for triethylamine sensing in an operation temperature around 200 °C (Liu et al. 2021), and the Pd-decorated nanoflowers, allowing the detection of methylamine even at room temperature (Bruce et al. 2022). Also, the Co-doping in nanostructured ZnO films grown by spray-pyrolysis for ozone (O_3) detection increases the responses one order of magnitude ($S_{rel} = 6.1$) even for low O_3 concentration level as 70 ppb (Onofre et al. 2019).

The strategy to combine materials in the nanometric scale are also explored adding other interfaces in the form of heterostructures, enabling the functionalization by the control of charge transport and adding specific characteristics that enhance the detection process. In the presence of two or more interfaces, the interactions with the sensing surface are modified, which can either inhibit or promote the adsorption of certain gas molecules, thus contributing to the optimization of the selectivity. When an n-type and a p-type oxides are combined in a p–n junction, each oxide has a characteristic surface chemistry and an opposite electrical response (increase or decrease in resistance) when interacting with the target gas, in addition to charge transfer between them due to the Fermi level equilibrium at the interface. This concept of opposite responses allows the modulation of the conduction channel width by adjusting the ratio of n-type and p-type oxides and favors the overall sensor response to a specific gas through preferential gas adsorption, compensating the response to one gas relative to another (Yin et al. 2020). This approach has been demonstrated to help counteract the adsorption of interfering gases and the effect of humidity (Walker et al. 2022). Additional combinations also demonstrated advantages for selectivity improvement, as in the case of SnO_2/ZnO, which is effective for acetone detection (Du et al. 2018), NiO/SnO_2 for ethanol (Wang et al. 2016), SnO_2/CuO for toluene (John and Ruban Kumar 2021), and Ag-modified In_2O_3/ZnO for formaldehyde (Fang et al. 2015).

Heterojunctions between conducting polymers and semiconducting metal oxides holds promising for gas sensor applications due to enhanced surface adsorption, redox property modulation, and electronic interactions. The common conducting polymers employed for this purpose include polyacetylene (PA), polyaniline (PANI), polypyrrole (PPy), and polythiophene (PT) (Liu et al. 2022; Wawrzyniak 2023). As p-type materials, these polymers can donate or accept electrons upon gas exposure, and due to their specific chemical affinities, especially with alkanes, conducting polymer-based heterojunctions have shown selective detection capabilities for volatile organic compounds (VOCs) such as 2-propanol (Schröder et al. 2023) and methanol (Sáaedi et al. 2019). They have also demonstrated selectivity for other gases, including NH_3 (Malook et al. 2019), CO_2 (Mandal et al. 2018), and H_2S (Shu et al. 2017). Other interesting approach is the heterostructures composed of carbon-based materials (Yin et al. 2021). The good performance is attributed to their ability to detect small resistance changes (sensitivity to low gas concentrations), catalytic activity, and chemical stability (Yin et al. 2021; Chakraborthy et al. 2022; Wawrzyniak 2023). Particularly, graphene-based heterostructures exhibit enhanced selectivity, attributed to the hydrophobic nature of graphene multilayers, which minimizes humidity interference and facilitates the absorption of organic molecules (Yin et al. 2021). Additionally, large surface area, high electron mobility, and electrical conductivity contribute to the highly specific detection of target gases (Chakraborthy et al. 2022; Wawrzyniak 2023). Specific detection toward VOC gases has been observed in the following material combinations: rGO/CuO (Modenes-Junior et al. 2019), rGO/TiO_2 (Ye et al. 2016), and SnO_2/rGO (Rong et al. 2018).

Table 10.2 Summary of semiconducting metal oxides-based sensing materials to volatile organic compounds detection

Sensing system	Analyte	Level (ppm)	Sensitivity	t_{resp} (s)	t_{rec} (s)	References
SnO_2 nanowires	Formaldehyde	10	$S_{rat} = 2.5$	90	150	Castro-Hurtado et al. (2016)
ZnO nanodisks	Acetone	100	$S_{rel} = 42$	2	4	Alenezi et al. (2013)
WO_3 nanoflowers	Acetylene	200	$S_{rat} = 32.3$	12	17	Wei et al. (2019)
In_3O_2 nanospheres	Triethylamine	200	$S_{rat} = 100$	1	85	Jin et al. (2021)
Au-doped ZnO nanorods	Acetone	100	$S_{rat} = 102$	8	5	Huang et al. (2020)
Co-doped ZnO	Acetaldehyde	5	$S_{rat} = 238$	<85	<85	Thomas et al. (2024)
Cr-doped WO_3 nanoflakes	Xylene	50	$S_{rat} = 6.4$	12	7	Parne et al. (2022)
NiO–SnO_2 nanofiber	Toluene	50	$S_{rat} = 11$	11.2	4	Liu et al. (2011)
CuO–SnO_2	Toluene	75	$S_{rat} = 540$	100	36	Hermawan et al. (2020)
Cr_2O_3/WO_3	Xylene	100	$S_{rat} = 26$	5	20	Li et al. (2015)
ZnO/PAni nanocomposites	Methanol	100	$S_{rel} = 19.2$	18.2	5	Sáaedi et al. (2019)
PV4D4/CuO	2-Propanol	100	$S_{rel} = 0.64$	9.5	35	Schröder et al. (2023)
rGO/Pt/SnO_2	Methanol	40	$S_{rel} = 61.6$	6	21	Peng et al. (2018)
GO-WO_3 hemitubes	Acetone	0.1	$S_{rat} = 1.7$	11	13.5	Choi et al. (2014)
SnO_2/rGO nanocomposite	Formaldehyde	100	$S_{rat} = 40$	<20	<90	Rong et al. (2018)

Table 10.2 exhibits some selected examples of semiconducting metal oxides employed to detect some VOCs and representative values of sensitivity and response times. This technology, still under development, shows compatible response time scale as compared with the data in Table 10.1 concerning human health.

Finally, thermal modulation is an approach that makes possible by adjusting the device operating temperature, optimizing the selectivity mechanism. This is achieved by the control of heater input voltage integrated in a single sensor to detect multiple gases at their respective optimal temperatures (Wawrzyniak 2023). This method allows a sensing that simulates the functionality of multiple sensors, each responding distinctly to different gas components. Dynamic shifts between high and low operating temperatures selectively activate or inhibit adsorption and redox reactions on the sensor surface, resulting in variations in sensor resistance (Schütze et al. 2017; Wawrzyniak 2023). The temperature-dependent resistance behavior produces distinct patterns that serve as unique identifiers for specific gases or gas mixtures, facilitating accurate gas composition analysis (Wawrzyniak 2023). Interpreting thermally modulated sensor responses requires advanced signal processing techniques to analyze these complex patterns and differentiate between gases that may appear indistinguishable in a static operation mode (Schütze et al. 2017). Oxides gas sensors have achieved high selectivity using thermal modulation; for instance, a

ZnO gas sensor reached 100% accuracy in identifying four different alcohol gases (Shi et al. 2022), and Au- and Pd-modified SnO_2 sensors demonstrated enhanced differentiation between methane and propane (Krivetskiy et al. 2021).

10.6 Conclusions and Perspectives

The fundamental and applied research of nanoscale science enables the development of fast and very sensitive devices to measure the presence of pollutants gas in atmosphere. The integration of these sensors in indoor and outdoor environments provides a suitable monitoring of air quality in an almost real time. With the employment of advanced IoT (Internet of Things) and AI (Artificial Intelligence) technologies associated to its simplicity of synthesis and low cost, semiconducting metal oxides sensors are a quantitative tool to provide monitoring of air quality and the development of sustainable society for the next generations.

References

American Industrial Hygiene Association-AIHA (1971) Anhydrous ammonia. In: hygienic guide series. Am Ind Hyg Assoc J 32:139–142

Alenezi MR, Henley SJ, Emerson NG, Silva SRP (2013) From 1D and 2D ZnO nanostructures to 3D hierarchical structures with enhanced gas sensing properties. Nanoscale 6(1):235–247. https://doi.org/10.1039/C3NR04519F

Andre RS, Sanfelice RC, Pavinatto A, Mattoso LHC, Correa DS (2018) Hybrid nanomaterials designed for volatile organic compounds sensors: a review. Mater Des 156:154–166. https://doi.org/10.1016/j.matdes.2018.06.041

Degler D, Weimar U, Barsan N (2019) Current understanding of the fundamental mechanisms of doped and loaded semiconducting metal-oxide-based gas sensing materials. ACS sensors 4(9):2228–2249. 10.1021/acssensors.9b00975

Barsan N, Weimar U (2001) Conduction model of metal oxide gas sensors. J Electroceram 7:143–167. https://doi.org/10.1023/A:1014405811371

Benamara M, Ly A, Soltani S, Essid M, Dahman H, Dhahri R, El Mir L, Debliquy M, Lahem D (2023) Enhanced detection of low concentration volatile organic compounds using advanced doped zinc oxide sensors. RSC Adv 13(43):30230–30242. https://doi.org/10.1039/D3RA03143H

Bochenkov VE, Sergeev GB (2010) Sensitivity, selectivity, and stability of gas-sensitive metal-oxide nanostructures. In: Metal oxide nanostructures and their applications, vol 3. American Scientific Publishers, Stevenson Ranch, pp 31–52

Bruce J, Bosnick K, Kamali Heidari E (2022) Pd-decorated ZnO nanoflowers as a promising gas sensor for the detection of meat spoilage. Sensors Actuators B Chem 355:131316. https://doi.org/10.1016/j.snb.2021.131316

Cali K, Tuccori E, Persaud KC (2020) Gravimetric biosensors. Methods Enzymol 642:435–468

Castro-Hurtado I, Gonzalez-Chávarri J, Morandi S, Samà J, Romano-Rodríguez A, Castaño E, Mandayo GG (2016) Formaldehyde sensing mechanism of SnO2 nanowires grown on-chip by sputtering techniques. RSC Adv 6(22):18558–18566. https://doi.org/10.1039/C5RA26105H

Chakraborthy A, Nuthalapati S, Nag A, Afsarimanesh N, Alahi MEE, Altinsoy ME (2022) A critical review of the use of graphene-based gas sensors. Chemosensors 10(9):355

Choi SJ, Fuchs F, Demadrille R, Grévin B, Jang BH, Lee SJ, Lee JH, Tuller HL, Kim ID (2014) Fast responding exhaled-breath sensors using WO3 hemitubes functionalized by graphene-based electronic sensitizers for diagnosis of diseases. ACS Appl Mater Interfaces 6(12):9061–9070. https://doi.org/10.1021/AM501394R/SUPPL_FILE/AM501394R_SI_001.PDF

Cui Y, Wei Q, Park H, Lieber CM (2001) Nanowire nanosensors for highly sensitive and selective detection of biological and chemical species. Science (80-) 293(5533):1289–1292. https://doi.org/10.1126/science.1062711

de Araújo EP, Amorim CA, Arantes AN, Chiquito AJ (2022) A label-free Acetone based SnO2 nanowire network sensor at room temperature. Appl Phys A Mater Sci Process 128(5):404. https://doi.org/10.1007/s00339-022-05540-x

Du H, Li X, Yao P, Wang J, Sun Y, Dong L (2018) Zinc oxide coated tin oxide nanofibers for improved selective acetone sensing. Nanomaterials 8(7):509. https://doi.org/10.3390/nano8070509

Fang F, Bai L, Song D, Yang H, Sun X, Sun H, Zhu J (2015) Ag-modified In2O3/ZnO nanobundles with high formaldehyde gas-sensing performance. Sensors 15(8):20086–20096. https://doi.org/10.3390/s150820086

Franco MA, Conti PP, Andre RS, Correa DS (2022) A review on chemiresistive ZnO gas sensors. Sens Actuators Rep 4:100100. https://doi.org/10.1016/j.snr.2022.100100

Garg A, Gupta NC (2019) A comprehensive study on spatio-temporal distribution, health risk assessment and ozone formation potential of BTEX emissions in ambient air of Delhi, India. Sci Total Environ 659:1090–1099. https://doi.org/10.1016/j.scitotenv.2018.12.426

Haidry AA, Yucheng W, Fatima Q, Raza A, Zhong L, Chen H, Mandebvu CR, Ghani F (2024) Synthesis and characterization of TiO2 nanomaterials for sensing environmental volatile compounds (VOCs): a review. TrAC Trends Anal Chem 170:117454. https://doi.org/10.1016/j.trac.2023.117454

Hanada T (2009) Basic properties of ZnO, GaN, and related materials. In: Oxide and nitride semiconductors. Advances in materials research, pp 1–19. https://doi.org/10.1007/978-3-540-88847-5_1

Hermawan A, Asakura Y, Inada M, Yin S (2020) A facile method for preparation of uniformly decorated-spherical SnO2 by CuO nanoparticles for highly responsive toluene detection at high temperature. J Mater Sci Technol 51:119–129. https://doi.org/10.1016/J.JMST.2020.02.041

Huang C, Chen CH, Li L, Cheng Z, Wang HL, Huang HY, Streets DG, Wang YJ, Zhang GF, Chen YR (2011) Emission inventory of anthropogenic air pollutants and VOC species in the Yangtze River Delta region, China. Atmos Chem Phys 11(9):4105–4120. https://doi.org/10.5194/acp-11-4105-2011

Huang J, Zhou J, Liu Z, Li X, Geng Y, Tian X, Du Y, Qian Z (2020) Enhanced acetone-sensing properties to ppb detection level using Au/Pd-doped ZnO nanorod. Sensors Actuators B Chem 310:127129. https://doi.org/10.1016/J.SNB.2019.127129

Jin X, Li Y, Zhang B, Xu X, Sun G, Wang Y (2021) Temperature-dependent dual selectivity of hierarchical porous In2O3 nanospheres for sensing ethanol and TEA. Sensors Actuators B Chem 330:129271. https://doi.org/10.1016/J.SNB.2020.129271

John RAB, Ruban Kumar A (2021) A review on resistive-based gas sensors for the detection of volatile organic compounds using metal-oxide nanostructures. Inorg Chem Commun 133:108893. https://doi.org/10.1016/j.inoche.2021.108893

Kaden DA, Mandin C, Nielsen GD, Wolkoff P (2010) Formaldehyde. In WHO Guidelines for indoor air quality: selected pollutants. World Health Organization, Geneva.

Khatib M, Haick H (2022) Sensors for volatile organic compounds. ACS Nano 16(5):7080–7115. https://doi.org/10.1021/acsnano.1c10827

Koppmann R (ed) (2007) Volatile organic compounds in the atmosphere. Blackwell, Oxford

Krivetskiy VV, Andreev MD, Efitorov AO, Gaskov AM (2021) Statistical shape analysis preprocessing of temperature modulated metal oxide gas sensor response for machine learning improved selectivity of gases detection in real atmospheric conditions. Sensors Actuators B Chem 329:129187. https://doi.org/10.1016/J.SNB.2020.129187

Kumar A, Singh D, Kumar K, Singh BB, Jain VK (2018) Distribution of VOCs in urban and rural atmospheres of subtropical India: temporal variation, source attribution, ratios, OFP and risk assessment. Sci Total Environ 613–614:492–501. https://doi.org/10.1016/j.scitotenv.2017.09.096

Li Y, Li F, Li C, Wei W, Jiang D, Zhu L, Sun D, Zhang X, Ruan S (2015) The preparation of Cr2O3@WO3 hierarchical nanostructures and their application in the detection of volatile organic compounds (VOCs). RSC Adv 5(76):61528–61534. https://doi.org/10.1039/C5RA06667K

Li Z, Askim JR, Suslick KS (2019) The optoelectronic nose: colorimetric and fluorometric sensor arrays. Chem Rev 119(1):231–292. https://doi.org/10.1021/acs.chemrev.8b00226

Li Z, Zeng W, Li Q (2022) SnO2 as a gas sensor in detection of volatile organic compounds: a review. Sensors Actuators A Phys 346:113845. https://doi.org/10.1016/j.sna.2022.113845

Li X, Fu L, Karimi-Maleh H, Chen F, Zhao S (2024) Innovations in WO3 gas sensors: nanostructure engineering, functionalization, and future perspectives. Heliyon 10(6):e27740. https://doi.org/10.1016/j.heliyon.2024.e27740

Liu L, Zhang Y, Wang G, Li S, Wang L, Han Y, Jiang X, Wei A (2011) High toluene sensing properties of NiO–SnO2 composite nanofiber sensors operating at 330 °C. Sensors Actuators B Chem 160(1):448–454. https://doi.org/10.1016/J.SNB.2011.08.007

Liu J, Zhang L, Fan J, Zhu B, Yu J (2021) Triethylamine gas sensor based on Pt-functionalized hierarchical ZnO microspheres. Sensors Actuators B Chem 331:129425. https://doi.org/10.1016/J.SNB.2020.129425

Liu X, Zheng W, Kumar R, Kumar M, Zhang J (2022) Conducting polymer-based nanostructures for gas sensors. Coord Chem Rev 462:214517. https://doi.org/10.1016/j.ccr.2022.214517

Malook K, Khan H, Shah M, Haque I (2019) Highly selective and sensitive response of Polypyrrole–MnO2 based composites towards ammonia gas. Polym Compos 40(4):1676–1683. https://doi.org/10.1002/pc.24917

Mandal B, Biswas A, Aaryashree, Sharma DS, Bhardwaj R, Das M, Rahman MdA, Kuriakose S, Bhaskaran M, Sriram S, Htay MT, Das AK, Mukherjee S (2018) π-Conjugated Amine–ZnO Nanohybrids for the Selective Detection of CO2 Gas at Room Temperature ACS Applied Nano Materials 1(12):6912–6921. https://doi.org/10.1021/acsanm.8b01731

Modenes-Junior MA, Zito CA, Perfecto TM, Volanti DP (2019) Ethanol detection using composite based on reduced graphene oxide and CuO hierarchical structure under wet atmosphere. Mater Sci Eng B 248:114385. https://doi.org/10.1016/j.mseb.2019.114385

Murrells T, Derwent RG (2007) Climate change consequences of VOC emission controls. In: The Department for Environment, Food and Rural Affairs, Welsh Assembly Government, the Scottish Executive and the Department of the Environment for Northern Ireland. AEA Energy & Environment, Ireland

National Institute for Occupational Safety and Health. Division of Standards Development and Technology Transfer (1992) NIOSH recommendations for occupational safety and health: compendium of policy documents and statements. NIOSH Numbered Publications, Cincinnati

Onofre YJ, Catto AC, Bernardini S, Fiorido T, Aguir K, Longo E, Mastelaro VR, da Silva LF, de Godoy MPF (2019) Highly selective ozone gas sensor based on nanocrystalline Zn0.95Co0.05O thin film obtained via spray pyrolysis technique. Appl Surf Sci 478:347–354. https://doi.org/10.1016/j.apsusc.2019.01.197

Parne SR, Pothukanuri N, Sriram SR, Joshi D, Edla DR (2022) Facile synthesis of pure and Cr-doped WO3 thin films for the detection of xylene at room temperature. ACS Omega 7(51):47796–47805. https://doi.org/10.1021/ACSOMEGA.2C05589/ASSET/IMAGES/LARGE/AO2C05589_0013.JPEG

Peng R, Chen J, Nie X, Li D, Si P, Feng J, Zhang L, Ci L (2018) Reduced graphene oxide decorated Pt activated SnO2 nanoparticles for enhancing methanol sensing performance. J Alloys Compd 762:8–15. https://doi.org/10.1016/J.JALLCOM.2018.05.177

Pinthong N, Thepanondh S, Kultan V, Keawboonchu J (2022) Characteristics and impact of VOCs on ozone formation potential in a petrochemical industrial area, Thailand. Atmosphere (Basel) 13(5):732. https://doi.org/10.3390/atmos13050732

Postica V, Vahl A, Strobel J, Santos-Carballal D, Lupan O, Cadi-Essadek A, de Leeuw NH, Schütt F, Polonskyi O, Strunskus T, Baum M, Kienle L, Adelung R, Faupel F (2018) Tuning doping and surface functionalization of columnar oxide films for volatile organic compound sensing: experiments and theory. J Mater Chem A 6(46):23669–23682. https://doi.org/10.1039/C8TA08985J

Postica V, Vahl A, Santos-Carballal D, Dankwort T, Kienle L, Hoppe M, Cadi-Essadek A, de Leeuw NH, Terasa M-I, Adelung R, Faupel F, Lupan O (2019) Tuning ZnO sensors reactivity toward volatile organic compounds via Ag doping and nanoparticle functionalization. ACS Appl Mater Interfaces 11(34):31452–31466. https://doi.org/10.1021/acsami.9b07275

Rong X, Chen D, Qu G, Li T, Zhang R, Sun J (2018) Effects of graphene on the microstructures of SnO2@rGO nanocomposites and their formaldehyde-sensing performance. Sensors Actuators B Chem 269:223–237. https://doi.org/10.1016/j.snb.2018.04.176

Sáaedi A, Shabani P, Yousefi R (2019) High performance of methanol gas sensing of ZnO/PAni nanocomposites synthesized under different magnetic field. J Alloys Compd 802:335–344. https://doi.org/10.1016/J.JALLCOM.2019.06.088

Schröder S, Ababii N, Brînză M, Magariu N, Zimoch L, Bodduluri MT, Strunskus T, Adelung R, Faupel F, Lupan O (2023) Tuning the selectivity of metal oxide gas sensors with vapor phase deposited ultrathin polymer thin films. Polymers (Basel) 15(3):524. https://doi.org/10.3390/POLYM15030524

Schütze A, Baur T, Leidinger M, Reimringer W, Jung R, Conrad T, Sauerwald T (2017) Highly sensitive and selective VOC sensor systems based on semiconductor gas sensors: how to? Environment 4(1):20. https://doi.org/10.3390/ENVIRONMENTS4010020

Scolfaro D, Onofre YJ, Teodoro MD, De Godoy MPF (2018) Atmosphere-dependent photoconductivity of ZnO in the Urbach tail. Int J Photoenergy 2018:1–8. https://doi.org/10.1155/2018/8607247

Shah S, Hussain S, Din STU, Shahid A, Amu-Darko JNO, Wang M, Tianyan Y, Liu G, Qiao G (2024) A review on In2O3 nanostructures for gas sensing applications. J Environ Chem Eng 12(3):112538. https://doi.org/10.1016/j.jece.2024.112538

Shi X, Zhang H, Ji H, Meng F (2022) Dynamic measurement of VOCs with multiple characteristic peaks based on temperature modulation of ZnO gas sensor. Chemosensors 10(6):226. https://doi.org/10.3390/CHEMOSENSORS10060226

Shu J, Qiu Z, Lv S, Zhang K, Tang D (2017) Cu2+-doped SnO2 nanograin/polypyrrole nanospheres with synergic enhanced properties for ultrasensitive room-temperature H2 S gas sensing. Anal Chem 89(20):11135–11142. https://doi.org/10.1021/acs.analchem.7b03491

Silva ALC, Rodrigues ADG, Jasinevicius RG, de Godoy MPF (2022) Tuning the photoconductivity of Co3O4 thin films by defect engineering. Appl Surf Sci 606:154943. https://doi.org/10.1016/j.apsusc.2022.154943

Sze SM, Ng KK (2006) Physics of semiconductor devices, 3rd edn. John Wiley & Sons, Inc, Hoboken

Thomas A, Thirumalaisamy L, Madanagurusamy S, Sivaperuman K (2024) Switching the selectivity of ZnO thin films for ultra-sensitive acetaldehyde gas sensors through Co doping. Sensors Actuators B Chem 401:135043. https://doi.org/10.1016/J.SNB.2023.135043

US Environmental Protection Agency (2024a) Technical overview of volatile organic compounds. https://www.epa.gov/indoor-air-quality-iaq/technical-overview-volatile-organic-compounds#8

US Environmental Protection Agency (2024b) Volatile organic compounds impact on indoor air quality. https://www.epa.gov/indoor-air-quality-iaq/volatile-organic-compounds-impact-indoor-air-quality#Health_Effects

US Environmental Protection Agency (2024c) NAAQS Table. https://www.epa.gov/criteria-air-pollutants/naaqs-table

Walker J, Karnati P, Akbar SA, Morris PA (2022) Selectivity mechanisms in resistive-type metal oxide heterostructural gas sensors. Sensors Actuators B Chem 355:131242. https://doi.org/10.1016/j.snb.2021.131242

Wang C (2020) Triethylamine sensing properties of ZnO nanostructures prepared by hydrothermal method at different pH values. Chem Phys Lett 749:137471. https://doi.org/10.1016/j.cplett.2020.137471

Wang L, Ma S, Xu X, Li J, Yang T, Cao P, Yun P, Wang S, Han T (2021) Oxygen vacancy-based Tb-doped SnO2 nanotubes as an ultra-sensitive sensor for ethanol detection. Sensors Actuators B Chem 344:130111. https://doi.org/10.1016/j.snb.2021.130111

Wang Y, Zhang H, Sun X (2016) Electrospun nanowebs of NiO/SnO2 p-n heterojunctions for enhanced gas sensing. Applied Surface Science 389:514–520. https://doi.org/10.1016/j.apsusc.2016.07.073

Wawrzyniak J (2023) Advancements in improving selectivity of metal oxide semiconductor gas sensors opening new perspectives for their application in food industry. Sensors 23(23):9548. https://doi.org/10.3390/s23239548

Wei Z, Zhou Q, Lu Z, Xu L, Gui Y, Tang C (2019) Morphology controllable synthesis of hierarchical WO3 nanostructures and C2H2 sensing properties. Phys E: Low-Dimens Syst Nanostructures 109:253–260. https://doi.org/10.1016/J.PHYSE.2019.01.006

World Health Organization (2021) WHO global air quality guidelines: Particulate matter (PM2.5 and PM10), ozone, nitrogen dioxide, sulfur dioxide and carbon monoxide, vol. 267, no. September. World Health Organization, Geneva.

World Health Organization. Regional Office for Europe (1989) Indoor air quality : organic pollutants. In: WHO meeting, Berlin, West, 23-27 August 1987. EURO reports and studies, vol 111. WHO Regional Office for Europe, Copenhagen

Yamazoe N (1991) New approaches for improving semiconductor gas sensors. Sensors Actuators B Chem 5(1–4):7–19. https://doi.org/10.1016/0925-4005(91)80213-4

Ye Z, Tai H, Xie T, Yuan Z, Liu C, Jiang Y (2016) Room temperature formaldehyde sensor with enhanced performance based on reduced graphene oxide/titanium dioxide. Sensors Actuators B Chem 223:149–156. https://doi.org/10.1016/j.snb.2015.09.102

Yin Z, Wang X, Sun F, Tong X, Zhu C, Lv Q, Ye D, Wang S, Luo W, Huang YA (2017) Aligned hierarchical Ag/ZnO nano-heterostructure arrays via electrohydrodynamic nanowire template for enhanced gas-sensing properties. Sci Rep 7(1):1–10. https://doi.org/10.1038/s41598-017-12553-7

Yin XT, Li J, Dastan D, Zhou WD, Garmestani H, Alamgir FM (2020) Ultra-high selectivity of H2 over CO with a p-n nanojunction based gas sensors and its mechanism. Sensors and Actuators B: Chemical 319:128330. https://doi.org/10.1016/j.snb.2020.128330

Yin F, Yue W, Li Y, Gao S, Zhang C, Kan H, Niu H, Wang W, Guo Y (2021) Carbon-based nanomaterials for the detection of volatile organic compounds: a review. Carbon 180:274–297. https://doi.org/10.1016/J.CARBON.2021.04.080

Yu PY, Cardona M (2010) Fundamentals of semiconductors. Springer Berlin Heidelberg, Berlin, Heidelberg

Zhang N, Gao C, Xiong Y (2019) Defect engineering: a versatile tool for tuning the activation of key molecules in photocatalytic reactions. J Energy Chem 37:43–57. https://doi.org/10.1016/j.jechem.2018.09.010

Zhou X, Zhou X, Wang C, Zhou H (2023) Environmental and human health impacts of volatile organic compounds: a perspective review. Chemosphere 313:137489. https://doi.org/10.1016/j.chemosphere.2022.137489

Chapter 11
Nanofibrous Membranes in Air Purification

Marwa Nabil, Hassan Shokry Hassan, and Marwa Farouk Elkady

Abstract Generally, the term pollution is derived from the production and consumption of resources, where the produced streams are very wasteful in their current state. Most of these waste streams cannot be reintegrated into the environment efficiently or economically. Air pollution resulting from ambient pollutants such as particulate matter, volatile organic matter, and/or gaseous contaminants represents one of the key environmental concerns of this century. There are many technologies developed for effectively treating air pollutants. However, the science and engineering area that is responsible for tomorrow's breakthroughs is membrane-based technology. Membrane is one of the promising technologies for air filtration due to its high efficiency, little cost, and simplicity of scale-up. Concerning the outstanding properties of high surface area, enormously porous structure, adaptable pore size and controllable morphology of electrospinning nanofiber membranes, it was an exceptional air-filter candidate. However, despite its exceptional characteristics, the main weakness of nanofibrous membranes are their poor mechanical properties during various applications, especially the separation processes. To overcome this limitation, electrospinning technology can offer an efficient one-step methodology to produce functionalized structures including core-shell and multilayered structures. Moreover, the incorporation of nanofillers onto the different electrospun nanofibers polymeric matrices can easily achieved using electrospinning technology to improve the mechanical and reactive characteristics of the composite fabricated nanofiber

M. Nabil (✉)
Advanced Technology and New Materials Research Institute, City of Scientific Research and Technology Applications (SRTA-City), New Borg El-Arab City, Alexandria, Egypt

H. S. Hassan
Environmental Engineering Department, Egypt-Japan University for Science and Technology (E-JUST), New Borg El-Arab City, Alexandria, Egypt
e-mail: hassan.shokry@ejust.edu.eg

M. F. Elkady
Chemical and Petrochemical Engineering Department, Egypt-Japan University for Science and Technology (E-JUST), New Borg El-Arab City, Alexandria, Egypt
e-mail: marwa.elkady@ejust.edu.eg

N. Ghosh et al. (eds.), *Nanotechnology in Air Quality Management*, Sustainable Development and Biodiversity 39, https://doi.org/10.1007/978-981-96-8339-0_11

membranes. So, this chapter will highlight the manipulation of nanofibrous membrane materials for controlling the air filtration performances of nanofibrous membranes either fabricated using single or blended polymeric matrices. The role of nanomaterials addition to the modification of nanofibrous membrane properties toward air filtration applications will be focused on. Finally, a comprehensive overview of the basic principles of the air pollutant separation mechanisms onto the nanofibrous membranes will be discussed.

Keywords Nanofibrous membrane · Air pollution · Nanomaterials · Electrospinning · Separation process

11.1 Introduction

Air pollution is the existence of chemical, biological, and physical contaminates in the Earth's atmosphere that alter the natural properties of the atmosphere. Recently, air pollution has become a pressing problem due to its dangerous effects on both environment and human health. So, effective means must be adapted to remove pollutants from the air (De Riccardis 2023).

Air pollution represents worldwide problem that has received particular awareness mainly, owing to difficulty to manage a pollutant which is released into the environment. Particulate matter (PM) is primary concern air pollutants, it represents complex combination of airborne particles including organic, inorganic, solid, and, liquid that able to transport various disease-causing microbes, such as saliva microparticles, viruses, bacteria, vehicle emissions gases, dust, and pollen (Abozeid et al. 2019).

Generally, air purification technologies rely on adsorption, photocatalytic oxidation, electrostatic precipitators, ionizers, and solid filtration using membrane technology. Membrane filtration represents the most efficient process to effectively remove PM from the air. Membranes act as a physical barrier during the air decontamination process and preserve exclusively the air contaminates and transport the purified air. However, the conventional air filters are based on micro-sized fibers that fabricated from synthetic materials such as fiberglass, polyester, and polypropylene, which cannot be retained finest PM than the range of 0.1–2.5 μm due to their larger pore size, which is risky for humans' health (Espinoza-Montero et al. 2023).

Recently, researchers and manufacturers developed advanced materials and novel membrane technologies to construct extremely efficient air filters. Recently, electrospun nanofiber membranes represent an attractive key tool to achieve effective air filtration performance. Among the different membrane fabrication technologies, the electrospinning technique allows nanofiber diameters engineered at broad range from 40 to 2000 nm through appropriate combination of polymers and solvents. This membrane fabrication technology is characterized by its versatility of electrospinnable materials, low cost, and the possibility of scaling up (Espinoza-Montero et al. 2023) besides its suitability for the production of environmentally sustainable filters (De Riccardis 2023).

Nanofiber-based air filters are characterized by their high area per unit mass, low weight and low-pressure drop, so they represent efficient media for air purification. Generally, electrospun membranes characterized by their outstanding porosity owing to interconnected structure, enhanced specific surface area, exceptional topography, and large diversity of chemical properties (Ali et al. 2020). This technique is very adaptable for suitable membrane fabrication due to its ability for modifications to perform a specific function. For instance, various fiber membranes can be developed through functionalization of nanofibers membranes with active species such as nanomaterials. These nanofiber air filters can also be assembled on a variety of matrices by stacking, alignment, or folding. Moreover, the polymer and solvent selected for nanofiber air filters fabrication have to be carefully selected due to their vital role for the membrane hydrophilicity/hydrophobicity properties (Espinoza-Montero et al. 2023).

Accordingly, a comprehensive overview of the electrospinning technology utilized for electrospun membrane fabrication for air filters will be discussed. The basics of air filtration mechanisms concerned with electrospun nanofibers membranes for air filtration will be explored. The recent aspects of development and challenges for air filtration and purification using polymeric electrospun membranes will be highlighted.

11.1.1 Air Pollution

Air pollution is the discharge of pollutants such as gases, particles, and biological molecules into the atmosphere in sufficient quantities and periods to harm all living things and the environment. Accordingly, the air pollutants may be classified as either particulate matter or gaseous contaminates. They are spread everywhere indoors and outdoors. Indoor air is typically two to five times more polluted than outdoor air. Indoor pollution with PM takes place due to the cooking, heating, smoking, and air conditioning activities (El Fawal et al. 2019). Outdoor industrial processes such as petroleum are responsible for emission PM and chemicals into the environment. Outdoor air contamination is considered a vital environmental health problem that concern humans because these activities degrade from the quality of healthy, breathable air quality, by critically contaminating the open air, which is a common public resource (Kadam et al. 2018). There is the greatest worry about the air pollution problem. Recently, it is scheduled as the most significant issue in the "2030 Agenda." So, the efficient decontamination of air pollutants from polluted air represents a hot research topic. Various strategies were developed for the efficient separation of air pollutants from polluted air (Elkady and Hassan 2021).

11.1.2 Air Pollution Control Using Filtration Strategy

Ventilation or filtration control strategy has been extensively applied for the efficient separation of both indoor and outdoor air contaminants. The selection of effective air filter is determined by its ability to capture various air contaminates. The efficiency of air filter is based on the nature of air pollutants, which is classified as shown in Fig. 11.1 and may be adjusted by the capturing mechanism of contaminates (Hearth and Sondarangalla 2021). As the fiber diameters decreased, the air filter's specific surface area improved, which enhanced the air filtration performance. As a result, nanofibers have augmented attention in air purification applications (Elkady et al. 2020).

11.2 Polymeric-Based Nanofibrous Membranes for Air Purification

Nanofiber membranes may be utilized as a common filter membrane that can be manufactured to capture fine particles and gaseous contaminates simultaneously. The air filtration efficiency of nanofiber membranes is increased compared to other filtration media due to two parameters, which are the effect of slip-flow with little airflow resistance, especially for nanofibers with diameter less than 500 nm (De Riccardis 2023): the specific surface area of nanofibers compared to microfibers (Ghasemiyeh et al. 2024).

Polymeric-based nanofibrous membrane technology represents important potential for different treatment challenges including air purification. Polymer membranes are mainly fabricated from solution-processable materials such as hollow fibers, fiber mats, and flat sheet membranes. The performance of the nanofibrous membrane can be tailored through the variation of polymer structure, polymer type, crosslinking degree, cross-linker, integration with nano-functional materials as

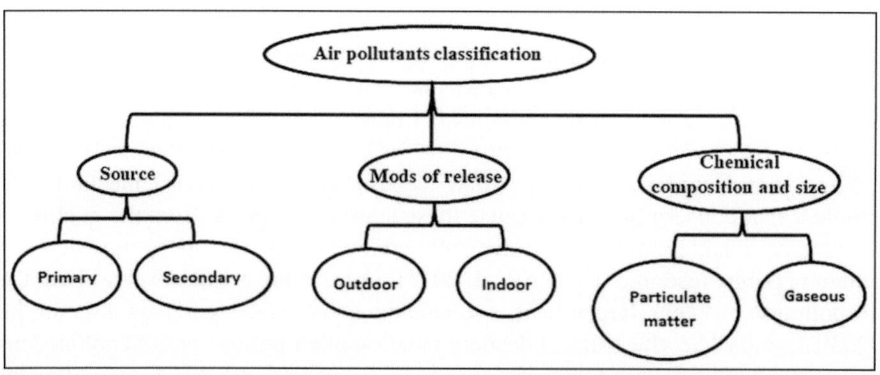

Fig. 11.1 Classification of sir pollutants

filler, and the parameter of membrane spinning processes for membrane fabrication (Shokry Hassan 2019).

11.2.1 General Techniques for Fabrication Nanofibrous Filters

Several techniques are available to fabricate nanofibers for the air purification process. These include chemical vapor deposition (Mokhena et al. 2022), sea-island technique (conjugate spinning) (Mamun et al. 2021), melt-blowing (Agrawal et al. 2021), phase separation (sol–gel process) (Anusiya and Jaiganesh 2022), and electrospinning (Islam et al. 2019). Notably, electrospinning represents adaptable and extensively accepted process for air filter production. Electrospun air filters may be fabricated from synthetic polymers or recently the trended materials are biopolymers. Lately, electrospun nanofibers with some additives have been used to capture gaseous contaminates simultaneously with PM filtration. The collected results on electrospun nanofibers to capture various air pollutants will be helpful to the scientific society as a reference for the development of advanced electrospun filters (Elkay et al. 2016).

11.2.2 Electrospun Nanofiber Air Filters

Electrospinning technology enables the design of nanofiber diameters at variable range from 40 to 2000 nm through proper combination of polymers and solvents, as shown in Fig. 11.2. Nanofiber filters that are produced via electrospun have unique characteristics, such as controllable morphology and connectivity, high surface-to-volume ratio, connectivity, and low-pressure drop. This uniqueness properties of nanofiber recorded excellent filtration performance for several sizes of air pollutants

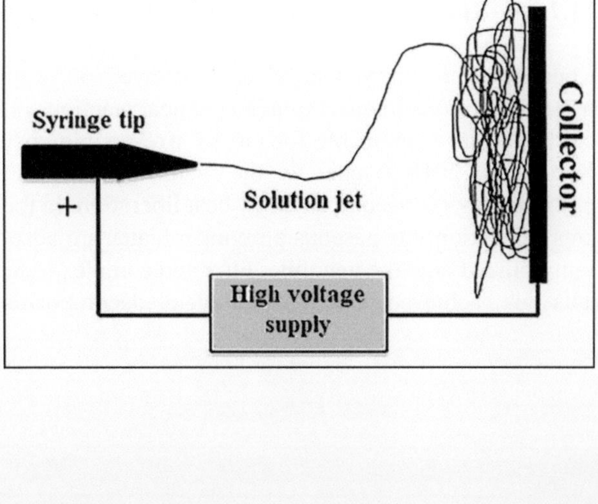

Fig. 11.2 Scheme of electrospinning system

(Saleh et al. 2023). The morphology and size of the fabricated fibers depend on several factors, i.e., environmental conditions, solution properties, and processing conditions. As an efficient factor in the production process of nanofiber, the solution concentration includes concentration, viscosity, surface tension, conductivity, molecular weight of the polymer, and the type of solvent (Essam et al. 2020).

11.2.3 Pollutants Capture Mechanism onto the Nanofibrous Filter

The air filtration mechanism is how pollutants are absorbed, bound, and captured by the air filter. It depends largely on the characteristics of the filtration medium and the nature of contaminates. The filtration mechanism for gaseous pollutants and fine particles is based upon the adsorption processes either the physical adsorption process that is predominated for fine particulate matters or chemical adsorption that may be included for the separation of gaseous pollutants (Kadam et al. 2018).

The adsorption process determines the variety of surface features of nanofibrous air filters. The adsorption of gaseous air pollutants is usually considered a multilayer process concerning the micro-porosity properties of nanofiber air filters besides the interactions between the air filters and air pollutants. Also, the adsorption process conditions are very important, including the gas pressure of the treated air and its temperature, the activity of the adsorptive surface energy distribution, and the surface area and porosity of the nanofiber (Al-Ghouti et al. 2023).

Generally, the adsorption (capture) mechanisms of the gaseous air pollutants include both the chemisorptions and physiosorption. The chemisorption mechanism depends on the conversion of pollutants to simple compounds due to chemical interactions such as catalytic or non-catalytic degradation. This mechanism has more selectivity concerning physisorption. For enhancement of its activity, the functionality and the chemical activity of the fiber must be improved (Gunawardene et al. 2022).

11.2.3.1 Gaseous Pollutants Capture Using the Chemisorption Process

Chemisorption filtration refers to the conversion of gaseous air contaminates into most simple constituents through chemical reactions in nanofiber filters. The chemical bonds are responsible for mostly irreversible chemisorption (Solis-Ibarra and Karunadasa 2014). Generally, the chemisorption process of gaseous air pollutants onto nanofiber is accompanied by heat liberation. In the case of catalyzed air pollutants separation, the gaseous air contaminates are adsorbed onto the active catalyst immobilized onto the nanofiber filter and complex reduction processes of air pollutants are achieved for the complete decomposition of air pollutants. The chemisorption process of gaseous air pollutants based on numerous factors such as the rate of adsorption, type of catalyst, capacity of gas adsorption onto the nanofiber filter, catalyst selectivity and thermal stability of the filter (Hassan et al. 2015).

The heterogeneous catalysis of gaseous air pollutants via solid catalysts incorporated into the nanofiber filters is directly or indirectly connected with adsorption. So, the chemisorption process impacts directly all classical concepts of catalysis, as will be discussed (Cao and Nørskov 2023). As evident from Fig. 11.3, the active centers on the catalyst surface immobilized onto nanofiber filters represent the points that have at least one of the air pollutants that are chemisorbed. These adsorbed pollutants are transferred to the intermediate compounds as a result of the catalytic reactions with active centers; also, the clusters as a form of intermediates are formed on catalytic centers (usually poly-nuclear) (Daligaux et al. 2021).

As the nanofiber air filter includes more than one active catalyst center onto the filter matrix, more than one atom of a catalytic center onto the air filter is shared in the interaction of reduction of gaseous air pollutants. The electronic structure of the gaseous air pollutants adsorbed onto the nanofibrous air filter is modified for the formed intermediates and transferred to unstable compounds that dissociated to simple compounds such as water vapors and carbon dioxide (Vasiliev et al. 2023).

Thus, this concept of catalytic reduction of gaseous air pollutants onto the nanofibrous air filters is considered the base of the quantum mechanical treatment of chemisorption. As a result of the geometric factor of the electrospun nanofiber air filters, the arrangement of active catalytic centers is responsible for the suitable chemisorption of gaseous air pollutants (Hassan et al. 2021).

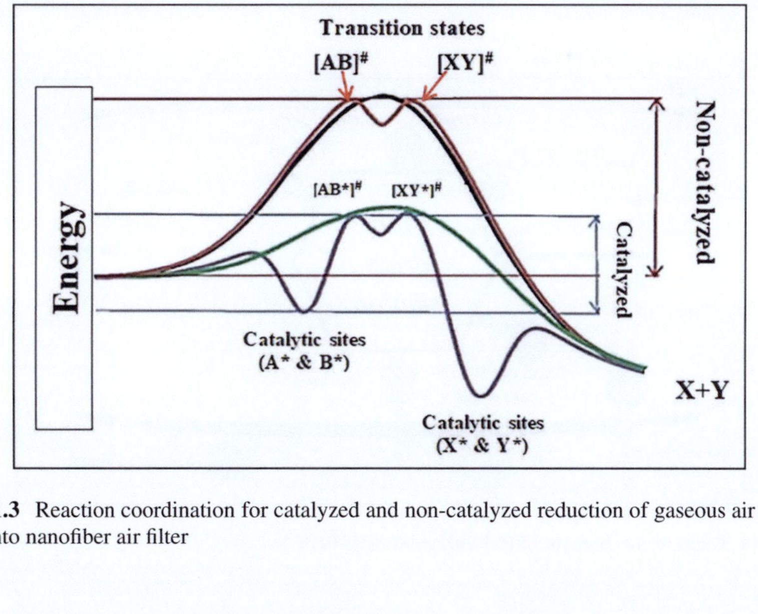

Fig. 11.3 Reaction coordination for catalyzed and non-catalyzed reduction of gaseous air pollutants onto nanofiber air filter

11.2.3.2 Particulate Matter (PM) Capture Using Physiosorption Process

PM filtration by nanofibers is dominated by physical parameters including the size of pollutants, the geometry of the filter, and air flow rate. The PM filtration performance is mainly controlled by physical separation mechanisms. It may include interception, inertial impaction, gravitation, diffusion, intermolecular interaction (Van der Waals), straining, and electrostatic interaction of contaminates with filter surface, as shown in Fig. 11.4.

11.2.3.2.1 Interception Separation

As known, the particles flow in the same direction of the air streamlines; in addition, it is intercepted during interaction with surface of nanofiber via the van der Waals forces, as shown in Fig. 11.5. This phenomenon is known as interception; it is identified as the principal filtration mechanism onto the nanofiber air filter. It is independent of the flow velocity of air. However, it is still efficient for particle capture in the range of 0.1–1 μm of diameter (Sanyal and Sinha-Ray 2021).

11.2.3.2.2 Inertial Impaction Separation

As a result of the nanofiber's random orientation of the electrospun membranes, the streamlined pathway of the air is usually twisted due to filtration process. The polluted contaminates diverge from the direction of air flow via inertia forces, and after that, it affects the nanofibers. So, this provokes an efficient filtration produced by the

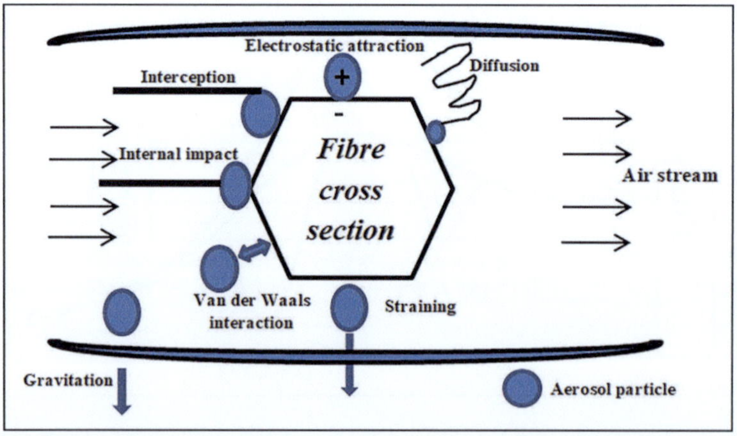

Fig. 11.4 Capture mechanism of PM via nanofiber filters

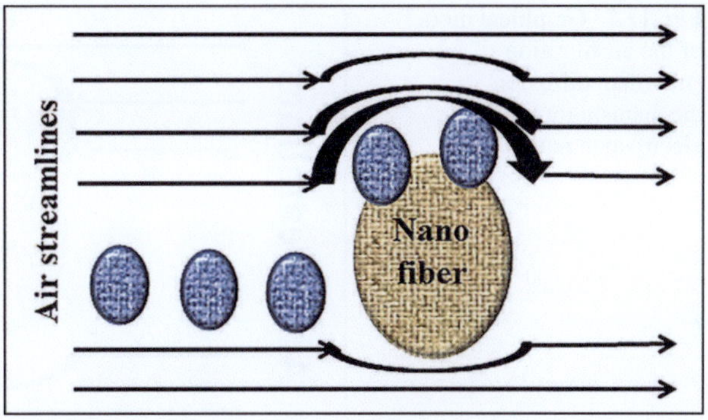

Fig. 11.5 Graphical model of the air filtration interception mechanism through electrospun nanofibers

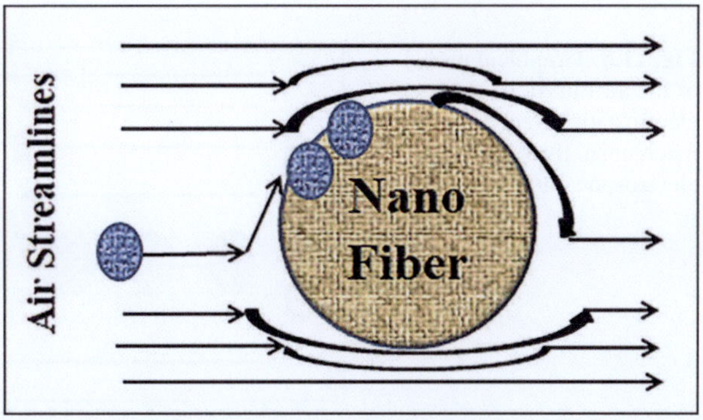

Fig. 11.6 Graphical model of the air filtration inertial impaction mechanism through electrospun nanofibers

inertial impaction. Smaller particles can be accompanied by the drag effect and transportation with the streamlines. However, the inertial momentum enhanced with the particle size and faces the drag force for large contaminates that cause divergence from the streamlines as a result of the impact onto nanofibers, as presented in Fig. 11.6. So, the inertial impaction efficiency improves with particle size aside from the velocity effect of air and the nanofibers orientations of the filter matrix (Ibrahim et al. 2021).

11.2.3.2.3 Brownian Diffusion Separation

Also, the Brownian motion causes the particles to deviate slightly from their original flow direction through the nanofiber air filters, as shown in Fig. 11.7. On the other hand, the particles diverge from the diffusion of the Brownian motion toward nanofiber, followed by interception (Chai et al. 2022). It typically takes place for small particles (lower than 1 μm).

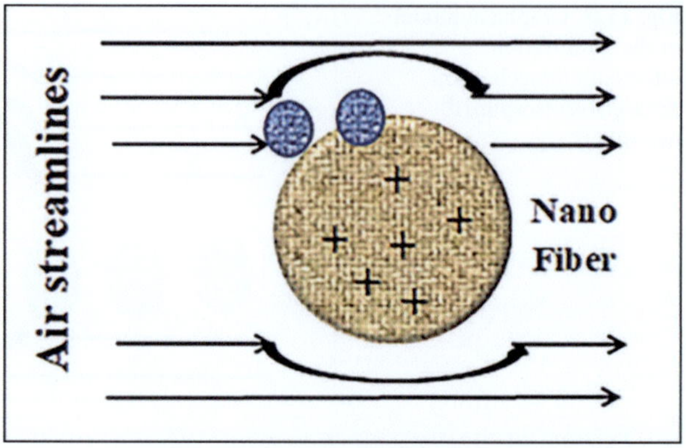

Fig. 11.7 Graphical model of the air filtration Brownian diffusion mechanism through electrospun nanofibers

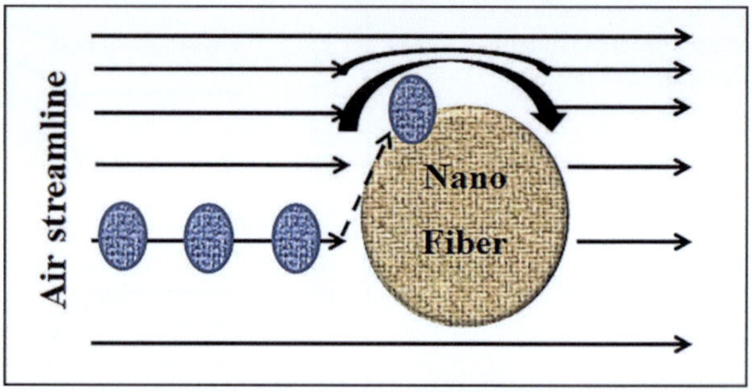

Fig. 11.8 Graphical model of the air filtration electrostatic effect mechanism through electrospun nanofibers

11.2.3.2.4 Electrostatic Separation Effect

The filtering efficiency is improved via the Coulombic interaction between charged contaminates and monopolar or bipolar charged nanofibers, as shown in Fig. 11.8. Also, the uncharged particles can be polarized due to either unipolar or bipolar charged nanofibers or dielectrophoretic forces produced by an external field (Joe et al. 2017). Finally, the gravity effect is negligible for the particles that have particle sizes smaller than 0.5 μm.

11.3 Enhanced Electrospun Nanofiber Membranes for Air Purification

Electrospun nanofibers can be functionalized through specific additives for improvement of air pollutants-capturing ability (Zhang et al. 2020). Also, the parameters affecting the electrospinning process may be controlled by altering fluid characteristics and reducing the fiber diameter. The type of electrospun material is a very important choice, it may include a single polymer, polymers blend, or a nanocomposite from nanofillers incorporated at the melt polymer (Al-Abduljabbar and Farooq 2023).

11.3.1 *Enhanced Nanofiber Air Filter Using Filler Additives*

Air filtration membranes produced from single polymer suffer from poor filtration efficiency, as well as weak mechanical characteristics such as ultra-lightness and elasticity. So, the integration of proper fillers to the base polymer improves the spinability and enhances jet stretching force. The mechanical properties of the polymeric materials represent the main factor that impacts the mechanical properties of the produced electrospun nanofiber. By incorporation proper fillers to the base polymer, the spinnability will be improved and enhancement of the jet stretching force. Then, the morphology of the fiber may be changed as a result of using some filler, also, the crystallinity of the fiber is enhanced. Then, it finally improves the mechanical strength of the electrospun nanofibrous filters.

11.3.1.1 Inorganic Nanofillers

Generally, inorganic nanomaterials are added to the nanofibers for improvement of the fiber's mechanical strength, uniformity, nano-size, and ability to separate and capture air pollutants. Types of inorganic fillers may include and not limited to nanoclays, nanocarbons, and nanometals. Generally, the light polymers are blended with specific inorganic nanoparticles that are characterized by their superior mechanical properties for improvement of the mechanical strength of bonded polymeric composite matrix as indicated in Fig. 11.9 (Shaoyun et al. 2019).

Moreover, the inorganic filler may enhance the size of the formulated nanofiber during the electrospinning process due to its electrical properties. For instance, the addition of anionic nanoclays as filler to the polymeric nanofibers reduces the average diameter of the electrospun polymeric composite matrix besides the enhancement of the tensile strength of composite nanofiber. In the case of using mixed multi-walled carbon nanotubes as filler with the polymeric fiber via ultrasonic polymerization technique, the produced composite nanofiber greatly enhanced in its

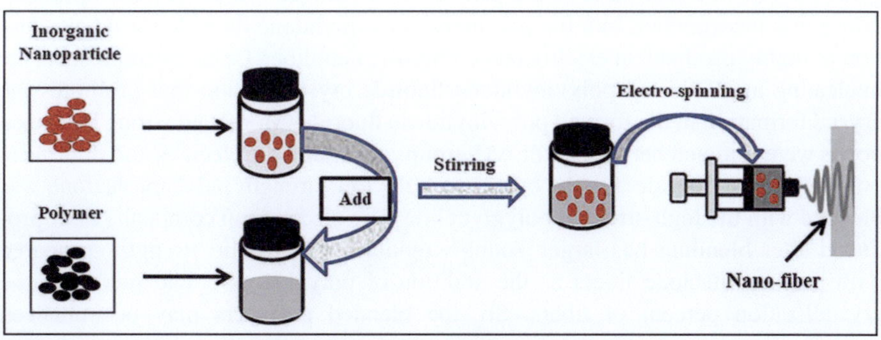

Fig. 11.9 Schematic diagram of nanofiber preparation using inorganic nanoparticles additives

uniformity and tensile strength compared with the polymeric matrix (Kong et al. 2012).

Moreover, using the precipitation method, Zn-doped calcium phosphate nanomaterials are added as filler to the polymer for the formation of electrospun fibers composite nanofiber. So, several properties are improved as the mechanical characteristics and thermal degradation rate compared with the free polymeric nanofiber (Shokry Hassan et al. 2014).

In addition, using metal oxide of nanoparticles such as manganese oxide (MnO_2) for composite nanofiber preparation may enhance the separation performance of air pollutants for the composite nanofiber. Also, the capturing effectiveness of metal oxide for suspended matters can be enhanced using magnetic nanoparticles incorporated at the nanofibers. For instance, the improvement in dust trapping performance of PVP nanofibers containing Fe_3O_4 magnetic nanoparticles is directly proportional to the enhancement of the magnetic nanoparticles' concentration in the polymeric mixture (Jialing et al. 2023; Bin Rashid et al. 2024).

11.3.1.2 Blended Polymers

In order to enhance the mechanical characteristics of the hybrid nanofiber, more than one polymer type may be blended during the electrospinning process. The mixing of strong and weak mechanical characteristics of the two polymer types enhances the mechanical characteristics of the overall yielded hybrid nanofiber. In addition, the blending of short molecular chain polymers with long chains decreases the entanglement of the blended mixture. Due to the controlling of entanglement in the polymer fluidity and reducing the crystals number, the excellent polymer fluidity can be provided via cold crystallization, in addition to the excellent crystallinity for the blended polymers. During the blending of more than polymer types, some phenomena can be occurred such as viscosity improvement of the blended polymers due to the enhancement of the attractive mutual force between the polymeric molecular chains. In addition, as a result of the restriction of the polymer fluidity due to the polymer blending, Young's modulus of the blended fiber is improved via the generation of new chemical bonds such as the hydrogen bonds. For example, cellulose was incorporated into the polymeric polyvinylidene fluoride for the production of highly mechanical electrospun composite nanofiber. Cellulose represents the nucleating agent during polyvinylidene fluoride crystallization that promotes the crystal formation in the formed polyvinylidene fluoride. Moreover, strong hydrogen bonds were initiated between hydroxyl groups at cellulose molecules and fluorine at polyvinylidene fluoride. On the other hand, the low-strength polycaprolactone was blended with the high-strength polyglycolide. The electrospun composite fiber produced after blending has larger Young's modulus and tensile strength compared with polycaprolactone fibers as the addition of polyglycolide acid increased the crystallization percent of fibers. So, the blended polymers may be enhanced regarding the crystallinity and toughness. Through the blending of one type of polymer via nucleating properties with another polymer to improve its crystallinity that leads to enhancement the overall mechanical characteristics of the blended polymeric nanofiber (Yueyang et al. 2023).

When the interaction between different materials occurs, excellent properties of the composite appears. The solvent is rapidly volatilized via the electrospinning process. The crosslinking between the obtained fibers does not occur; also, it is independent of each other. Thus, at stretching process of the fiber membrane, a single fiber is hard to resist the sizeable tensile strength, so it is easily to be pulled away (Medeiros et al. 2022). Finally, the gradual destruction of the fiber membrane is occurred. The posttreatment method contains freeze-drying, immersion, ultrasonic adsorption, and plasma treatment; also, certain unique materials including metal coatings, graphene oxide, and chitosan are inserted in the electrospun fiber membrane (Kanjwal and Ghaferi 2022).

11.3.2 Posttreatment Techniques

For mechanical properties enhancement of the polymeric fiber membrane, the posttreatment method was applied extensively. The posttreatment method includes annealing, twisting, crosslinking, post-stretching, and solvent steam. It improves the fiber's mechanical properties by enhancing the fiber morphology, strengthening the connection between the fibers, and enhancing chemical bonding between the polymeric chains (Li et al. 2023).

11.3.2.1 Annealing

The crystallization process for producing different electrospun fibers is still not explained. So, most fibers are partially amorphous that affects their mechanical properties. The annealing process of the fiber produces a partial transformation from an amorphous into crystalline structure through the increment at the annealing temperature. This annealing temperature may increase the crystallinity of some irregular molecular chains, so the molecular chains at amorphous materials are easily moving toward crystalline regions (Fonseca et al. 2021).

There is a direct proportion between the temperature value of the annealing process and the relaxing process of the polymeric materials by annealing process. So, the molecular segments change into opposite charges. Thus, the motion of different chains is provided and forms covalent bonds. For some unique materials, at high temperatures, the thermodynamic movement of the chain segment and the new chemical bonds formed are intensified.

11.3.2.2 Stretching

As a result of the short time for fibers production, the nanofibers have low crystal orientation. So, this fiber amorphous structures impact its mechanical strength. After stretching of the fiber in the water bath, the macromolecular chains enhance in the orientation along the stretching direction axis. During the fibers stretched at the water bath, the superior the stretching ratio, the greater the orientation of the macromolecular chain, which increase fiber crystallinity. Then, both tensile strength and initial fiber modulus improved after the stretching process (Demir et al. 2006).

11.3.2.3 Twisting

The straight connection and cohesion of fibers are improved due to twisting treatment that enhanced the friction between fibers (Hosseinali and Thomasson 2018). Thus, the fiber's uniformity was enhanced. So, the fiber's mechanical properties are strong. In addition, there is a direct proportionality between the number of twisted fibers and their mechanical properties. The twisted fiber characterized by its superior specific surface area and higher surface roughness compared with straight nanofibers. Twisting can be divided into core twisting and noncore twisting. The core bundle mechanical properties are low due to the damage at core bundle uniformity of the fiber.

11.3.2.4 Solvent Steam Treatment

The development of spherical and large α-phase crystal produced after fibers' crystallization enhances the crystals more brittle and decreases the fiber's mechanical properties (Dashan et al. 2019). The treatment of fiber with ethanol adapted its crystal structure from α-helical structure to β-sheet through enhancing the crystallization of β-phase that enhances the fiber's mechanical strength.

11.3.2.5 Crosslinking

Mainly, the crosslinking is aimed at the enhancement of the fiber membranes' loose overlap as well as random accumulation (Benselfelt et al. 2023). Additionally, chemical crosslinking via crosslinking agent and physical crosslinking contain unique pros. The physical treatment including hot pressing, cold curing, and ultraviolet irradiation enhances the mechanical strength of fiber filter and eliminates the retained crosslinking agent.

11.4 Controlling Nanofiber Properties Affecting the Air Filtration Process

Air filtration via nanofiber materials is controlled by controlling pore size, fiber diameter, surface area, mass coverage, fiber thickness, and filter construction. These properties impact air pollutant filtration efficiency using nanofiber air fitters.

11.4.1 Pore Size

Nanofibers consist of clogged pores, blind pores, and structure pores. The closed pores are hardly penetrated by the air pollutants as it has closed terminate internally. On the other hand, the open pores are suitable to be used for air filtration and purification processes. The size of the pore diameter is very important characteristic of through pores (Grigoriev et al. 2020). The smaller one produces efficient filtration but unfavorably impacts both permeability and pressure drop through the air filter. Membranes characterized with large and small pore size combination have the best air filtration performance compared with membranes with little pore size at the same thickness (Wang et al. 2022).

11.4.2 Fiber Diameter

The diameter of fiber has great effect on the filtration performance of air filter. The large diameters nanofiber up to 300 nm has about 99% of the filtration efficiency for different air pollutants. Also, the decline at fiber diameter (300–120 nm) enhances the filtration performance of air pollutants (Frising et al. 2003). The filtration efficiency of the nanofiber is inversely proportional to the size of fiber diameter.

11.4.3 Surface Area of Nanofiber

The nanofiber surface area improves the particle deposition probability onto fiber surface and delays the filter saturation that enhances filter life. Owing to the high specific surface area of the membrane, it recorded high air filtration efficiency and suitable for the treatment all air pollutants including biological contaminants. In addition, the deposition probability may be improved by functionalization with various nanofillers.

11.4.4 Membrane Thickness

The membrane thickness is directly proportional to the time of the electrospinning process. Also, it is inversely proportional to air permeability directly and proportional to pressure drop (Selatile et al. 2018). The case of a thin layer of a little micron restricts the airflow of the membrane to a large extent. The filtration process is improved if thin nanofiber membranes are arranged at layers as an alternative of thick layers (Zhang et al. 2010). However, it continues to increase the pressure drop and thus the quality factor deteriorates.

Also, the values of mass coverage and membrane thickness balance each other at the nanofilter. A low mass coverage of nanofibers can have a significant influence on both the efficiency of filtration and air pressure drop.

11.4.5 Construction of Nanofiber Membrane for Air Purification

There are several promising structures for the fabrication of efficient nanofiber filters such as hybrid materials (Sanchis-Gual et al. 2023), bead-on-a-string, and multilayered (Ahmadi Soufivand et al. 2023) that have been studied in previous researches to maximize quality factor for nanofibers filters. At composite coarse fine filter construction, fine fibers enhance the filtration efficiency, while coarser fiber leads to rigidity. Then, the sandwich configuration of composite nanofiber may have versatile applications. On the other hand, the beads that are prepared via electrospinning form pores in the fiber structure due to the little viscosity and deficient stretching of the solution. This specific structure improves the distance between nanofibers and decreases the membrane volume fraction that facilitates the air flow through the air filter. The bead-on-chain structure showed high filtration efficiency and low-pressure drop. So, the manifold thin layers of nanofibers, fabricated at shorter times with similar weight, enhance the quality factor of the filter. Furthermore, the core-shell, hollow, and porous nanofiber filter structures are suitable for producing ultrafiltration media for the separation of PM and gaseous air contaminants (Wang et al. 2023).

11.5 Conclusion

Recently, air pollution is a challenging problem; so, in this chapter, we discussed how to use nanofibrous membranes in air purification. The nanofibers prepared using the electrospinning techniques have many advantages compared with other filter fabrication techniques. The synthesized fiber membranes need various requirements for the air filtration process such as high specific surface area, high porosity, good connectivity of pore channels, and small diameter to fabricate highly efficient air filters suitable for the separation of different pollutants.

This chapter presented an overview of the potential trapping mechanisms of air pollutants that occur during the pollutants separation onto the electrospun membrane air filters. Electrospun nanofiber membranes are effective for air pollutant separation mainly due to electrostatic attraction. Previous research showed the effectiveness of electrospun nanofiber filter for PM separation. The separation of gaseous air pollutants filtration was shortly investigated. It has a serious environmental impact of complex air pollutants on human health. Although many achievements have been recorded in this field, there are a lot of challenges that still require potential solutions. Regrettably, the bottlenecks of electrospun membranes are mainly due to their poor mechanical stability. This chapter presented the basic concepts and solutions suggested to overcome the main limitation of nanofiber air filters to enhance their mechanical properties.

Nowadays, the most enhanced nanofiber filters are fabricated either by blended polymeric or composite materials. Also, the fiber diameter, pore size, and nanofiber membrane thickness are optimized mathematically and empirically to achieve maximum quality factor for air pollutant separation.

References

Abozeid MA, Hassan HS, Morsi I, Kashyout AB (2019) Development of nano-WO_3 doped with NiO for wireless gas sensors. Arab J Sci Eng 44:647–654. https://doi.org/10.1007/s13369-018-3608-1

Agrawal S, Ranjan R, Lal B, Rahman A, Singh SP, Selvaratnam T, Nawaz T (2021) Synthesis and water treatment applications of nanofibers by electrospinning. Processes 9:1779. https://doi.org/10.3390/pr9101779

Ahmadi Soufivand A, Faber J, Hinrichsen J et al (2023) Multilayer 3D bioprinting and complex mechanical properties of alginate-gelatin mesostructures. Sci Rep 13:11253. https://doi.org/10.1038/s41598-023-38323-2

Al-Abduljabbar A, Farooq I (2023) Electrospun polymer nanofibers: processing, properties, and applications. Polymers 15:65. https://doi.org/10.3390/polym15010065

Al-Ghouti MA, Ashfaq MY, Khan M, Al Disi Z, Da'na DA, Shoshaa R (2023) State-of-the-art adsorption and adsorptive filtration based technologies for the removal of trace elements: a critical review. Sci Total Environ 895:164854. https://doi.org/10.1016/j.scitotenv.2023.164854

Ali I, Kashyout AE-HB, Tayel M, Shokry Hassan H, Rizk M (2020) Ruthenium (Ru) doped zinc oxide nanostructure-based radio frequency identification (RFID) gas sensors for NH_3 detection. J Mater Res Technol 9:15693–15704. https://doi.org/10.1016/j.jmrt.2020.11.033

Anusiya G, Jaiganesh R (2022) A review on fabrication methods of nanofibers and a special focus on application of cellulose nanofibers. Carbohydr Polym Technol Appl 4:100262. https://doi.org/10.1016/j.carpta.2022.100262

Benselfelt T, Kummer N, Nordenström M, Fall AB, Nyström G, Wågberg L (2023) The colloidal properties of nanocellulose. ChemSusChem 16:e202201955. https://doi.org/10.1002/cssc.202201955

Bin Rashid A, Haque M, Islam SMM, Labib KMRU (2024) Nanotechnology-enhanced fiber-reinforced polymer composites: recent advancements on processing techniques and applications. Heliyon 10:e24692. https://doi.org/10.1016/j.heliyon.2024.e24692

Cao A, Nørskov JK (2023) Spin effects in chemisorption and catalysis. ACS Catal 13:3456–3462. https://doi.org/10.1021/acscatal.2c06319
Chai Z, Anthony C, Busnaina AA (2022) Directed assembly of nanomaterials for making nanoscale devices and structures: mechanisms and applications. ACS Nano 16(11):17641–17686. https://doi.org/10.1021/acsnano.2c07910
Daligaux V, Richard R, Manero M-H (2021) Deactivation and regeneration of zeolite catalysts used in pyrolysis of plastic wastes—a process and analytical review. Catalysts 11:770. https://doi.org/10.3390/catal11070770
Dashan M, Yingxiong W, Maja K, Laurens D, Yixin J, Ludwig C, Jie Z, Kim R (2019) Effects of phase morphology on mechanical properties: oriented/unoriented PP crystal combination with spherical/microfibrillar PET phase. Polymers 11:248. https://doi.org/10.3390/polym11020248
De Riccardis MF (2023) Electrospun nanofibrous membranes for air filtration: a critical review. Compounds 3:390–410. https://doi.org/10.3390/compounds3030030
Demir H, Atikler U, Balköse D, Tıhmınlıoğlu F (2006) The effect of fiber surface treatments on the tensile and water sorption properties of polypropylene–luffa fiber composites. Compos A: Appl Sci Manuf 37(3):447–456. https://doi.org/10.1016/j.compositesa.2005.05.036
El Fawal GF, Hassan HS, El-Aassar MR, Elkady MF (2019) Electrospun polyvinyl alcohol nanofibers containing titanium dioxide for gas sensor applications. Arab J Sci Eng 44:251–257. https://doi.org/10.1007/s13369-018-3529-z
Elkady MF, Hassan HS (2021) Photocatalytic degradation of malachite green dye from aqueous solution using environmentally compatible Ag/ZnO polymeric nanofibers. Polymers 13:2033. https://doi.org/10.3390/polym13132033
Elkady M, Salama E, Amer WA, Ebeid EM, Ayad MM, Shokry H (2020) Novel eco-friendly electrospun nanomagnetic zinc oxide hybridized PVA/alginate/chitosan nanofibers for enhanced phenol decontamination. Environ Sci Pollut Res 27:43077–43092. https://doi.org/10.1007/s11356-020-10247-8
Elkay MF, Shokry Hassan H, Salama E (2016) Sorption profile of phosphorus ions onto ZnO nanorods synthesized via sonic technique. J Eng 2016:2308560. https://doi.org/10.1155/2016/2308560
Espinoza-Montero PJ, Montero-Jiménez M, Rojas-Quishpe S, Alcívar León CD, Heredia-Moya J, Rosero-Chanalata A, Orbea-Hinojosa C, Piñeiros JL (2023) Nude and modified electrospun nanofibers, application to air purification. Nanomaterials 13:593–622. https://doi.org/10.3390/nano13030593
Essam A, El-Shazly A, Shokry H (2020) Piezoelectric pvdf nanofibrous membrane electrospinning optimization parameters. Key Eng Mater 897:71–76. https://doi.org/10.4028/www.scientific.net/KEM.897.71
Fonseca LM, El Halal SLM, Dias ARG, Zavareze ER (2021) Physical modification of starch by heat-moisture treatment and annealing and their applications: a review. Carbohydr Polym 15:118665. https://doi.org/10.1016/j.carbpol.2021.118665
Frising T, Thomas D, Contal P, Bémer D, Leclerc D (2003) Influence of filter fibre size distribution on filter efficiency calculations. Chem Eng Res Des 81(9):1179–1184. https://doi.org/10.1205/026387603770866353
Ghasemiyeh P, Mohammadi-Samani S, Nokhodchi A (2024) An overview of nanofibers and microfibers for improved oral delivery of medicines: challenges and advances. J Drug Deliv Sci Technol:105235. https://doi.org/10.1016/j.jddst.2023.105235
Grigoriev MV, Dyachkova IG, Buzmakov AV, Povolotsky MA, Kokhand VV, Chukalin MV, Uvarov VI (2020) μCT Analysis of porous cermet membranes with the use of enhanced filtration and binarization algorithms. J Surf Invest: X-Ray Synchrotron Neutron Tech 14:1294–1303. https://doi.org/10.1134/S1027451020060312
Gunawardene OHP, Gunathilake CA, Vikrant K, Amaraweera SM (2022) Carbon dioxide capture through physical and chemical adsorption using porous carbon materials: a review. Atmosphere 13:397. https://doi.org/10.3390/atmos13030397

Hassan HS, Kashyout AB, Morsi I, Nasser AAA, Raafat A (2015) Fabrication and characterization of nano-gas sensor arrays. AIP Conf Proc 1653:020042. https://doi.org/10.1063/1.4914233

Hassan HS, Elkady MF, Serour NM (2021) Intelligent nanosensors (INS) for environmental applications. In: Handbook of nanomaterials for sensing applications. Elsevier, pp 321–344. https://doi.org/10.1016/B978-0-12-820783-3.00017-8

Hearth HMRG, Sondarangalla SDA (2021) Enhancing air quality and controlling indoor air pollution by using low-cost air purification strategy, fostering opportunities for techno-preneurship in the new normal. International research symposium. https://www.researchgate.net/publication/357299975

Hosseinali F, Thomasson J (2018) Multiscale frictional properties of cotton fibers: a review. Fibers 6:49. https://doi.org/10.3390/fib6030049

Ibrahim Y, Kashyout AB, Morsi I, Shokry Hassan H (2021) Development of nano-SnO_2 and SnO_2:V_2O_5 thin films for selective gas sensor devices. Arab J Sci Eng 46:669–686. https://doi.org/10.1007/s13369-020-04735-9

Islam MS, Ang BC, Andriyana A, Afifi AM (2019) A review on fabrication of nanofbers via electrospinning and their applications. SN Appl Sci 1:1248. https://doi.org/10.1007/s42452-019-1288-4

Jialing S, Xuanhao L, Liang YE, Sam FYL, Manhong H (2023) A review on electrospinning as versatile supports for diverse nanofbers and their applications in environmental sensing. Adv Fiber Mater 5:429–460. https://doi.org/10.1007/s42765-022-00237-5

Joe Y-H, Shim J, Shin WG, Park H-S (2017) Effects of an external electric field on the collection efficiency of air filters: filtration mechanisms with an external E-field. Aerosol Sci Technol 51:1409–1418. https://doi.org/10.1080/02786826.2017.1361014

Kadam VV, Wang L, Padhye R (2018) Electrospun nanofiber materials to filter air pollutants – a review. J Ind Text 47(8):2253–2280. https://doi.org/10.1177/1528083716676812

Kanjwal MA, Ghaferi AA (2022) Graphene incorporated electrospun nanofiber for electrochemical sensing and biomedical applications: a critical review. Sensors 22:8661. https://doi.org/10.3390/s22228661

Kong Y, Yuan J, Qiu J (2012) Preparation and characterization of aligned carbon nanotubes/polylactic acid composite fibers. Phys B Condens Matter 407(13):2451–2457. https://doi.org/10.1016/j.physb.2012.03.045

Li H, Li Z, Wang N, Peng Y, Jiang Z, Zhang Q (2023) Improving the mechanical properties of CCFRPLA by enhancing the interface binding energy and strengthening the anti-separation ability of a PLA matrix. Polymers 15:2554. https://doi.org/10.3390/polym15112554

Mamun A, Blachowicz T, Sabantina L (2021) Electrospun nanofiber mats for filtering applications— technology, structure and materials. Polymers 13:1368. https://doi.org/10.3390/polym13091368

Medeiros GB, Lima FA, de Almeida DS, Guerra VG, Aguiar ML (2022) Modification and functionalization of fibers formed by electrospinning: a review. Membranes 12:861. https://doi.org/10.3390/membranes12090861

Mokhena TC, Matabola KP, Mokhothu TH, Mtibe A, Mochane MJ, Ndlovu G, Andrew JE (2022) Electrospun carbon nanofibres: preparation, characterization and application for adsorption of pollutants from water and air. Sep Purif Technol 288:120666. https://doi.org/10.1016/j.seppur.2022.120666

Saleh WM, Ahmad MI, Yahya EB, Khalil HPSA (2023) Nanostructured bioaerogels as a potential solution for particulate matter pollution. Gels 9:575. https://doi.org/10.3390/gels9070575

Sanchis-Gual R, Coronado-Puchau M, Mallah T, Coronado E (2023) Hybrid nanostructures based on gold nanoparticles and functional coordination polymers: chemistry, physics and applications in biomedicine, catalysis and magnetism. Coord Chem Rev 480:215025. https://doi.org/10.1016/j.ccr.2023.215025

Sanyal A, Sinha-Ray S (2021) Ultrafine PVDF nanofibers for filtration of air-borne particulate matters: a comprehensive review. Polymers 13:1864. https://doi.org/10.3390/polym13111864

Selatile MK, Ray SS, Ojijo V, Sadiku R (2018) Recent developments in polymeric electrospun nanofibrous membranes for seawater desalination. RSC Adv 8:37915–37938. https://doi.org/10.1039/C8RA07489E
Shaoyun F, Zheng S, Pei H, Yuanqing L, Ning H (2019) Some basic aspects of polymer nanocomposites: a critical review. Nano Mater Sci 1:2–30. https://doi.org/10.1016/j.nanoms.2019.02.006
Shokry Hassan H (2019) Role of preparation technique in the morphological structures of innovative nano-cation exchange. J Mater Res Technol 8:2854–2864. https://doi.org/10.1016/j.jmrt.2019.04.023
Shokry Hassan H, Kashyout AB, Morsi I, Nasser AAA, Raafat A (2014) Fabrication and characterization of gas sensor micro-arrays. Sens Bio-Sens Res 1:34–40. https://doi.org/10.1016/j.sbsr.2014.04.001
Solis-Ibarra D, Karunadasa HI (2014) Reversible and irreversible chemisorption in nonporous-crystalline hybrids. Angew Chem Int Ed 53(4):1039–1042. https://doi.org/10.1002/anie.201309786
Vasiliev A, Shaposhnik A, Moskalev P, Kul O (2023) Kinetics of chemisorption on the surface of nanodispersed SnO_2–PdOx and selective determination of CO and H_2 in air. Sensors 23:3730. https://doi.org/10.3390/s23073730
Wang H, Bao Y, Yang X, Lan X, Guo J, Pan Y, Huang W, Tang L, Luo Z, Zhou B et al (2022) Study on filtration performance of PVDF/PUL composite air filtration membrane based on far-field electrospinning. Polymers 14:3294. https://doi.org/10.3390/polym14163294
Wang P, Lv H, Cao X, Liu Y, Yu D-G (2023) Recent progress of the preparation and application of electrospun porous nanofibers. Polymers 15:921. https://doi.org/10.3390/polym15040921
Yueyang M, Wen S, Shang W, Xuemei G, Fen A, Yuanlan N, Yali L, Zhiming L (2023) Electrospun polymers: using devices to enhance their potential for biomedical applications. React Funct Polym 186:105568. https://doi.org/10.1016/j.reactfunctpolym.2023.105568
Zhang Q, Welch J, Park H, Wu C-Y, Sigmund W, Marijnissen JCM (2010) Improvement in nanofiber filtration by multiple thin layers of nanofiber mats. J Aerosol Sci 41(2):230–236. https://doi.org/10.1016/j.jaerosci.2009.10.001
Zhang W, He Z, Han Y, Jiang Q, Zhan C, Zhang K, Li Z, Zhang R (2020) Structural design and environmental applications of electro-spun nanofibers. Compos Part A Appl Sci Manuf 137:106. https://doi.org/10.1016/j.compositesa.2020.106009

MIX
Papier aus verantwortungsvollen Quellen
Paper from responsible sources
FSC® C105338

If you have any concerns about our products,
you can contact us on
ProductSafety@springernature.com

In case Publisher is established outside the EU,
the EU authorized representative is:
Springer Nature Customer Service Center GmbH
Europaplatz 3, 69115 Heidelberg, Germany

Printed by Libri Plureos GmbH
in Hamburg, Germany